Computational Multiphase Geomechanics

Computational Multiphase Geomechanics

Fusao Oka and Sayuri Kimoto

CRC Press
Taylor & Francis Group
Boca Raton London New York

CRC Press is an imprint of the
Taylor & Francis Group, an **informa** business

First edition published 2022
by CRC Press
2 Park Square, Milton Park, Abingdon, Oxon, OX14 4RN

and by CRC Press
6000 Broken Sound Parkway NW, Suite 300, Boca Raton, FL 33487-2742

CRC Press is an imprint of Informa UK Limited

British Library Cataloguing-in-Publication Data
A catalogue record for this book is available from the British Library

Library of Congress Cataloging-in-Publication Data

Names: Oka, Fusao, author. | Kimoto, Sayuri, author.
Title: Computational multiphase geomechanics / Fusao Oka, Sayuri Kimoto.
Description: First edition. | Abingdon, Oxon ; Boca Raton, FL : CRC Press, [2022] | Includes bibliographical references and index.
Identifiers: LCCN 2021024103 (print) | LCCN 2021024104 (ebook) | ISBN 9781032059556 (hbk) | ISBN 9781032059570 (pbk) | ISBN 9781003200031 (ebk)
Subjects: LCSH: Geotechnical engineering--Mathematics. | Soil mechanics--Mathematical models.
Classification: LCC TA705 .O388 2022 (print) | LCC TA705 (ebook) | DDC 624.1/51--dc23
LC record available at https://lccn.loc.gov/2021024103
LC ebook record available at https://lccn.loc.gov/2021024104

ISBN: 978-1-032-05955-6 (hbk)
ISBN: 978-1-032-05957-0 (pbk)
ISBN: 978-1-003-20003-1 (ebk)

DOI: 10.1201/9781003200031

Typeset in Sabon
by Deanta Global Publishing Services, Chennai, India

Contents

Preface

It is well known that numerical methods play a very important role in geotechnical engineering and in a related activity called "computational geotechnics". The area that is covered by computational geotechnics is growing, in particular that which relates to the multiphase nature of geomaterials. This book is the second volume of *Computational Modeling of Multiphase Geomaterials* that was published in 2012 and provides recent progress in this area, i.e., the basic concept of the air–water–soil mixture, cyclic constitutive models, anisotropic models, non-coaxial models, gradient models, strain localization including compaction bands, dynamic strain localization, and the instability of unsaturated soils. In addition, this book includes the application of computational modeling to slope stability, large-scale excavation of ground, liquefaction analysis of levees during earthquakes, a methane hydrate development-related problem, and the shielding of contamination by bentonite. Furthermore, the erosion problem of geomaterials due to seepage flow is presented. In the analyses presented in this book, we used the computer codes "CONVI2D" for the two-dimensional problem and "CONVI3D" for the three-dimensional problem, and several extended codes such as CONVI2D-DY for dynamic analysis, etc. These codes have been developed by the authors and their colleagues in the geomechanics laboratory of Kyoto University.

Chapter 1 deals with the formulation of gas–liquid–solid three-phase porous media with the fundamental concept. The effective stress and skeleton stresses are derived based on the theory of porous media in the context of the mixture theory of continuum mechanics.

Chapter 2 presents various constitutive equations, i.e., the cyclic elastoplastic model, the cyclic viscoplastic model, the anisotropic model, unsaturated soil models, the non-coaxial model, the strain gradient-dependent non-local model, and the strain-softening model. This chapter is essential for the numerical analysis of various geomaterials.

Chapter 3 deals with the governing equations of multiphase geomaterials and the dynamic finite element formulation with the updated Lagrangian formulation for large deformation analysis.

In Chapter 4, the special topic of the strain localization problems of geomaterials, i.e., two typical localization modes, is discussed. In particular, compaction bands are described in detail and discussed for mudstone. In addition, the dynamic localization problems of unsaturated soil are numerically presented in comparison with the response of saturated soil.

In Chapter 5, the stability of a water infiltration problem in an unsaturated soil and a numerical simulation are presented. The results of a linear stability analysis of elasto-viscoplastic unsaturated material show that the parameters of the suction–saturation relation play a significant role in the onset of the instability of an unsaturated viscoplastic material subjected to water infiltration. The parametric studies of a numerical simulation of the problem with special reference to the suction–saturation relation are consistent with the linear instability analysis. Comparisons of the numerical simulation and the infiltration test results are shown.

Chapter 6 presents a study to clarify the effect of rainfall infiltration on the deformation of unsaturated river embankments considering seepage flow. From the numerical results, it is seen that the deformation of the embankment significantly depends on the water permeability of the soil and it is localized on the slope surface at the river side.

In Chapter 7, the large deformation dynamic response of river embankments subjected to seismic excitations is presented using a u–p formulation developed in the context of the porous theory. Efforts are made to model the failure modes and damage patterns of river embankments observed in the 2011 Tohoku earthquake. The effects of the ground profile, water table, and the earthquake motion on the seismic response and the damage pattern of river embankments are particularly emphasized.

In Chapter 8, large-scale excavation in construction is numerically back analyzed using a soil-water coupled finite element method with an elasto-viscoplastic model that considers strain-induced degradation. A comparison between the numerical results and the measurements of the excavation shows that the simulation method can capture the overall deformation of the soft ground and the earth retaining walls including the time-dependent behaviour.

Chapter 9 presents an elasto-viscoplastic model for unsaturated expansive soil. An evolutional equation is adopted to describe the absorption of water into the interlayer of clay platelets. The internal compaction effect caused by swelling of the clay unit is expressed with the expansion of the overconsolidation boundary surface and the static yield surface. A finite element analysis is conducted to study the development of the swelling pressure. Comparing the experimental results and the simulated results, it is found that the proposed model can reproduce the effects of dry density and the initial water content on the swelling behavior.

In Chapter 10, a numerical simulation of the gas production process using the depressurizing method is presented. The simulation is conducted for the seabed ground model with a hydrate-bearing layer to investigate the mechanical behavior during dissociation. The method has been developed based on the chemo-thermo-mechanically coupled analysis, taking into account the phase changes from solids to fluids, i.e., water and gas, the flow of fluids, heat transfer, and the ground deformation. An extended elasto-viscoplastic model for unsaturated soils considering the effect of hydrate bonding is used for the hydrate-bearing sediments. The numerical results of the seabed ground during the dissociation of methane hydrate are discussed.

In Chapter 11, we present the formulation of the constitutive equations of internal erosion with the erosion criteria and the rate equation of mass transfer. The driving force for erosion is assumed to be given by the interaction term, i.e., the relative velocity between two phases in the equation of motion for a two-phase mixture. Then, field equations to simulate the hydro-mechanical behavior due to internal erosion are derived in the framework of the multiphase porous theory. Laboratory erosion tests using a gap-graded sandy soil are simulated by the proposed model and the validity is discussed with respect to the rate of eroded soil mass and the particle size distribution measured after the erosion test.

Acknowledgment

During the writing and preparation of this book, the authors became indebted to many researchers and students. In particular, we wish to express our sincere thanks to the late Professors K. Akai and T. Adachi of Kyoto University, and the late Professor I.G. Vardoulakis of the National Technical University of Athens. Professor H.-B. Mühlhaus of the University of Queensland, Professor B. Loret of the Institute National Polytechnique University of Grenoble, Professor A. Yashima of Gifu University, Professor T. Kodaka of Meijo University, Professors R. Uzuoka and Y. Higo of Kyoto University, Professor Jidong Zhao of Hong Kong University of Science and Technology, Dr H. Iwai, Dr K.B. Shahbodagh, Dr E.F. Garcia, Dr S. Boonlert, Dr Y.-S. Kim, Dr H. Feng, Dr H. Sadeghi, Dr N. Takada, Dr T. Akaki, and all the graduate students of the geomechanics laboratory of Kyoto University for their contributions and discussions. Many thanks go to the members of the TC-34 and TC-103 of ISSMGE and the LIQCA Liquefaction Geo Research Institute for their discussions and supports. I would also like to thank Dr Shahbodagh Khan Babak of the University of New South Wales for his invaluable comments on the manuscript. Finally, we would like to dedicate this book to our families, in particular, O. Keiko and K. Keiko.

Many thanks are also due to the following researchers and organizations for permissions to use data: Mr Chikamatsu; and for the reuse of the figures and tables: the Institution of Civil Engineers (London), ASCE, the South East Asia Geotechnical Society and the Association of Geotechnical Societies in South East Asia, John Wiley & Sons, Springer Nature.

Fusao Oka

Sayuri Kimoto

Authors

Fusao Oka is a professor emeritus of Kyoto University and president of the LIQCA Liquefaction Geo Research Institute. Until 1997, he was a professor in the Department of Civil and Earth Resources Engineering of Kyoto University. He specializes in computational geomechanics with particular regard to constitutive equations, consolidation, liquefaction, and strain localization analyses. He organized several international conferences and workshops, such as the Fourth International Workshop on Localization and Bifurcation Theory for Soils and Rocks (1997), the ISSMGE International Symposium on Deformation and Progressive Failure in Geomaterials (1997), the International Symposium on Prediction and Simulation Methods for Geohazard Mitigation (2009), and the 14th International Conference of IACMAG (2014). He has published more than 300 papers on geomechanics and is a co-author of *Computational Modeling of Multiphase Geomaterials* (CRC Press, 2012).

Sayuri Kimoto had been working on geomechanics as an associate professor at Kyoto University for years. She is presently a professor of Osaka Sangyo University. She specializes in the elasto-viscoplastic constitutive equations of soils and the numerical analysis of multiphase geomaterials, such as the behavior analysis of seabed ground due to the production of methane gas. She is a co-author of *Computational Modeling of Multiphase Geomaterials* (CRC Press, 2012).

Chapter 1

Fundamental equations of multiphase geomaterials

1.1 COMPOSITION PARAMETERS AND FUNDAMENTAL PHASE RELATIONS OF POROUS MATERIAL

1.1.1 Basic quantities

Geomaterials such as soils are composed of the solid skeleton (soil particles), pore water, and pore gas. Let us consider the porous material composed of solid, water, and air. Common parameters for the description of the porous material used in this chapter are as follows (see Figure 1.1):

$V = \Sigma_{\alpha=S,W,G} V^{\alpha}$: total volume, apparent volume of the solid skeleton ($=V_b$)
V^S: volume of the solid skeleton (soil particles)
V^v: volume of void; V^W: volume of liquid (water); V^G: volume of gas (air)
M: total mass; M^S: mass of solid substance; M^W: mass of liquid; M^G: mass of gas
W: total weight; W^S: weight of the solid substance; W^W: weight of liquid; W^G: weight of gas

The following parameters are defined for the arbitrary volume of porous material as

$e = V^v/V^S$: void ratio; $f = 1 + e$: specific volume; $n = V^v/V$: porosity
$S_r = V^W/V^v$: saturation; $w = M^W/M^S$: water content when the liquid is water

1.1.2 Volumes of air–water–solid three phases and volumetric strains

Using the volumes of the solid substance V^S, pore water V^W, air V^G, and the porosity n, the apparent volumes of the solid skeleton V_b, water V_f, and air V_g are described as follows (see Figures 1.1 and 1.2):

DOI: 10.1201/9781003200031-1

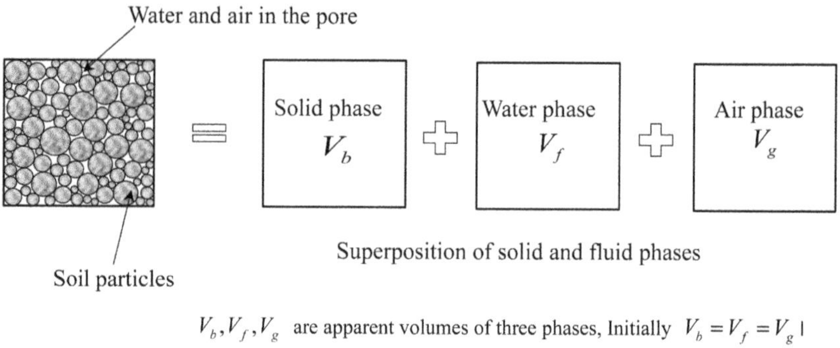

Figure 1.1 Superposition of three phases.

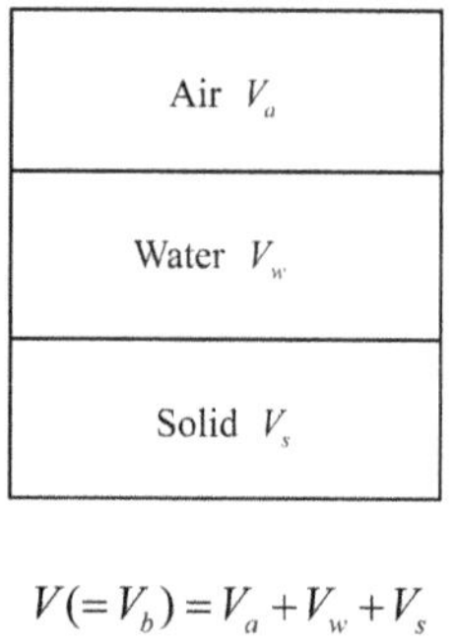

Figure 1.2 Constitution of porous media.

$$V_b = \frac{V^S}{1-n}, \quad V_f = \frac{V^W}{nS_r}, \quad V_g = \frac{V^G}{(1-S_r)n} \tag{1.1}$$

where S_r is the saturation.

Equation 1.1 indicates that V^W is the actual volume of the water in the skeleton and V_f is the extended apparent (phase) volume of the distributed pore water in the skeleton to satisfy the assumption of the superposition of three phases for porous media. Similar definitions are used for V^G and V_g. Hence, $V_f = V_g = V_b$ at the initial state. Differentiating Equation 1.1, the following differential equations are obtained:

$$\frac{dV_b}{V_b} = \frac{dV^S}{V^S} + \frac{dn}{(1-n)} \tag{1.2}$$

$$\frac{dV_f}{V_f} = \frac{dV^W}{V^W} - \frac{dn}{n} - \frac{dS_r}{S_r} \tag{1.3}$$

$$\frac{dV_g}{V_g} = \frac{dV^G}{V^G} - \frac{dn}{n} + \frac{dS_r}{(1-S_r)} \tag{1.4}$$

Assuming that the void volume is equal to the pore volume, V_p:

$$V_b = V^S + V_p \tag{1.5}$$

We have an apparent volumetric strain increment of the solid skeleton:

$$d\varepsilon^{(s)} = d\varepsilon^s_{kk} = de_b = \frac{dV_b}{V_b} = (1-n)\frac{dV^S}{V^S} + n\frac{dV_p}{V_p} \tag{1.6}$$

Using Equations 1.1 and 1.6, we obtain a strain of the pore:

$$de_p = \frac{dV_p}{V_p} = \frac{dn}{n(1-n)} + \frac{dV^S}{V^S} \tag{1.7}$$

Using Equation 1.7, Equations 1.3 and 1.4 give the following relations for the apparent volumetric strain increment of the liquid and gas phases:

$$d\varepsilon^{(w)} = d\varepsilon^w_{kk} = \frac{dV_f}{V_f} = \frac{dV^W}{V^W} - (1-n)\left[\frac{dV_p}{V_p} - \frac{dV^S}{V^S}\right] - \frac{dS_r}{S_r} \tag{1.8}$$

$$d\varepsilon^{(g)} = d\varepsilon^g_{kk} = \frac{dV_g}{V_g} = \frac{dV^G}{V^G} - (1-n)\left[\frac{dV_p}{V_p} - \frac{dV^S}{V^S}\right] + \frac{dS_r}{(1-S_r)} \tag{1.9}$$

Substituting the definitions of strains $de_S = dV^S / V^S$ (strain of the solid skeleton), $de_W = dV^W / V^W$ (strain of liquid [water]), and $de_G = dV^G / V^G$ (strain of gas [air]) into Equations 1.6, 1.8, and 1.9, we obtain the relations among the strains:

$$d\varepsilon^{(s)} = de_b = (1-n)de_S + nde_p \tag{1.10}$$

$$d\varepsilon^{(w)} = de_W + (1-n)(de_S - de_p) - dS_r / S_r \tag{1.11}$$

$$d\varepsilon^{(g)} = de_G + (1-n)(de_S - de_p) + dS_r / (1-S_r) \tag{1.12}$$

Assuming the isotropic deformation, $de_p = de_S$ in the unjacketed test.

Alternatively, using the volume fractions, Equations 1.10–1.12 can be written as

$$d\varepsilon^{(s)} = de_b = n^S de_S + n^W de_W + n^G de_G \tag{1.13}$$

$$d\varepsilon^{(w)} = n^S de_S + \left(1 - \frac{n^S n^W}{n}\right) de_W - \frac{n^S n^G}{n} de_G - dS_r / S_r \tag{1.14}$$

$$d\varepsilon^{(g)} = n^S de_S - \frac{n^S n^W}{n} de_W + \left(1 - \frac{n^S n^G}{n}\right) de_G + dS_r / (1 - S_r) \tag{1.15}$$

where $n^S = V^S / V_b$, $n^W = V^W / V_b$, and $n^G = V^G / V_b$ are volume fractions.

Equation 1.10 can be derived from the mass conservation law (e.g., Vardoulakis and Sulem 1995).

1.2 FORMULATION OF THE CONSTITUTIVE EQUATIONS FOR THREE-PHASE POROUS MATERIALS

1.2.1 Introduction

This section presents the formulation of the constitutive model for three-phase geomaterials based on continuum thermodynamics with internal variables (Oka and Kimoto 2019, 2020). A linear constitutive model is derived for three-phase porous materials such as soils and rocks, which is an extension of the Biot model (1957). It is well known that Terzaghi's effective stress and the effective stress with the Biot coefficient have been used for geomaterials. For partially saturated soils, several stress concepts have been used such as the net stress (Fredlund et al. 1978; Alonso et al. 1990), the Bishop stress (Bishop 1959) and the generalized Bishop stress (Khalili and Khabbaz 1998; Kohgo et al. 1993; Bolzon et al. 1996; Wheeler and Sivakumar 1995; Ehlers et al. 2004; Nuth and Lalouis 2008), and the skeleton stress (Jommi 2000; Nuth and Lalouis 2008; Oka et al. 2006, 2008, 2010; Zhao et al. 2016). For the elastic behavior of water-saturated geomaterials, Biot (1957, 1962) developed an elastic water-saturated model with the effective stress having the so-called Biot constant or the Biot coefficient. Kimoto et al. (2017) showed that the Biot-type two-phase theory is applicable to the elastic behavior of dry soil, i.e., air-filled sand, and that the classical effective stress cannot be applied to the air-filled dry soil. On the other hand, compared with the water-saturated porous material, for the three-phase porous material such as partially saturated soil, the skeleton stress with the Biot coefficient has not been well discussed. This section shows that the skeleton stress with the Biot coefficient can be derived for three-phase materials. In addition, the choice of stress variables is discussed. For inelastic strain rates, however, the choice of stress variables is still an open problem.

Many researchers (e.g., Coussy 1995; de Boer and Ehlers 1990; de Boer 1996; Schrefler 2002; Lalouis et al. 2003; Borja 2004) studied the general formulation of multiphase materials based on continuum mechanics with thermodynamics in the context of the mixture theory (e.g., Bowen 1976, 1982). The classical mixture theory deals with the multiphase properties but does not consider the interfacial properties such as the surface tension or surface stress (Gurtin and Murdoch 1975), although the capillary pressure by surface tension is essential for the three-phase unsaturated soil. Hassanizadeh and Gray (1990) advocated the effect of the interface boundaries between different phases, i.e., they explicitly provided the functional dependence on the capillary pressure. Duriez et al. (2017) have derived relations between internal surface stresses and the surface tension based on the micromechanical considerations for three-phase granular media. It has been recommended that the choice of stress variables should be based on the micromechanical considerations of the interaction between three phases such as the capillary stress and experimental evidence of the interactions. It is reasonable to explicitly include the real value of the surface tension with the evolutional equation of the surface tension associated with the change of solid structure. It is, however, currently not so easy to explicitly introduce the surface tension into the continuum model. Hence, herein, the effect of surface tension is implicitly included in the constitutive equations through the matric suction (e.g., Laplace relation), as has been used in geomechanics. The constitutive equations with interactions between fluids and the solid skeleton are derived in the frame of macroscopic continuum mechanics. The model includes the matric suction–saturation relation as well as the usual stress–strain relations.

1.2.2 General setting for the theory of three-phase porous media

The porous media theory has been developed based on the mixture theory of an immiscible mixture of solid and fluids (e.g., Bowen 1976; Bedford and Drumheller 1983; Passman et al. 1984; De Boer 1997). The mixture is defined as the superposition of the continua of multiphases and each point of the mixture is occupied by two or three continua, e.g., liquid and solid, or solid, liquid, and gas continua. We assume that the material points of all continua for constituents occupy the same spatial point of the actual configuration simultaneously (e.g., Sec 158, Truesdell and Toupin 1960). The independent governing equation such as the equation of motion with interaction is assigned for each continuum of constituents. The material to be modeled is composed of three phases, namely, solid, liquid, and gas, which are continuously distributed throughout space. In the present study, the gas, liquid, and solid phases are set to air, water and solid skeleton, respectively.

The particles of the multiphase material under consideration occupy the same position at time t:

$$x_i^\beta = \hat{x}_i^\beta(X_j^\beta, t); \ \beta = S(\text{solid}),\ W(\text{liquid}),\ G\,(\text{gas}) \tag{1.16}$$

where X_j^β is the position of the particle of β phase at the reference state and x_i^β is the position of the particle of β phase at time t. (Note: The reference paper [Oka and Kimoto 2020] used $\beta = s, w, a$, which are represented by $\beta = S, W, G$.)

At time t:

$$x_i = x_i^S(t) = x_i^W(t) = x_i^G(t) \tag{1.17}$$

We now consider the infinitesimal strain field, i.e., a small deformation gradient without loss of generality. Using the following, the strains for the three phases ε_{ij}^β are obtained:

$$\varepsilon_{ij}^\beta = \frac{1}{2}\left(\frac{\partial u_i^\beta}{\partial x_j} + \frac{\partial u_j^\beta}{\partial x_i}\right) \tag{1.18}$$

where u_i^β is the displacement vector for β phase continua.

The velocity is defined as the time rate of the displacement.

The total volume V is obtained from the sum of the partial volumes of the constituents; solid (S), water (W), and air (G) (Figures 1.1 and 1.2), namely,

$$\sum_\beta V^\beta = V \quad (\beta = S, W, G) \tag{1.19}$$

The volume of void or pore V_p, which is composed of water and air, is given as follows:

$$\sum_\beta V^\beta = V_p \quad (\beta = W, G) \tag{1.20}$$

The volume fraction n^β is defined as the local ratio of the volume element with respect to the total volume, namely,

$$n^\beta = \frac{V^\beta}{V} \tag{1.21}$$

$$\sum_\beta n^\beta = 1 \quad (\beta = S, W, G) \tag{1.22}$$

Equation 1.22 is an internal constraint of the porous media.

The volume fraction of the void, i.e., porosity n, is written as

$$n = \sum_\beta n^\beta = \frac{V_p}{V} = \frac{V - V^S}{V} = 1 - n^S \quad (\beta = W, G) \tag{1.23}$$

The volume fraction concept was adopted to construct the theory of mixture (e.g., Mills 1967; Morland 1972). The historical development of the volume fraction theory was well discussed by de Boer (1996, 1998).

In addition, the water saturation is required in the model, namely,

$$S_r = \frac{V^W}{V^W + V^G} = \frac{n^W}{n^W + n^G} \tag{1.24}$$

The corresponding partial Cauchy stress tensors are symmetric and defined by σ_{ij}^{β}, $\beta = S, W, G$; solid (S), liquid (W), and gas (G).

Considering that the area fraction is statistically proportional to the volume fraction (e.g., Biot 1962; de Boer 1998; Borja 2006; Oka and Kimoto 2012), we assume that

$$\sigma_{ij} = \sigma_{ij}^S + \sigma_{ij}^W + \sigma_{ij}^G \tag{1.25}$$

$$\sigma_{ij}^W = -n^W P^W \delta_{ij} \tag{1.26}$$

$$\sigma_{ij}^G = -n^G P^G \delta_{ij} \tag{1.27}$$

where σ_{ij} is the total Cauchy stress, and P^G and P^W are the pore gas and pore liquid pressures, respectively. We adopt the convention of tension being positive throughout this section.

1.2.3 Thermodynamic formulation for multiphase porous materials

The existence of energy functions is crucial for the stability of the materials. It will be shown that the present model is consistent with the restriction of thermodynamics.

Herein, a formulation in the small strain field is adopted without a loss in generality. It is assumed that the temperatures for the three phases are equal. The first law of thermodynamics is given as

$$\dot{u} = \sigma_{ij}^{\beta} \dot{\varepsilon}_{ij}^{\beta} + r - q_{i,i} \tag{1.28}$$

where the temperature is assumed to be equal for two phases; u is the internal energy density per unit volume; σ_{ij}^{β} is the stress tensor; $\dot{\varepsilon}_{ij}^{\beta}$ is the strain rate tensor for the β phase; q_i is the heat flux vector; and r is the other energy supply rate such as radiation or body heating. The summation convention is used for the indices β, i, j; the dot denotes the time differentiation; and "i" indicates the spatial differentiation with respect to x_i.

The local form of the Clausius–Duhem inequality is given by

$$\dot{\eta} \geq \frac{r}{\theta} - \left(\frac{q_i}{\theta}\right)_{,i} \tag{1.29}$$

where η is the entropy density per unit volume and θ is the temperature.

Using the first law of thermodynamics, i.e., the energy balance equation (Equation 1.28), we obtain

$$\dot{\eta}\theta + \sigma_{ij}^{\beta}\dot{\varepsilon}_{ij}^{\beta} - \dot{u} - \frac{1}{\theta}q_i\frac{\partial\theta}{\partial x_i} \geq 0 \tag{1.30}$$

Using the free energy function per unit volume, namely,

$$\Phi\left(\varepsilon_{ij}^{\beta}, \xi_{ij}^{s}, \theta\right) = u - \eta\theta \tag{1.31}$$

we have

$$-(\dot{\Phi} + \eta\dot{\theta}) + \sigma_{ij}^{\beta}\dot{\varepsilon}_{ij}^{\beta} - \frac{1}{\theta}q_i\frac{\partial\theta}{\partial x_i} \geq 0 \tag{1.32}$$

where ξ_{ij}^{s} is an internal variable for the inelastic strain–like variables of the solid skeleton.

Then, substituting the complementary energy, i.e., the Legendre transformation of the free energy as

$$\Psi\left(\sigma_{ij}^{\beta}, \xi_{ij}^{s}, \theta\right) = \sigma_{ij}^{\beta}\varepsilon_{ij}^{\beta} - \Phi \tag{1.33}$$

into Equation 1.32 yields

$$\left(\frac{\partial\Psi}{\partial\theta} - \eta\right)\dot{\theta} + \left(\frac{\partial\Psi}{\partial\sigma_{ij}^{\beta}} - \varepsilon_{ij}^{\beta}\right)\dot{\sigma}_{ij}^{\alpha} + \frac{\partial\Psi}{\partial\xi_{ij}^{s}}\dot{\xi}_{ij}^{s} - \frac{1}{\theta}q_i\frac{\partial\theta}{\partial x_i} \geq 0 \tag{1.34}$$

Using Coleman's method (Coleman and Gurtin 1967), in which the stress rates and temperature rate are arbitrary, we obtain

$$\eta = \frac{\partial\Psi}{\partial\theta}, \quad \varepsilon_{ij}^{\beta} = \frac{\partial\Psi}{\partial\sigma_{ij}^{\beta}} \tag{1.35}$$

and the local dissipation as

$$\frac{\partial\Psi}{\partial\xi_{ij}^{s}}\dot{\xi}_{ij}^{s} - \frac{1}{\theta}q_i\frac{\partial\theta}{\partial x_i} \geq 0 \tag{1.36}$$

Upon the time differentiation of Equation 1.35(2), the strain rates for the solid skeleton are obtained as

$$\dot{\varepsilon}_{ij}^S = \frac{\partial^2 \Psi}{\partial \sigma_{ij}^S \partial \sigma_{kl}^\beta} \dot{\sigma}_{kl}^\beta + \frac{\partial^2 \Psi}{\partial \sigma_{ij}^S \partial \xi_{kl}^s} \dot{\xi}_{kl}^s + \frac{\partial^2 \Psi}{\partial \sigma_{ij}^S \partial \theta} \dot{\theta} \tag{1.37}$$

in which the internal variable for the solid phase is only considered as ξ_{ij}^s.

By setting,

$$\dot{\varepsilon}_{ij}^{S(el)} = \frac{\partial^2 \Psi(\sigma_{kl}^\chi, \xi_{kl}^\chi, \theta)}{\partial \sigma_{ij}^S \partial \sigma_{mn}^\beta} \dot{\sigma}_{mn}^\beta \tag{1.38}$$

$$\dot{\varepsilon}_{ij}^{S(inel)} = \frac{\partial^2 \Psi(\sigma_{kl}^\chi, \xi_{kl}^s, \theta)}{\partial \sigma_{ij}^S \partial \xi_{mn}^s} \dot{\xi}_{mn}^s \tag{1.39}$$

the strain rates for the solid skeleton are given under isothermal conditions as

$$\dot{\varepsilon}_{ij}^S = \dot{\varepsilon}_{ij}^{S(el)} + \dot{\varepsilon}_{ij}^{S(inel)} \tag{1.40}$$

where $\dot{\varepsilon}_{ij}^{S(el)}$ is the quasi-elastic strain rates and $\dot{\varepsilon}_{ij}^{S(inel)}$ is the inelastic strain rates.

In the following sections, the isothermal conditions are assumed.

1.2.4 Constitutive equations for three-phase porous material

Let us discuss the air–water–soil three-phase porous materials in the infinitesimal strain field.

There are several methods for the energy functions. One is that the energy function is composed of functions for single constituents (e.g., Green and Steel 1966; Bowen 1976) and another method adopts the energy function for the whole porous media (e.g., Green and Steel 1966; Coussy 1995; Borja 2006). For the complementary energy function, we assume that the energy function for the whole porous media is dependent on the partial stresses and the internal variable. Hence, in order to discuss the linear model in the next section, the following quadratic form of the complementary energy is assumed:

$$\begin{aligned} \Psi = {} & A_{mnkl}\sigma_{mn}^S\sigma_{kl}^S + B_{mnkl}\sigma_{mn}^S\sigma_{kl}^W + C_{mnkl}\sigma_{mn}^W\sigma_{kl}^W + D_{mnkl}\sigma_{mn}^S\sigma_{kl}^G \\ & + E_{mnkl}\sigma_{mn}^G\sigma_{kl}^G + F_{mnkl}\sigma_{mn}^W\sigma_{kl}^G + H_{mnkl}\sigma_{mn}^S\xi_{kl}^s \end{aligned} \tag{1.41}$$

the stress–strain relations are obtained as

$$\begin{aligned} \varepsilon_{ij}^S = \frac{\partial \Psi}{\partial \sigma_{ij}^S} = {} & A_{mnkl}\delta_{mi}\delta_{nj}\sigma_{kl}^S + A_{mnkl}\delta_{ki}\delta_{lj}\sigma_{mn}^S \\ & + B_{mnkl}\delta_{mi}\delta_{nj}\sigma_{kl}^W + D_{mnkl}\delta_{mi}\delta_{nj}\sigma_{kl}^G + H_{mnkl}\delta_{mi}\delta_{nj}\xi_{kl}^s \end{aligned} \tag{1.42}$$

$$\varepsilon_{ij}^{W} = \frac{\partial \Psi}{\partial \sigma_{ij}^{W}} = B_{mnkl}\delta_{ki}\delta_{lj}\sigma_{mn}^{S} + C_{mnkl}\delta_{mi}\delta_{nj}\sigma_{kl}^{W} + C_{mnkl}\delta_{ki}\delta_{lj}\sigma_{mn}^{W} + F_{mnkl}\delta_{mi}\delta_{nj}\sigma_{kl}^{G} \tag{1.43}$$

$$\varepsilon_{ij}^{G} = \frac{\partial \Psi}{\partial \sigma_{ij}^{G}} = D_{mnkl}\delta_{ki}\delta_{lj}\sigma_{mn}^{S} + E_{mnkl}\delta_{mi}\delta_{nj}\sigma_{kl}^{G} + E_{mnkl}\delta_{ki}\delta_{lj}\sigma_{mn}^{G} + F_{mnkl}\delta_{ki}\delta_{lj}\sigma_{mn}^{W} \tag{1.44}$$

where δ_{ij} is Kronecker's delta: $\delta_{ij}\{=1 \text{ if } i=j,\ =0 \text{ if } i \neq j\}$.

It is assumed that A_{mnkl}, B_{mnkl}, C_{mnkl}, D_{mnkl}, E_{mnkl}, F_{mnkl}, and H_{mnkl} are the fourth-order isotropic tensors for having the isotropic solid and fluids, namely,

$$\begin{aligned} &A_{mnkl} = a_1\delta_{mn}\delta_{kl} + a_2(\delta_{mk}\delta_{nl} + \delta_{ml}\delta_{nk}) \quad B_{mnkl} = b\delta_{mn}\delta_{kl} \\ &C_{mnkl} = c\delta_{mn}\delta_{kl} \quad D_{mnkl} = d\delta_{mn}\delta_{kl} \quad E_{mnkl} = e\delta_{mn}\delta_{kl} \quad F_{mnkl} = f\delta_{mn}\delta_{kl} \\ &H_{mnkl} = h(\delta_{mk}\delta_{nl} + \delta_{ml}\delta_{nk}) \end{aligned} \tag{1.45}$$

where a, b, c, d, e, f, and h are the scalar parameters.

Equations 1.42–1.45 yield the following stress–strain relations:

$$\varepsilon_{ij}^{S} = 2a_1\delta_{ij}\sigma_{kk}^{S} + 4a_2\sigma_{ij}^{S} + b\delta_{ij}\sigma_{kk}^{W} + d\delta_{ij}\sigma_{kk}^{G} + 2h\xi_{ij}^{S} \tag{1.46}$$

$$\varepsilon_{ij}^{W} = b\delta_{ij}\sigma_{kk}^{S} + 2c\delta_{ij}\sigma_{kk}^{W} + f\delta_{ij}\sigma_{kk}^{G} \tag{1.47}$$

$$\varepsilon_{ij}^{G} = d\delta_{ij}\sigma_{kk}^{S} + 2e\delta_{ij}\sigma_{kk}^{G} + f\delta_{ij}\sigma_{kk}^{W} \tag{1.48}$$

In Equation 1.46, an evolutional equation is needed for the internal variables ξ_{ij}^{s} that may be given by the plastic flow rule or the viscoplastic flow rule, etc., as the inelastic strain rate.

In the above equations, the fluid strains are isotropic because of Equation 1.45.

1.2.5 Elastic constitutive model for three-phase porous materials with interaction between fluids and solid

In order to include the non-linearity of the material model, we discuss the incremental form of the model. In the following incremental formulation, assuming the incremental linearity with respect to the volume fraction, Equations 1.26 and 1.27 lead to $d\sigma_{ij}^{W} = -n^{W}dP^{W}\delta_{ij}$ and $d\sigma_{ij}^{G} = -n^{G}dP^{G}\delta_{ij}$.

From Equation 1.42, the incremental stress–strain relation is obtained for the solid phase of the three-phase porous material, including the inelastic strain equivalent internal variable ξ_{ij}^{s}, as

$$
\begin{aligned}
d\varepsilon_{ij}^{S} &= 2a_1\delta_{ij}d\sigma_{kk}^{S} + 2a_2\left(\delta_{ik}\delta_{jl} + \delta_{il}\delta_{jk}\right)d\sigma_{kl}^{S} + b\delta_{ij}d\sigma_{kk}^{W} + d\delta_{ij}d\sigma_{kk}^{G} + 2hd\xi_{ij}^{s} \\
&= \frac{1}{2\mu}d\sigma_{ij}^{S} - \frac{\lambda}{2\mu(3\lambda+2\mu)}\delta_{ij}d\sigma_{kk}^{S} + b\delta_{ij}d\sigma_{kk}^{W} + d\delta_{ij}d\sigma_{kk}^{G} + 2hd\xi_{ij}^{s}
\end{aligned}
\tag{1.49}
$$

where λ and μ are Lamé's elastic constants.

The evolutional equation is necessary for the internal variable in Equation 1.49. When $2h = 1.0$ is set in Equation 1.49, the internal variable can be taken as the inelastic strain that corresponds to Equation 1.40. Considering Equations 1.41 and 1.45, the first term on the left-hand side of Equation 1.36 becomes $\sigma_{ij}^{S}\dot{\varepsilon}_{ij}^{S(inel)}$. Herein, the additional evolutional equations can be specified such as the flow rule and hardening equations, e.g., Oka et al. (2006, 2008) proposed an elasto-viscoplastic constitutive model with the skeleton stress although the solid constituent compressibility is assumed to be zero.

1.2.6 Volumetric linear elastic model for three-phase materials

In this section, in order to derive a skeleton stress with the Biot coefficient, we will discuss the air–water–solid elastic porous media with the interaction between the solid skeleton and immiscible fluids. Relations between the volumetric strains and the volume fractions are shown in Section 1.1. In this section, we used s, w, a instead of S, W, G for solid, water, and air, respectively, and $d\sigma_{ij}^{(w)} = -n^{w}du_{w}\delta_{ij}$ and $d\sigma_{ij}^{(a)} = -n^{a}du_{a}\delta_{ij}$.

From Equations 1.46–1.48, the incremental elastic mean stress–volumetric strain relations can be written as follows:

$$d\sigma^{(s)} = \alpha_b d\varepsilon^{(s)} + \alpha_c d\varepsilon^{(w)} + \alpha_d d\varepsilon^{(a)} \tag{1.50a}$$

$$d\sigma^{(w)} = \alpha_c d\varepsilon^{(s)} + k_c d\varepsilon^{(w)} + h_c d\varepsilon^{(a)} \tag{1.50b}$$

$$d\sigma^{(a)} = \alpha_d d\varepsilon^{(s)} + h_c d\varepsilon^{(w)} + k_d d\varepsilon^{(a)} \tag{1.50c}$$

where $d\sigma^{(\beta)} = d\sigma_{kk}^{\beta}/3$, $d\varepsilon^{(\beta)} = d\varepsilon_{kk}^{\beta}$; parameters α_b, k_c, and k_d specify the volume change characteristics of the constituents of the solid–water–air three-phase material; and parameters α_c, α_d, and h_c indicate the volume change interaction of the solid–water–air three-phase material.

For the two-phase water-saturated case, Equation 1.50 becomes

$$\begin{aligned} d\sigma^{(s)} &= \alpha_b d\varepsilon^{(s)} + \alpha_c d\varepsilon^{(w)} \\ d\sigma^{(w)} &= \alpha_c d\varepsilon^{(s)} + k_c d\varepsilon_l^{(w)} \end{aligned} \tag{1.51}$$

The coefficients of the above model have been used by Ishihara (1965), Oka (1996), and Kimoto et al. (2017) etc., and are equivalent to those by Biot (1957). Kimoto et al. (2017) clearly studied this type of model and they derived Terzaghi-type effective stress with the Biot coefficient for the water-saturated soil. The soil–water two-phase case will be discussed in Section 1.2 in detail.

In the macroscopic constitutive theory based on the continuum mixture theory, it is not so easy to include the surface tension. As is known, the meniscus between the soil particles of unsaturated soil is related to the matric suction (pore air pressure–pore water pressure) and the radius of the meniscus by the Laplace relation: $(u_a - u_w) = s_t(\text{surface tension}) / R(\text{radius of menisucus})$. Additionally, the suction greatly influences the deformability of unsaturated soils (e.g., Alonso et al. 1990; Wheeler et al. 1995; Oka et al. 2010). Hence, the interaction parameter, h_c, between the gas (air) and the liquid (water) phases manifests the surface tension. The relationship between parameters α_c, k_c, and α_b and the compressibility parameters in the case of water-saturated porous material were discussed by Kimoto et al. (2017) in detail.

Substituting Equations 1.50b and 1.50c into Equation 1.50a, we have

$$\begin{aligned} d\sigma^{(s)} &= \left(\alpha_b - \frac{\alpha_c(\alpha_c k_d - \alpha_d h_c) + \alpha_d(\alpha_d k_c - \alpha_c h_c)}{k_c k_d - h_c^2} \right) d\varepsilon^{(s)} \\ &\quad + \frac{\alpha_c k_d - \alpha_d h_c}{k_c k_d - h_c^2} d\sigma^{(w)} + \frac{\alpha_d k_c - \alpha_c h_c}{k_c k_d - h_c^2} d\sigma^{(a)} \\ &= A_1 d\varepsilon^{(s)} + B_1 d\sigma^{(w)} + C_1 d\sigma^{(a)} \end{aligned} \tag{1.52}$$

where

$$\begin{aligned} A_1 &= \alpha_b - \frac{\alpha_c(\alpha_c k_d - \alpha_d h_c) + \alpha_d(\alpha_d k_c - \alpha_c h_c)}{k_c k_d - h_c^2} \\ B_1 &= \frac{\alpha_c k_d - \alpha_d h_c}{k_c k_d - h_c^2} \quad C_1 = \frac{\alpha_d k_c - \alpha_c h_c}{k_c k_d - h_c^2} \end{aligned} \tag{1.53}$$

Equation 1.50 can be written as

$$d\varepsilon^{(s)} = \frac{1}{A_1}\left[d\sigma^{(s)} - B_1 d\sigma^{(w)} - C_1 d\sigma^{(a)} \right] \tag{1.54a}$$

$$d\varepsilon^{(w)} = -\frac{B_1}{A_1}d\sigma^{(s)} + \left(\frac{k_d}{k_c k_d - h_c^2} + \frac{B_1^2}{A_1}\right)d\sigma^{(w)} - \left(\frac{h_c}{k_c k_d - h_c^2} - \frac{B_1 C_1}{A_1}\right)d\sigma^{(a)} \tag{1.54b}$$

$$d\varepsilon^{(a)} = -\frac{C_1}{A_1}d\sigma^{(s)} + \left(\frac{k_c}{k_c k_d - h_c^2} + \frac{C_1^2}{A_1}\right)d\sigma^{(a)} - \left(\frac{h_c}{k_c k_d - h_c^2} - \frac{B_1 C_1}{A_1}\right)d\sigma^{(w)} \tag{1.54c}$$

Let us discuss the meaning of the parameters.

Under the drained conditions, $d\sigma^{(w)} = d\sigma^{(a)} = 0$:

$$d\varepsilon^{(s)} = \frac{1}{A_1}d\sigma^{(s)} = C_b d\sigma^{(s)} \tag{1.55}$$

The compressibility of the solid skeleton, C_b, is given by

$$C_b = \frac{1}{A_1} \tag{1.56}$$

Under the unjacketed compression test in the water (the unjacketed test is one in which the porous material is submerged in a fluid chamber and the load is applied by increasing the fluid pressure in the chamber (e.g., Cheng 2016)):

$$d\sigma^{(w)} = -ndu_w,\ d\sigma^{(s)} = -(1-n)du_w,\ d\sigma^{(a)} = 0$$

Considering the unjacketed condition $de_p = de_s$, de_p and de_s are defined in Section 1.1, Equations 1.21–1.23, 1.54a, and 1.54b yields

$$d\varepsilon^{(s)} = -\frac{1}{A_1}\left((1-n) - B_1 n\right)du_w = -C_s du_w \tag{1.57}$$

$$d\varepsilon^{(w)} = -\left(-\frac{B_1}{A_1}(1-n) + \left(\frac{k_d}{k_c k_d - h_c^2} + \frac{B_1^2}{A_1}\right)n\right)du_w = -C_w du_w \tag{1.58}$$

where C_s and C_w are the compressibilities of solid and water, respectively.

Similarly, in the unjacketed compression test in air:

$$d\sigma^{(a)} = -ndu_a,\ d\sigma^{(s)} = -(1-n)du_a,\ d\sigma^{(w)} = 0$$

From Equations 1.6, 1.7, 1.9, 1.54a, and 1.54c, we obtain

$$d\varepsilon^{(s)} = -\frac{1}{A_1}\left((1-n) - C_1 n\right)du_a = -C_s du_a \tag{1.59}$$

$$d\varepsilon^{(a)} = -\left(-\frac{C_1}{A_1}(1-n) + \left(\frac{k_c}{k_c k_d - h_c^2} + \frac{C_1^2}{A_1}\right)n\right)du_a = -C_a du_a \tag{1.60}$$

where C_a is the compressibility of the pore air.

From Equations 1.57 and 1.59, it follows that

$$\frac{\alpha_c}{\alpha_d} = \frac{k_c + h_c}{k_d + h_c} \tag{1.61}$$

Finally, the relationship between parameters is obtained as

$$C_s = C_b\left((1-n) - B_1 n\right) \tag{1.62}$$

$$C_w = \left(-\frac{B_1}{A_1}(1-n) + \left(\frac{k_d}{k_c k_d - h_c^2} + \frac{B_1^2}{A_1}\right)n\right) \tag{1.63}$$

$$C_a = \left(-\frac{C_1}{A_1}(1-n) + \left(\frac{k_c}{k_c k_d - h_c^2} + \frac{C_1^2}{A_1}\right)n\right) \tag{1.64}$$

Although the number of material parameters of Equation 1.50 is six, the independent parameters are five since Equation 1.61 holds. Hence, the parameters $\alpha_b, \alpha_c, \alpha_d, k_c, k_d, h_c$ can be determined by the physically meaningful parameters C_b, C_s, C_w, C_a and the slope of the suction–saturation curve. In the following, the suction–saturation relation is discussed using Equation 1.70. Measurement of the volumetric strains under these conditions is recommended.

1.2.7 Skeleton stress

From Equation 1.54a, the partial stress for the solid skeleton is given by

$$d\sigma^{(s)} = A_1 d\varepsilon^{(s)} + B_1 d\sigma^{(w)} + C_1 d\sigma^{(a)} \tag{1.65}$$

The partial stresses for the water and air phases are given by

$$d\sigma^{(w)} = -S_r n du_w \quad \text{and} \quad d\sigma^{(a)} = -(1-S_r) n du_a \tag{1.66}$$

where S_r is the saturation; n is the porosity; du_w is the pore water pressure increment; and du_a is the pore air pressure increment.

From Equations 1.57 and 1.59, we have $B_1 = C_1 = 1/n - 1 - C_s/(nC_b)$.

Hence, the total stress increment becomes

$$d\sigma = d\sigma^{(s)} + d\sigma^{(w)} + d\sigma^{(a)} = \frac{1}{C_b}d\varepsilon^{(s)} - B_1 S_r n du_w$$

$$-C_1(1-S_r)ndu_a - S_r n du_w - (1-S_r)ndu_a$$

$$= \frac{1}{C_b}d\varepsilon^{(s)} - \left(1-\frac{C_s}{C_b}\right)S_r du_w - \left(1-\frac{C_s}{C_b}\right)(1-S_r)du_a \tag{1.67}$$

$$= \frac{1}{C_b}d\varepsilon^{(s)} - \left(1-\frac{C_s}{C_b}\right)\left(S_r du_w + (1-S_r)du_a\right)$$

$$= \frac{1}{C_b}d\varepsilon^{(s)} - \left(1-\frac{C_s}{C_b}\right)dP_f$$

where $dP_f = S_r du_w + (1-S_r)du_a$ is the average pore pressure increment.

From Equation 1.67, the increment of the skeleton stress with the Biot coefficient ($\alpha = 1 - C_s/C_b$) is obtained:

$$d\sigma' = d\sigma + \left(1-\frac{C_s}{C_b}\right)dP_f = d\sigma + \alpha dP_f = \frac{1}{C_b}d\varepsilon^{(s)} \tag{1.68}$$

where $d\sigma'$ is the increment of the skeleton stress and dP_f is the average pore pressure increment.

Since C_s/C_b is smaller than unity for soils (e.g., 1.5×10^{-3} to 2.7×10^{-4}; Skempton 1954), many researchers (e.g., Jommi 2000; Wheeler and Sivakumar 1995; Schrefter 2002; Oka et al. 2010) are successfully using the skeleton stress or the equivalent stress with the unit value of the Biot coefficient. On the other hand, the Biot coefficient is not negligible and has been used for multiphase rocks such as water-saturated hard rock (e.g., Makhnenko and Sulem 2016). Gray and Schrefler (2007) derived the Biot coefficient for three-phase material but they assumed that the solid phase stress tensor can be decomposed to the effective stress and the additional stress. It is worth noting that the same expression for the skeleton stress as Equation 1.68 can be derived for the fully linear stress–strain relation as Equation 1.50.

1.2.8 Suction–saturation relation

Let us discuss the parameter of the suction–saturation relationship. As mentioned above, the effect of surface tension is described by the suction through the Laplace relation. Since $d\sigma^{(w)} = -nS_r du_w$ and $d\sigma^{(a)} = -n(1-S_r)du_a$, from Equations 1.50b and 1.50c and Equations 1.10–1.12, the following equations can be obtained:

$$nd(u_a - u_w) = \left(\frac{\alpha_c}{S_r} - \frac{\alpha_d}{1-S_r}\right)d\varepsilon^{(s)} + \left(\frac{k_c}{S_r} - \frac{h_c}{1-S_r}\right)d\varepsilon^{(w)} + \left(\frac{h_c}{S_r} - \frac{k_d}{1-S_r}\right)d\varepsilon^{(a)}$$

$$= Qde_b + R\left(de_w + \frac{1-n}{n}(de_s - de_b)\right) + T\left(de_a + \frac{1-n}{n}(de_s - de_b)\right)$$

$$+ \left(T\frac{dS_r}{1-S_r} - R\frac{dS_r}{S_r}\right) \tag{1.69}$$

where

$$Q = \frac{\alpha_c}{S_r} - \frac{\alpha_d}{1-S_r},\ R = \frac{k_c}{S_r} - \frac{h_c}{1-S_r},\ T = \frac{h_c}{S_r} - \frac{k_d}{1-S_r}$$

Since $de_b/dt = \dot{e}_b$, $de_w/dt = \dot{e}_w$, $de_a/dt = \dot{e}_a$, and the rate of the matric suction $\dot{P}^c = (\dot{u}_a - \dot{u}_w)$, the rate form is obtained as

$$n\dot{P}^C = Q\dot{e}_b + R\left(\dot{e}_w + \frac{1-n}{n}(\dot{e}_s - \dot{e}_b)\right) + T\left(\dot{e}_a + \frac{1-n}{n}(\dot{e}_s - \dot{e}_b)\right)$$

$$+ \left(\frac{T}{1-S_r} - \frac{R}{S_r}\right)\dot{S}_r \tag{1.70}$$

$$= Q\dot{e}_b + R\left(\dot{e}_w + \frac{1-n}{n}(\dot{e}_s - \dot{e}_b)\right) + T\left(\dot{e}_a + \frac{1-n}{n}(\dot{e}_s - \dot{e}_b)\right) + Z\dot{S}_r$$

$$Z = \left(\frac{2h_c}{S_r(1-S_\gamma)} - \frac{k_d}{(1-S_r)^2} - \frac{k_c}{S_r^2}\right)$$

In Equation 1.70, Z/n is given by the slope of the suction–saturation curve. Parameter h_c could manifest effects such as surface tension on the saturation–suction relation and the interaction through the interface between water and air. Equation 1.70 is precisely equivalent to the suction–saturation relation when the volumes of pore water, pore air, and the solid skeleton constituent are constant. It is worth noting that the value for Z in Equation 1.70 can be negative if $h_c < 0$, as is quantitatively consistent with the experimental results (e.g., van Genuchten 1980; Kimoto et al. 2011).

In fact, using the suction, van Genuchten's model (1980) of the suction–saturation relation is given by

$$S_{re} = (1 + (\alpha_{VG}P^c)^{n'})^{-m},\ m = 1 - 1/n'$$

where S_{re} is the effective saturation ($S_{re} = (S_r - S_{ri})/(S_{r\max} - S_{ri})$, $S_{r\max}$ is the maximum value of the saturation, $(S_r - S_{ri})$ is the saturation reduced to the part contributing to flow, and $(1 - S_{ri})$ is the maximum value of the saturation that contributes to flow, e.g., Pinder and Gray 2008); $P^c = u_a - u_w$ is the suction; and $\alpha_{VG} > 0$, $n' > 1.0$, and $m = (1 - 1/n') > 0$ are the constitutive parameters. Differentiating the van Genuchten model with respect to the suction, we obtain

$$\frac{\partial S_{re}}{\partial P^c} = -m\alpha_{VG}n'\left(1 + (\alpha_{VG}P^c)^{n'}\right)^{-(m+1)}\left(\alpha_{VG}P^c\right)^{n'-1}$$

Consequently, the gradient of the suction–saturation is negative.

Using the relationships between the volumetric fractions shown in Section 1.1 (Equations 1.10–1.12) and the partial stresses for the liquid (water) and gas (air) phases, $\sigma_{ij}^w = -nS_r u_w \delta_{ij}$ and $\sigma_{ij}^a = -n(1 - S_r)u_a\delta_{ij}$, the stress power in the unit volume of the energy balance yields

$$\begin{aligned}
\sigma_{ij}^{\beta}\dot{\varepsilon}_{ij}^{\beta} &= \sigma_{ij}^{s}\dot{\varepsilon}_{ij}^{s} + \sigma_{ij}^{w}\dot{\varepsilon}_{ij}^{w} + \sigma_{ij}^{a}\dot{\varepsilon}_{ij}^{a} \\
&= \sigma_{m}^{s}\dot{\varepsilon}_{kk}^{s} + s_{ij}^{s}\dot{e}_{ij}^{s} \\
&\quad - nS_r u_w\left(\dot{e}_w + \frac{1-n}{n}\left(\dot{e}_s - \dot{e}_b\right) - \frac{\dot{S}_r}{S_r}\right) \\
&\quad - n\left(1 - S_\gamma\right)u_a\left(\dot{e}_a + \frac{1-n}{n}\left(\dot{e}_s - \dot{e}_b\right) + \frac{\dot{S}_r}{\left(1 - S_r\right)}\right) \\
&= \sigma_{m}^{s}\dot{\varepsilon}_{kk}^{s} + s_{ij}^{s}\dot{e}_{ij}^{s} - \dot{S}_r n\left(u_a - u_w\right) - nS_r u_w\dot{e}_w \\
&\quad - n\left(1 - S_r\right)u_a\dot{e}_a - (1-n)\left(\dot{e}_s - \dot{e}_b\right)\left(S_r u_w + \left(1 - S_r\right)u_a\right)
\end{aligned} \tag{1.71}$$

where σ_m^s and s_{ij}^s are the mean partial stress and the deviatoric partial stress tensor for the solid phase, respectively, and $\dot{e}_{ij}^s$ is the deviatoric strain rate tensor.

The third term, $-\dot{S}_r n(u_a - u_w)$, in the last line on the right-hand side of Equation 1.71 indicates that $n(u_a - u_w)$, i.e., the suction multiplied by the porosity between the air pressure and the water pressure conjugates with the saturation rate. This was first addressed by Houlsby (1997) and then confirmed by Borja (2006). The inter-particle forces through the menisci of the pore water and the particle surface cause the suction. The saturation–suction relation is one of the constitutive equations for unsaturated soils, and the suction dependency of the constitutive parameters needs to be considered in order to express the suction-dependent behavior of unsaturated soil. This shows that it is preferable to express the stress in Equation 1.68

as the "skeleton stress (increment)" instead of the effective stress (increment) because constitutive parameters are required to be dependent on the suction. As discussed above, Equation 1.50 indicates the full linear formulation for three-phase porous materials including the suction–saturation constitutive equation. This is consistent with the fact that the surface tension is related to the suction (Laplace relation), i.e., the interaction between fluids and solid.

1.2.9 Usage of skeleton and effective stresses

Equation 1.68 indicates that the skeleton stress controls the strain of the solid skeleton. Under fully undrained tests for unsaturated soils, in which the strain of the solid skeleton is equal to the strain of fluids, i.e., $\varepsilon_{kk}^{s} = \varepsilon_{kk}^{a} = \varepsilon_{kk}^{w}, \varepsilon^{(s)} = \varepsilon^{(a)} = \varepsilon^{(w)}$(e.g., Oka et al. 2010; Kimoto et al. 2011), the volumetric strain is finite and not negligible because of the high compressibility of air. Relations between the volumetric strains are given by Equations 1.10–1.12. These equations show that the strains of fluids may change without the change in saturation.

On the other hand, for the water-saturated soil, the effective stress is applicable to both the drained behavior and the undrained behavior because the volumetric strain is negligible under the undrained conditions by the low compressibility of water compared to that of the soil skeleton (Kimoto et al. 2017).

In addition, the experimental evidence shows that the compressibility of the skeleton depends on the menisci suction (e.g., Alonso et al. 1990; Kahlili 1998; Wheeler and Sivakumar 1995; Oka et al. 2010) and the stress power includes the suction–saturation-related term. Hence, the suction dependency has to be considered through the material parameters. The above consideration indicates that it is preferable to use the skeleton stress rather than the effective stress as a technical term.

Up to the present time, it has not yet been established that the effective stress and the skeleton stress with the Biot coefficient must be used in the prediction of inelastic strains for both the fully and partially saturated hard rocks. From a theoretical point of view, the choice of stress variables for the prediction of inelastic strain, i.e., classical effective stress (skeleton stress) or the effective stress with the Biot coefficient, is still an unresolved problem. Rice (1977) proposed the use of the effective stress with the Biot coefficient for the elastic strain and the Terzaghi stress for the inelastic strain for fissured rock with a point contact structure. Borja (2004) suggested that the effective stress with the Biot coefficient would be applicable for inelastic solids when the bulk compressibility in the inelastic strain range is used.

Makhnenko and Labuz (2016) used the effective stress with the Biot coefficient for the elastic behavior, and Terzaghi's effective stress for the inelastic behavior of rock materials. It is recommended that the stress variables are chosen based on the experimental evidence and micromechanical considerations as well as the theoretical considerations.

For the validation of the present model, the authors recommend confirming the constitutive model with the skeleton stress (Equation 1.68) by analyzing the fully undrained isotropic compression tests for partially saturated geomaterial with the measurement of the volumetric strain of the solid skeleton, the pore water and pore air pressures, and saturation. The fully undrained test (Kimoto et al. 2011) can be performed by interrupting the drainage of water and air from the specimen.

1.3 TWO-PHASE THEORY OF POROUS MEDIA AND THE EFFECTIVE STRESS

1.3.1 Introduction

Many researchers have discussed the effective stress concept for water-saturated geomaterials (e.g., Fillunger 1913; Terzaghi 1936; Bishop 1959, 1973; Skempton 1960; de Boer and Ehlers 1990; Oka 1980, 1988a, 1988b, 1996; Coussy 1995; Lade and de Boer 1997; Borja 2006). Moreover, the effective stress has been generalized for two- and three-phase materials. The generalization for two-phase materials is known as Biot's effective stress. Biot (1962) discussed the effective stress for two-phase materials and showed the relationship between his theory and the effective stress used in soil mechanics. Biot (1962) stated that the only instance when the pore pressure does not appear in the stress–strain relations is when the particle compressibility is zero and the pore fluid is incompressible. Many researchers have emphasized the compressibility of soil grains (e.g., Skempton 1960) and proposed generalized equations, including one for the compressibility of soil particles that is similar to Biot's equation (e.g., Lade and de Boer 1997). However, the pore fluid compressibility is not included in the equation, even though Biot (1962) stated that near incompressibility is an important factor for the above-mentioned effective stress. In the field of soil mechanics concerning the compressibility of pore fluid, Bishop's B-value is considered to be an indication of the saturation.

The relative compressibility of the pore fluids to that of the soil skeleton plays an important role in the stress variables for saturated and unsaturated soils. Oka (1996) discussed the validity of Terzaghi's effective stress for porous materials from this viewpoint, and calculated the m values, determined by $\Delta\sigma = m\Delta u_w$, where $\Delta\sigma$ is the increment in the total stress and Δu_w is the increment in the pore water pressure under undrained conditions for various soils based on Biot's theory for porous media (e.g., Biot 1962; Biot and Willis 1957; Detournary and Cheng 1993). He concluded that Terzaghi's effective stress concept is valid when both the ratio of the compressibility of a soil particle to that of the soil skeleton and the ratio of the fluid compressibility to that of the soil skeleton are very small. Coussy (1995) studied the effective stress through the pioneering work of porous continua based on thermodynamics and derived the Biot coefficient and Bishop's pore pressure coefficient. Lade and de Boer (1997) studied the effective stress and concluded that Biot's effective stress equation is not applicable to the behavior of porous materials with very soft skeletons or to soil behavior under high confining pressures.

In rock mechanics, some laboratory data have been reported for Biot's coefficient α, which is obtained from the ratio between an unjacketed and a jacketed bulk modulus. For example, Makhnenko and Labuz (2013) performed unjacketed and jacketed compression tests for sandstone and discussed the difference between the jacketed bulk modulus and the unjacketed bulk modulus of quartz.

The effective stress concept for solid–fluid two-phase media is revisited in this section following the work by Kimoto et al. (2017). In particular, the effects of the compressibility of both the pore fluid and the soil particle are discussed under three different conditions, i.e., undrained, drained, and unjacketed conditions based on a Biot-type theory for two-phase porous media. It is confirmed that Terzaghi's effective stress holds at the moment when soil grains are assumed to be incompressible and when the compressibility of the pore fluid is small enough compared to that of the soil skeleton. Then, isotropic compression tests for dry sand under undrained conditions are analyzed with the triaxial apparatus in which the changes in the pore air pressure can be measured. The ratio of the increment in the cell pressure to the increment in the pore air pressure, m, corresponds to the inverse of the B-value by Bishop and was obtained during the step loading of the cell pressure. In addition, the m values are evaluated by comparing them with theoretically obtained values based on the solid–fluid two-phase mixture theory. The experimental m values were close to the theoretical values, as they were in the range of approximately 40–185, depending on the cell pressure. Finally, it is found that the soil material with a highly compressible pore fluid, such as air, must be analyzed with the multiphase porous mixture theory. However, Terzaghi's effective stress is practically applicable when the compressibilities of both the soil particles and the pore fluid are small enough compared to that of the soil skeleton.

1.3.2 Theory of solid–fluid two-phase porous media

In this section, the linear solid–fluid two-phase theory for porous materials by Biot (1962) is reviewed and the generalized effective stress equation with Biot's parameter is derived. In this derivation, for the sake of simplicity, only the volumetric deformation is considered.

1.3.2.1 Material parameters of Biot's solid–fluid two-phase theory

As discussed in Section 1.2, the linear stress–strain relations for the volumetric component of the solid phase and the fluid phase can be written using the parameters by Ishihara (1965, 1968, 1970) and Nagumo (1965) as

$$d\sigma^{(s)} = \alpha_b d\varepsilon^{(s)} + \alpha_c d\varepsilon^{(f)} \tag{1.72}$$

$$d\sigma^{(f)} = \alpha_c d\varepsilon^{(s)} + k_c d\varepsilon^{(f)} \tag{1.73}$$

where $d\sigma^{(i)} = d\sigma_{kk}^{(i)}/3$ $(i=s, f)$ is the isotropic part of the partial stress increment for the solid and fluid phases; $d\sigma = d\sigma^{(s)} + d\sigma^{(f)}$ is the definition of the total stress increment; $d\varepsilon^{(i)}(i=s,f)$ are the volumetric parts of the strain increment tensor for the solid and fluid phases, respectively; and $d\varepsilon^{(i)} = d\varepsilon_{kk}^{(i)}$, α_b, α_c, and k_c are the material constants. The stress and strain increments are defined as compression is positive in this section. Equations 1.72 and 1.73 can be rewritten as

$$d\varepsilon^{(s)} = \frac{k_c d\sigma^{(s)} - \alpha_c d\sigma^{(f)}}{\alpha_b k_c - \alpha_c^2} \tag{1.74}$$

$$d\varepsilon^{(f)} = \frac{\alpha_b d\sigma^{(f)} - \alpha_c d\sigma^{(s)}}{\alpha_b k_c - \alpha_c^2} \tag{1.75}$$

It is worth noting that the Biot-type constitutive equation includes the parameter α_c, which features the interaction between the solid and pore fluid phases.

Relationships between several strain increments are useful to discuss the geometrical and kinematical aspects of the problem. From Equations 1.6 and 1.8 with $S_r = 0$ and the index w replaced by f, i.e., $de_w = de_f$, strains are given by

$$d\varepsilon^{(s)} = n de_p + (1-n) de_s \tag{1.76}$$

$$d\varepsilon^{(f)} = de_f + (1-n)(de_s - de_p) \tag{1.77}$$

where de_p is the ratio of the change in the volume of the pore space to the reference pore volume (Geertsma 1957; Bishop 1973; Ishihara 1965, 1967; Oka 1996); de_s is the volumetric strain increment of the soil particles; de_f is the volumetric strain increment of the pore fluid; and n is the porosity.

Following the previous studies (Ishihara 1965, 1967; Oka 1996), we can obtain the relationship between material constants α_b, α_c, and k_c and the compressibilities of the soil particles, the soil skeleton, and the pore fluid, considering the following two specific conditions. Let us consider the two problems under isotropic stress conditions.

First, in unjacketed tests, the specimen, which is not covered with a membrane, is immersed in fluid, to which pressure u is applied. In this case, the pressure acts on both the solid portion, 1 – n, and the fluid portion, n, of the surfaces of the specimen:

$$d\sigma^{(s)} = (1-n)du \tag{1.78}$$

$$d\sigma^{(f)} = ndu \tag{1.79}$$

From the physical view of the unjacketed conditions, $de_p = de_s$ and Equation 1.76, and we have the strain:

$$d\varepsilon^{(s)} = de_s = de_p \tag{1.80}$$

Equation 1.80 holds when the shape of the particles is similar to the initial shape after the pore pressure was applied under unjacketed conditions. Substituting Equations 1.78 and 1.79 into Equations 1.74 and 1.75, we have

$$d\varepsilon^{(s)} = de_s = \frac{(1-n)k_c - n\alpha_c}{\alpha_b k_c - \alpha_c^2} du = C_s du \tag{1.81}$$

$$d\varepsilon^{(f)} = de_f = \frac{(n-1)\alpha_c + n\alpha_b}{\alpha_b k_c - \alpha_c^2} du = C_f du \tag{1.82}$$

where C_s is the compressibility of the soil particle and C_f is the compressibility of the pore fluid.

Second, in drained tests, the specimen is enclosed in a thin impermeable jacket and then subjected to an external pressure. In this case, the pore pressure remains constant, i.e., $d\sigma^{(f)} = 0$. From Equations 1.74 and 1.75, we have

$$d\varepsilon^{(s)} = \frac{k_c}{\alpha_b k_c - \alpha_c^2} d\sigma^{(s)} = C_b d\sigma^{(s)} \tag{1.83}$$

$$d\varepsilon^{(f)} = \frac{-\alpha_c}{\alpha_b k_c - \alpha_c^2} d\sigma^{(s)} \tag{1.84}$$

In addition, under drained conditions, $de_f = 0$, hence from Equations 1.75–1.77, we have

$$de_p = \frac{\alpha_c + k_c}{\alpha_b k_c - \alpha_c^2} d\sigma^{(s)} = C_p d\sigma^{(s)}, \ C_p = \frac{\alpha_c + k_c}{\alpha_b k_c - \alpha_c^2} \tag{1.85}$$

Equation 1.83 defines the compressibility of the solid skeleton, C_b, and Equation 1.84 expresses the strain of the pore fluid in this drained case, which should be related to the expelled pore water volume that may include a contribution from the membrane penetration for coarse sands. Equations 1.81–1.83 yield the relations between the constitutive constants for the solid and fluid phases and the coefficients of the compressibility, which have a clear physical meaning. The pore compressibility is C_p.

To obtain the partial stresses under undrained conditions, the macroscopic volumetric strains for the two phases are considered to be equal, that is, $d\varepsilon^{(s)} = d\varepsilon^{(f)}$ (e.g., Ishihara 1968). Therefore, the following relation is obtained from Equations 1.74 and 1.75, giving a relationship between total stress and partial stress of $d\sigma = d\sigma^{(s)} + d\sigma^{(f)}$.

$$d\sigma^{(f)} = \frac{\alpha_c + k_c}{k_c + 2\alpha_c + \alpha_b} d\sigma \tag{1.86}$$

Considering Equations 1.81–1.83, the partial stresses acting on the solid and the fluid phases under undrained conditions are

$$d\sigma^{(f)} = \frac{C_b - C_s}{(C_b/n) - (1 + 1/n)C_s + C_f} d\sigma \tag{1.87}$$

$$d\sigma^{(s)} = \frac{(1/n - 1)C_b - (C_s/n) + C_f}{(C_b/n) - (1 + 1/n)C_s + C_f} d\sigma \tag{1.88}$$

The parameters in the Biot-type equation of Equations 1.72 and 1.73 can be expressed by the parameters with the physical meaning as

$$\alpha_c = \frac{-n\left[(n-1)C_p + C_s\right]}{C_b(C_f - C_s) + C_p C_s} \tag{1.89}$$

$$k_c = \frac{nC_b}{C_b(C_f - C_s) + C_p C_s} \tag{1.90}$$

$$\alpha_b = \frac{(n-1)^2 C_p + (n-1)C_s + C_f}{C_b(C_f - C_s) + C_p C_s} \tag{1.91}$$

where C_b, C_f, and C_s are the compliance of the solid skeleton, the compliance of the pore fluid, and the compliance of the solid particles, respectively, and C_p is the pore compressibility:

$$C_p = \frac{C_b - C_s}{n} \tag{1.92}$$

Inversely, we have

$$C_b = \frac{k_c}{\alpha_b k_c - \alpha_c{}^2} \tag{1.93}$$

$$C_f = \frac{(n-1)\alpha_c + n\alpha_b}{\alpha_b k_c - \alpha_c^2} \tag{1.94}$$

$$C_s = -\frac{(n-1)k_c + n\alpha_c}{\alpha_b k_c - \alpha_c{}^2} \tag{1.95}$$

1.3.2.2 Derivation of the effective stress

The incremental strain caused by the incremental loads is now considered. From Equation (1.74):

$$d\sigma^{(s)} = \frac{\alpha_b k_c - \alpha_c^2}{k_c} d\varepsilon^{(s)} + \frac{\alpha_c}{k_c} d\sigma^{(f)} \tag{1.96}$$

Substituting Equations 1.89–1.91 into Equation 1.96, the constitutive equation for the solid phase is obtained:

$$d\sigma^{(s)} = \frac{1}{C_b} d\varepsilon^{(s)} + \left\{ \frac{1}{n}\left(1 - \frac{C_s}{C_b}\right) - 1 \right\} d\sigma^{(f)} \tag{1.97}$$

In the Biot-type theory, the partial stress increment for the fluid phase is given by

$$d\sigma^{(f)} = ndu \tag{1.98}$$

where n is the porosity and u is the pore fluid pressure. Using $d\sigma^{(f)} = ndu$, Equation 1.97 becomes

$$d\sigma^{(s)} = \frac{1}{C_b} d\varepsilon^{(s)} + \left(1 - \frac{C_s}{C_b}\right) du - d\sigma^{(f)} \tag{1.99}$$

With the definition of the total stress, $d\sigma = d\sigma^{(s)} + d\sigma^{(f)}$, Equation 1.99 becomes

$$d\sigma = \frac{1}{C_b} d\varepsilon^{(s)} + \left(1 - \frac{C_s}{C_b}\right) du \tag{1.100}$$

Since the effective stress only controls the deformation of the soil, herein called the volume change, the increment in the effective stress is given by $d\sigma' = d\varepsilon^{(s)}/C_b$. Hence, a well-known effective stress is obtained from Equation 1.100 in incremental form with the so-called Biot coefficient, $\alpha = 1 - C_s/C_b$:

$$d\sigma = d\sigma' + \left(1 - \frac{C_s}{C_b}\right)du \text{ or } d\sigma' = d\sigma - \left(1 - \frac{C_s}{C_b}\right)du \tag{1.101}$$

It is worth noting that the total stress and the pore fluid pressure are independent in Equation 1.101.

The same generalized Equation 1.101 was discussed by Skempton (1960) and Bishop (1973). Attention should be paid to the fact that Skempton (1960) derived Equation 1.101 by assuming that the deformation is caused by a change in the pore pressure, du, and by a net pressure increment, $d\sigma - du$. From a physical perspective, the drained test with back pressure is the only possible test to perform the above two processes in the same test. In other words, it can be said that Skempton (1960) generalized Terzaghi's equation by considering the compression of the soil particles.

For saturated soils, three isotropic loading conditions are representative, i.e., undrained, drained, and unjacketed. In the following sections, Equations 1.100 and 1.101 will be examined for the three conditions.

1.3.2.3 Undrained conditions

Under the undrained conditions, the pore pressure and the total stress are not independent. Since $d\varepsilon^{(s)} = d\varepsilon^{(f)}$ for undrained conditions (e.g., Ishihara 1967; Oka 1996), Equations 1.74 and 1.75 give the following relation between the partial stress increments and $d\sigma^{(f)}$ as

$$d\sigma^{(s)} = \frac{\alpha_c + \alpha_b}{k_c + \alpha_c} d\sigma^{(f)} = \left(\frac{(1/n - 1)(C_b - C_s) + C_f - C_s}{C_b - C_s}\right) d\sigma^{(f)} \tag{1.102}$$

Hence, the total stress increment is given by

$$d\sigma = d\sigma^{(s)} + d\sigma^{(f)} = \left(\frac{(C_b - C_s) + n(C_f - C_s)}{C_b - C_s}\right) du = m du \tag{1.103}$$

where m is the reciprocal of Bishop's B-value:

$$m = \frac{(C_b - C_s) + n(C_f - C_s)}{C_b - C_s} \tag{1.104}$$

Substituting Equation 1.103 into Equation 1.101 yields

$$d\sigma' = \left\{ m - \left(1 - \frac{C_s}{C_b} \right) \right\} du \tag{1.105}$$

Then, the strain increment in the soil skeleton is obtained since the effective stress induces the strain of the soil skeleton:

$$d\varepsilon^{(s)} = C_b d\sigma' = C_b \left\{ m - \left(1 - \frac{C_s}{C_b} \right) \right\} du = \frac{C_b}{m} \left\{ m - \left(1 - \frac{C_s}{C_b} \right) \right\} d\sigma \tag{1.106}$$

It follows from the above equations that the increment in the effective stress becomes zero when $m=1$ and $C_s/C_b=0$ under undrained conditions and that the strain of the skeleton is zero.

On the other hand, if $m \neq 1$ and/or $C_s/C_b \neq 0$, Terzaghi's effective stress equation $d\sigma' = d\sigma - du$ does not hold exactly because the volumetric strain of the soil skeleton is not zero in Equation 1.106 under undrained conditions. It is worth noting that the mean effective stress increment is zero under undrained isotropic compression conditions and the associated strain is zero from the perspective of Terzaghi's effective stress concept.

Substituting Equation 1.101 into the first equation of Equation 1.106, we have

$$d\varepsilon^{(s)} = C_b d\sigma' = C_b \left(d\sigma - du \right) + C_s du \tag{1.107}$$

This alternative expression is stated by Skempton (1960), Bishop (1973), Lade and de Boer (1997), and Molenkamp et al. (2014). It is noted that Equation 1.96 is derived based on Biot's simple two-phase theory for porous media, i.e., Equations 1.72 and 1.73 in the present formulation, while Skempton (1960) derived Equation 1.96 based on the physical consideration that the volumetric strain is induced by $d\sigma - du$ and du independently.

Next, the soil under unjacketed conditions, whereby the soil is immersed in water without the cover of a membrane, will be discussed.

1.3.2.4 Unjacketed conditions

Under unjacketed conditions, $d\sigma = du$ holds because the soil is immersed in water with no membrane. From Equations 1.101 and 1.106, we have

$$d\varepsilon^{(s)} = C_b \left\{ d\sigma - \left(1 - \frac{C_s}{C_b} \right) du \right\} = C_b \left\{ 1 - \left(1 - \frac{C_s}{C_b} \right) \right\} du = C_s du = de_s \tag{1.108}$$

The condition $d\varepsilon^{(s)} = de_s$ (the strain increment of the soil particle) holds when the entire surface of each particle is only subjected to the pore water pressure increment and the soil particles are similarly compressed due to the change in the pore pressure. However, this condition generally does not hold, as discussed by Lade and de Boer (1997). In addition, Equations 1.76 and 1.108 imply $de_p = de_s$. Herein, the simplified condition that Ishihara (1965, 1968) and Nagumo (1965) used is given. In the unjacketed tests, the zero-volumetric strain increment holds when $C_s = 0$ and, in that case, Terzaghi's effective stress equation holds.

1.3.2.5 Drained conditions

Finally, the soil under drained conditions is considered. For drained conditions, $d\sigma = d\sigma'$ and $du = 0$. Under these conditions, the effective stress concept trivially holds because of the zero-pore pressure.

From the above discussions, the conditions that $m = 1$ and $C_s = 0$ are sufficient for Terzaghi's effective stress equation for water-saturated soil because the effective stress must hold for drained, undrained, and unjacketed conditions. This indicates that the effective stress for water-saturated soils by Terzaghi correctly holds as

$$\begin{aligned} d\sigma' &= d\sigma - du \text{ when } m = 1 \text{ and } C_s = 0, \text{ or} \\ d\sigma' &= d\sigma - du \text{ when } C_f / C_b = 0 \text{ and } C_s = 0 \end{aligned} \tag{1.109}$$

1.3.3 Analysis of the isotropic compression tests on dry sand

To study the effect of the pore fluid compressibility on the formulation of the two-phase mixtures, Kimoto et al. (2017) conducted isotropic compression tests for dry sand by measuring the pore air pressure. Herein, we will reproduce the test results using the two-phase porous theory mentioned above.

1.3.3.1 Isotropic compression tests on dry sandy soil

The isotropic compression tests were performed on a dry (air saturated), sandy material under jacketed conditions, i.e., undrained for the air in the pores. The tests were conducted using the triaxial apparatus, which was modified to precisely measure the pore air pressure by Oka et al. (2010). The ratio m of the increment in pore air pressure to that of the cell pressure corresponds to the inverse of the Skempton coefficient (B-value) and was obtained during the step loading of the cell pressure. In addition, the m values were compared with the theoretical values obtained from Equation 1.104 based on Biot's theory for porous material.

After preparation, the specimens were set in the triaxial cell and an isotropic confining pressure of 20 kPa was first applied. Then, the cell pressure was isotropically increased to 200 kPa in steps of 50 kPa under fully

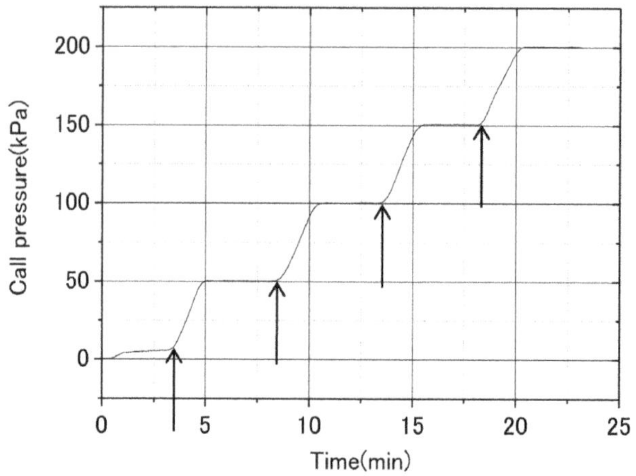

Figure 1.3 Time history of the cell pressure. (After Kimoto, S., Oka, F. and Morimoto, Y. 2017. The effective stress concept and the evaluation of changes in pore air pressure under jacketed isotropic compression tests for dry sand based on 2-phase porous theory. *Int. J. Numer. Anal. Meth. Geomech.* 41(18): 1894–1907.)

undrained conditions for both air and water. That is, both the air-operated valve on the top and the valve on the bottom were closed. The time history of the cell pressure is shown in Figure 1.3. The arrows in the figure indicate the points at which the cell pressure increased. During the tests, the pore air pressure and the volume change were continuously measured. The time history of the pore air pressure is shown in Figure 1.4. The pore air pressure showed stepwise increases along with the increasing cell pressure. The ratio of the increment in the pore air pressure to that of the cell pressure is calculated as

$$m = \frac{d\sigma}{du_a} \tag{1.110}$$

which corresponds to the inverse of the B-value for the water-saturated soil. In Equation 1.110, du is replaced by du_a from Equation 1.103, which is the increment in the pore air pressure and $d\sigma$ is the increment in the cell pressure at each step. The measured values of m are given in Table 1.1. The increment in the pore air pressure, du_a, decreased with the increasing value of the cell pressure. The time history of the volumetric strain is shown in Figure 1.5.

1.3.3.2 Evaluation of pore air pressure

Since the pore fluid for dry sand is highly compressible air, the pore air pressure of the dry sand was analyzed using the theory of two-phase porous media instead of Terzaghi's effective stress, i.e., the compressive

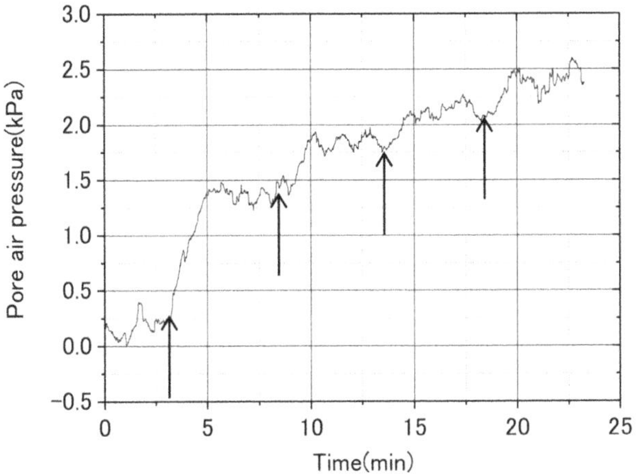

Figure 1.4 Time history of the pore air pressure during isotropic undrained compression. (After Kimoto, S., Oka, F. and Morimoto, Y. 2017. The effective stress concept and the evaluation of changes in pore air pressure under jacketed isotropic compression tests for dry sand based on 2-phase porous theory. *Int. J. Numer. Anal. Meth. Geomech.* 41(18): 1894–1907.)

Table 1.1 Measured *m* values

Cell pressure σ (kPa)	*$\Delta\sigma$ (kPa)*	*Δu_a (kPa)*	*m (measured) (=$\Delta\sigma/\Delta u_a$)*
6.57–50	43.43	1.081	40.18
50–100	50.0	0.460	108.7
100–150	50.0	0.291	171.8
150–200	50.0	0.272	183.8

Source: Data from Kimoto, S., Oka, F. and Morimoto, Y. 2017. The effective stress concept and the evaluation of changes in pore air pressure under jacketed isotropic compression tests for dry sand based on 2-phase porous theory. *Int. J. Numer. Anal. Meth. Geomech.* 41(18): 1894–1907.

deformation of the pore air cannot be ignored. The relationship between the applied total stress and the pore pressure for the solid–fluid two-phase material under undrained conditions is discussed in Section 1.3.2.

The values of m, as calculated from Equation 1.104 for the tested soil, are given in Table 1.2 with the measured values. A typical value for sand particles is used for the compressibility of the soil particle, C_s. The compressibility of the pore fluid, or the compressibility of air, C_f, for the dry material, is obtained by the ideal gas law as

$$C_f = 1/(p_a + u_a) \tag{1.111}$$

where $p_a = 101.325$ (kPa) is the atmospheric pressure. The bulk compressibility of the soil skeleton, C_b, is obtained from the isotropic drained

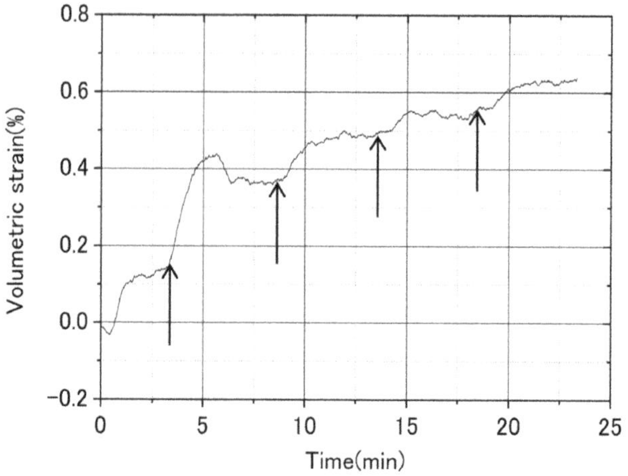

Figure 1.5 Time history of volumetric strain of the soil skeleton during isotropic undrained compression. (After Kimoto, S., Oka, F. and Morimoto, Y. 2017. The effective stress concept and the evaluation of changes in pore air pressure under jacketed isotropic compression tests for dry sand based on 2-phase porous theory. *Int. J. Numer. Anal. Meth. Geomech.* 41(18): 1894–1907.)

Table 1.2 The *m* values calculated from Equation 1.104

Cell pressure σ *(kPa)*	*n*	C_b *(1/kPa)*	C_s *(1/kPa)*	C_f *(1/kPa)*	*m* *(Equation 1.104)*	*m* *(measured)*
0–50	0.3874	9.63×10^{-5}	2.76×10^{-8}	9.74×10^{-3}	40.17	40.18
50–100	0.3868	3.29×10^{-5}	2.76×10^{-8}	9.69×10^{-3}	114.9	108.7
100–150	0.3865	2.56×10^{-5}	2.76×10^{-8}	9.67×10^{-3}	147.3	171.8
150–200	0.3859	2.23×10^{-5}	2.76×10^{-8}	9.64×10^{-3}	168.0	183.8

Source: Data from Kimoto, S., Oka, F. and Morimoto, Y. 2017. The effective stress concept and the evaluation of changes in pore air pressure under jacketed isotropic compression tests for dry sand based on 2-phase porous theory. *Int. J. Numer. Anal. Meth. Geomech.* 41(18): 1894–1907.

compression tests performed separately. The $e-\log p'$ relation from the isotropic drained compression test is shown in Figure 1.6. The theoretical values for m, shown in Table 1.2, agree well with the measured values for the tested material. The m values are larger for the larger cell pressure. This is mainly because the bulk compressibility of the soil skeleton, C_b, is smaller for the larger cell pressure. In addition, $m-(1-C_s/C_b)>>1.0$, i.e., Terzaghi's effective stress concept does not hold.

1.3.3.3 Evaluation of the bulk compressibility

The bulk compressibility under undrained conditions, C_u, is given by Ishihara (1965) as

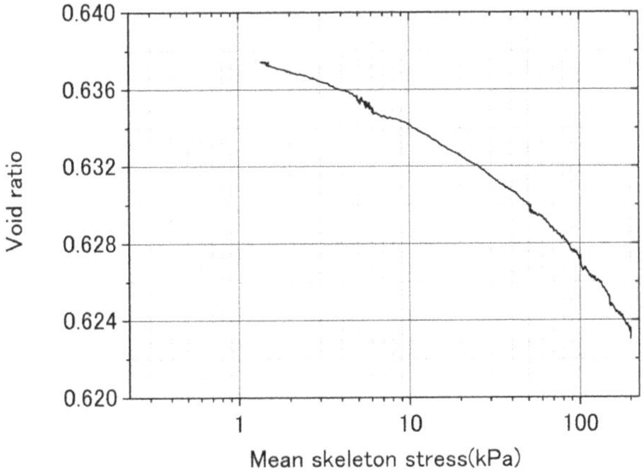

Figure 1.6 e–log p' relation from the isotropic drained compression tests. (After Kimoto, S., Oka, F. and Morimoto, Y. 2017. The effective stress concept and the evaluation of changes in pore air pressure under jacketed isotropic compression tests for dry sand based on 2-phase porous theory. *Int. J. Numer. Anal. Meth. Geomech.* 41(18): 1894–1907.)

Table 1.3 Bulk compressibility under undrained conditions

Cell pressure σ *(kPa)*	*n*	$\Delta\varepsilon_v$	C_u *(Equation 1.112) (1/kPa)*	C_u *(measured) (1/kPa)*
0–50	0.3874	0.287×10^{-2}	9.39×10^{-5}	6.61×10^{-5}
50–100	0.3868	0.103×10^{-2}	3.27×10^{-5}	2.06×10^{-5}
100–150	0.3865	0.054×10^{-2}	2.54×10^{-5}	1.08×10^{-5}
150–200	0.3859	0.067×10^{-2}	2.22×10^{-5}	1.34×10^{-5}

Source: Data from Kimoto, S., Oka, F. and Morimoto, Y. 2017. The effective stress concept and the evaluation of changes in pore air pressure under jacketed isotropic compression tests for dry sand based on 2-phase porous theory. *Int. J. Numer. Anal. Meth. Geomech.* 41(18): 1894–1907.

$$C_u = \frac{C_b(C_f - C_s) + C_p C_s}{C_p - C_s + C_f} \tag{1.112}$$

where C_p is the pore compressibility given by Equation 1.92.

The measured values for the bulk compressibility $C_u(= d\varepsilon_v/d\sigma)$ in which the volumetric strain increment $d\varepsilon_v$ of the solid skeleton is obtained from Figure 1.5, are compared with the values calculated by Equation 1.112, as shown in Table 1.3. The magnitude of the measured values is within 60%–70% of the theoretical values.

1.3.3.4 Discussion of experimental results

In the case of the water-saturated sandy soil, the compressibility of the pore fluid, C_f, is 4.8×10^{-7} (1/kPa), the compressibility of the soil particle, C_s, is 2.0×10^{-8} (1/kPa), and the bulk compressibility of the soil skeleton, C_b, is between 2.0×10^{-5} and 60×10^{-5} (1/kPa). In this case, it is assumed that $C_f/C_b \cong 0$ and $C_s/C_b \cong 0$ in Equations 1.87, 1.88, and 1.104. Consequently, $m-(1-C_s/C_b)\cong 0$ and the effective stress condition does hold. From this assumption, Equations 1.87, 1.88, and 1.104 yield

$$d\sigma^{(f)} = nd\sigma \tag{1.113}$$

$$d\sigma^{(s)} = (1-n)d\sigma \tag{1.114}$$

$$m \cong 1 \tag{1.115}$$

The m value calculated by Equation 1.104 becomes almost 1.0 for the water-saturated sandy soil. This means that the applied total stress is balanced with the pore water pressure.

For the partially saturated soil, the compressibility of the fluid, C_f, is approximately 1.0×10^{-2} (1/kPa), which is significantly larger than C_b from 2.0×10^{-5} to 9.0×10^{-5} (1/kPa) for the tested material. In this case, it can be assumed that $C_s/C_b \cong 0$, while the term C_f/C_b cannot be disregarded. From this assumption, Equation 1.104 yields

$$m \cong 1 + n\frac{C_f}{C_b} \tag{1.116}$$

The experiment of isotropic compression for dry sand provided values of $m \cong 40$ under cell pressure levels of 0–50 kPa and $m \cong 160 \sim 180$ under cell pressure levels of 150–200 kPa. The large values for m in dry sandy soil come from the higher compressibility of the pore fluid compared to the bulk compressibility of the soil skeleton.

For comparison, C_f=4.8×10^{-7} (1/kPa) and $C_b \cong 5.9\times10^{-8}$ (1/kPa) for water-saturated sandstone, and the calculated m value is 4.3 when $n=0.4$, i.e., $m-(1-C_s/C_b)>0$.

REFERENCES

Alonso, E.E., Gens, A. and Josa, A. 1990. A constitutive model for partially saturated soils. *Geotechnique* 40(3): 405–430.

Bedford, A. and Drumheller, D.S. 1983. Theories of immiscible and structured mixtures. *Int. J. Eng. Sci.* 21(8): 863–960.

Biot, M.A. 1962. Mechanics of deformation and acoustic propagation in porous media. *J. Appl. Phys.* 33(4): 1482–1498.

Biot, M.A. and Willis, D.G. 1957. The elastic coefficients of the theory of consolidation. *J. Appl. Mech.* 24(4): 594–601.

Bishop, A.W. 1959. The principle of effective stress. *Tekn. Ukeblad* 39: 859–863.

Bishop, A.W. 1973. The influence of an undrained change in stress on the pore pressure in porous media of low compressibility. *Geotechnique* 23(3): 435–442.
Bolzon, G., Schrefler, B.A. and Zienkiewicz, O.C. 1996. Elastoplastic soil constitutive laws generalized to partially saturated states. *Geotechnique* 46(2): 279–289.
Borja, R.I. 2004. Cam-clay plasticity part V: A mathematical framework for three-phase deformation and strain localization analysis of partially saturated porous media. *Com. Meth. Appl. Mech. Engng.* 193(48–51): 5301–5338.
Borja, R.I. 2006. On the mechanical energy and effective stress in saturated and unsaturated porous continua. *Int. J. Solids Struct.* 43(6): 1764–1786.
Bowen, R.M. 1976. Theory of mixtures. In *Continuum Physics*, Vol. III, ed. A.C. Eringen, 1–127. Academic Press, New York.
Bowen, R.M. 1982. Compressible porous media models by use of the theory of mixtures. *Int. J. Eng. Sci.* 20(6): 697–735.
Cheng, A.H.-D. 2016. *Poroelasticity.* Springer, New York.
Coleman, B.D. and Gurtin, M.E. 1967. Thermodynamics with internal state variables. *J. Chem. Phys.* 47(2): 597–613.
Coussy, O. 1995. *Mechanics of Porous Continua.* Wiley, Hoboken, NJ (originally published in French as *Mécanique de Milieux Poreux* 1991 Editions Technip).
de Boer, R. 1996. Highlights in the development of the theory of porous media: Toward a consistent macroscopic theory. *Appl. Mech. Rev.* 49(4): 201–262.
de Boer, R. 1997. The volume fraction concept in the porous media theory. *ZAMM* 77(8): 563–577.
de Boer, R. 1998. Theory of porous media-past and present. *ZAMM* 78(7): 441–466.
de Boer, R. and Ehlers, W. 1990. The development of the concept of effective stresses. *Acta Mech.* 83(1–2): 77–92.
Detournary, E. and Cheng, A.H.D. 1993. Fundamentals of poroelasticity. In *Comprehensive Rock Engineering*, 2nd ed., eds. C. Fairhurst, 113–171. Pergamon Press, Oxford.
Duriez, J., Eghbalian, M., Wan, R. and Darve, F. 2017. The micromechanical nature of stresses in triphasic granular media with interfaces. *J. Mech. Phys. Solids* 99: 495–511.
Ehlers, W., Graf, T. and Ammann, M. 2004. Deformation and localization analysis of partially saturated soil. *Com. Meth. Appl. Mech. Eng.* 193(27–29): 2885–2910.
Fillunger, P. 1913. Der Auftrieb in Talsperren, *Österr. Wochenschr. off. Baudienst* 19: 532–556, 567–570.
Fredlund, D.G., Morgenstern, N.R. and Widger, R.A. 1978. The shear strength of unsaturated soils. *Can. Geotech. J.* 15(3): 313–321.
Geertsma, J. 1957. The effect of fluid pressure decline on volumetric changes of porous rocks. *Petrol. Trans. AIME* 210(1): 331–340.
Gray, W.G. and Schrefler, B.A. 2007. Analysis of the solid phase stress tensor in multiphase porous media. *Int. J. Numer. Anal. Meth. Geomech.* 31(4): 541–581.
Green, A.E. and Steel, T.R. 1966. Constitutive equations for interacting continua. *Int. J. Eng. Sci.* 4(4): 483–500.
Gurtin, M.E. and Murdoch, A.I. 1975. A continuum theory of elastic material surfaces. *Arch. Rat Mech. Anal.* 57: 292–323.

Hassanizadeh, S.M. and Gray, W.G. 1990. Mechanics and thermodynamics of multiphase flow in porous media including interphase boundaries. *Adv. Water Resour.* 13(4): 169–186.

Houlsby, G.T. 1997. The work input to an unsaturated granular material. *Geotechnique* 47(1): 193–196.

Ishihara, K. 1965. Theory of consolidation of a porous material with heat effect based on the irreversible thermodynamics. *Trans. JSCE* 113(113): 28–42.

Ishihara, K. 1968. Propagation of compressional waves in a saturated soil. In *Proceedings of the International Symposium Wave Propagation and Dynamic Properties of Earth Materials*, ed. G.E. Triandafilidis, 195–206. University of New Mexico Press, Albuquerque, NM.

Ishihara, K. 1970. Approximate forms of wave equations for water-saturated porous materials and related dynamic modulus. *Soils Found.* 10(4): 10–38.

Jommi, C. 2000. Remarks on the constitutive modelling of unsaturated soils. In *Experimental Evidence and Theoretical Approaches in Unsaturated Soils*, eds. A. Tarantio and C. Mancuso, 139–153. Balkema, Rotterdam.

Khalili, N. and Khabbaz, M.H. 1998. A unique relationship for the determination of the shear strength of unsaturated soils. *Geotechnique* 48(5): 681–687.

Kimoto, S., Oka, F., Fukutani, J., Yabuki, T. and Nakashima, K. 2011. Monotonic and cyclic behavior of unsaturated sandy soil under drained and fully undrained conditions. *Soils Found.* 51(4): 663–681.

Kimoto, S., Oka, F. and Morimoto, Y. 2017. The effective stress concept and the evaluation of changes in pore air pressure under jacketed isotropic compression tests for dry sand based on 2-phase porous theory. *Int. J. Numer. Anal. Meth. Geomech.* 41(18): 1894–1907.

Kohgo, Y., Nakano, M. and Miyazaki, T. 1993. Theoretical aspects of constitutive modelling for unsaturated soils. *Soil. Mech. Found. Eng.* 33(4): 49–63 (in Japanese).

Lade, P. and de Boer, R. 1997. The concept of effective stress for soil, concrete and rock. *Geotechnique* 47(1): 61–78.

Laloui, L., Klubertanz, G. and Vulliet, L. 2003. Solid-liquid-air coupling in multiphase porous media. *Int. J. Numer. Anal. Meth. Geomech.* 27(3): 183–206.

Makhnenko, R.M. and Labuz, J.F. 2013. Unjacketed bulk compressibility of sandstone in laboratory experiments. In *Proceedings of the Fifth Biot Conference of Poromechanics*, ed. C. Hellmich, B. Picher and D. Adam, 481–488. ASCE,New York.

Makhnenko, R.Y. and Labuz, J.F. 2016. Elastic and inelastic deformation of fluid-saturated rock. *Philos. Trans. R. Soc. A* A374: 20150422.

Mills, N. 1967. On a theory of multi-component mixtures. *J. Mech. Appl. Math.* 20: 449–508.

Molenkamp, F., de Jager, R.R. and Mathijssen, F.A.J.M. 2014. Stress measures affecting deformation of granular materials. *Vadose Zone J.* 13(5): 1–17.

Morland, L.W. 1972. A simple constitutive theory for a fluid-saturated porous solid. *J. Geophys. Res.* 77(5): 890–900.

Nagumo, S. 1965. Effect of pore for deformation and failure of porous media. *Bull. Earthquake Res. Inst. Univ. Tokyo* 43:317–338.

Nuth, M. and Laloui, L. 2008. Effective stress concept in unsaturated soils: Clarification and validation of a unified framework. *Int. J. Numer. Anal. Meth. Geomech.* 32(7): 771–801.

Oka, F. 1980. On the definition of effective stress by the theory of 2-phase mixture. *Proc. Jpn. Soc. Civil Eng.* 299(299): 59–64 (in Japanese).

Oka, F. 1988a. Principle of effective stress in soil mechanics *Tsuchi to kiso. The Japanese Geotechnical Society* 36(6): 11–17 (in Japanese).

Oka, F. 1988b. The validity of the effective stress concept in soil mechanics. In *Micromechanics of Granular Materials. Studies in Applied Mechanics*, 20 (Proc. U.S./Japan Seminar on the Micromechanics of Granular Materials., Sendai, Japan, October 1987) eds. M. Satake and J.T. Jenkins, 207–214. Elsevier Sci. Pub., Amsterdam.

Oka, F. 1996. Validity and limits of the effective stress concept in geomechanics. *Mech. Cohesive-Frictional Mater.* 1(2): 219–234.

Oka, F. and Kimoto, S. 2012. *Computational Modeling of Multiphase Geomaterials.* CRC Press/Taylor & Francis, Boca Raton, FL.

Oka, F. and Kimoto, S. 2019. On the formulation of the multiphase porous geomaterials. In *Desiderata Geotechnica*, Springer Series in Geomechanics and Geoengineering, ed. Wei Wu, 143–146. Springer, New York.

Oka, F. and Kimoto, S. 2020. Formulation of a multiphase constitutive model for three-phase porous materials and the skeleton stress. *Soils Found.* 60: 1000–1010.

Oka, F., Kodaka, T., Kimoto, S., Kim, Y.-S. and Yamasaki, N. 2006. An elasto-viscoplastic model and multiphase coupled FE analysis for unsaturated soil. In *Proceedings of the Unsaturated Soils Conference*, April 2–6 2006, ASCE, Carefree Arizona, Geotechnical Special Publication, 147, ASCE 2: 2039–2050. ASCE, Reston, VI.

Oka, F., Feng, H., Kimoto, S., Kodaka, T. and Suzuki, H. 2008. A numerical simulation of triaxial tests of unsaturated soil at constant water and air content by using an elasto-viscoplastic model. In *Proceedings of the 1st European Conference on Unsaturated Soils*, E-UNSAT 2008, Durham, UK, July 2–4 2008, eds. D.G. Toll, C.E. Augarde, D. Gallipoli and S.J. Wheeler, 735–741. CRC Press, Boca Raton, FL.

Oka, F., Kodaka, T., Suzuki, H., Kim, Y.-S., Nishimatsu, N. and Kimoto, S. 2010. Experimental study on the behavior of unsaturated compacted silt under triaxial compression. *Soils Found.* 50(1): 27–44.

Passman, S.L., Nunziato, J.W. and Walsh, E.K. 1984. A theory of multiphase mixtures. In *Rational Thermodynamics*, ed. C. Truesdell, 286–325. Springer, New York.

Pinder, G.F. and Gray, W.G. 2008 *Essentials of Multi-Phase Flow and Transport in Porous Media.* Wiley, Hoboken, NJ.

Rice, J.R. 1977. Pore pressure effect in inelastic constitutive formulation for fissured rock masses. In *Proceedings of the 2nd ASCE Engineering Mechanics Speciality Conference*, Raleigh, NC, ASCE, New York, 295–297.

Schrefler, B.A. 2002. Mechanics and thermodynamics of saturated/unsaturated porous materials and quantitative solutions. *Appl. Mech. Rev.* 55(4): 351–388.

Skempton, A.W. 1954. The pore pressure coefficients A and B. *Geotechnique* 4(4): 143–147.

Skempton, A.W. 1960. Effective stress in soils, concrete and rocks. In *Proceedings of the Conference Pore Pressure and Suction in Soils*, 4–16. Butterworths, London.

Terzaghi, K. 1936. The shearing resistance of saturated soils and the angle between the planes of shear. In *Proceedings of the First International Conference on Soil Mechanics and Foundation Engineering*, 1, 54–56. Harvard University, Cambridge.

Truesdell, C. and Toupin, R. 1960. The classical field teoried. In *Encyclopedia in physics*, ed. S. Flugge, Vol.III/1, Springer-Verlag, Berlin: 223–793.

Valdoulakis, I. and Sulem, J. 1995 *Bifurcation Analysis in Geomechanics*. Blackie, Glasgow.

van Genuchten, M.Th. 1980. A closed-form equation for predicting the hydraulic conductivity of unsaturated soils. *Soil Sci. Soc. Am. J.* 44(5): 892–898.

Wheeler, S.J. and Sivakumar, V. 1995. An elasto-plastic critical state framework for unsaturated soil. *Geotechnique* 45(1): 35–53.

Zhao, C., Liu, Z., Shi, P., Li, J., Cai, G. and Wei, C. 2016. Average soil skeleton stress for unsaturated soils and discussion on effective stress. *Int. J. Geomech.* 16(6): D4015006.

Chapter 2

Constitutive models of geomaterials

2.1 CYCLIC ELASTO-PLASTIC CONSTITUTIVE MODEL

2.1.1 Introduction

A large number of constitutive models for geomaterials, such as soils and rocks, have been proposed over the last three decades. Cyclic models play a very important role in solving the dynamic behavior of geomaterials, e.g. behavior during earthquakes. Success has been found in constructing cyclic constitutive models for geomaterials with several theories such as elasto-plasticity, elasto-viscoplasticity, hypo-plasticity, and the endochronic theory. Within the framework of plasticity, models have been developed by Gaboussi and Momen (1979), Pastro et al. (1990), Prevost and Kean (1978), Dafalias et al. (1981), Adachi and Oka (1984a,b), Ishihara and Kabilamany (1990), Hashiguchi and Chen (1998), Oka et al. (1999), Pestana and Whittle (1999), Asaoka et al. (2002), and Chiaro et al. (2013). Oka et al. (1999) proposed a model based on the generalized flow rule using the non-linear kinematic hardening rule which was originally advocated by Armstrong and Frederick (1966). Then, Oka and Kimoto (2016, 2018) generalized this model incorporating structural degradation and the single yield surface. Niemunis and Herle (1997) proposed a hypo-plastic model and Valanis and Read (1982) developed a model without yield surface. As shown above, although many models have been proposed, Wichtmann and Triantafyllidis (2015) showed that it is still difficult to reproduce all aspects related to the cyclic behavior of soils quantitatively and qualitatively.

This section presents a cyclic constitutive model for sandy soils by Oka and Kimoto (2018), which is based on the non-linear kinematic hardening rule advocated by Armstrong and Frederick (1966) with strain-induced degradation. The model has been developed with a single yield function which includes changes in the stress ratio and in the mean effective stress, while the model by Oka et al. (1999) used two different yield surfaces for the changes in the stress ratio and the mean effective stress. In addition, the

DOI: 10.1201/9781003200031-2

model includes the degradation of the overconsolidation boundary surface with respect to the plastic strain. The introduction of degradation can bring about a strain-softening effect on deformation. Then, in order to control the non-associativity that comes from the friction of soils, a non-associativity parameter has been introduced (e.g., Bigoni 2012).

From the numerical simulation under undrained simple shear conditions, the cyclic behavior of sandy soils can be successfully described by the proposed model with the newly introduced material parameters for dilatancy, degradation, and non-associativity. The instability problem associated with the non-associative flow rule was discussed by Lade (1992) for the behavior under monotonic loading conditions. However, the effect of non-associativity on the cyclic behavior of soil has not yet been sufficiently studied. The simulated results show that non-associativity plays an important role in controlling the decrease in the mean effective stress under undrained cyclic conditions, which is significant for the simulation of liquefaction. The stronger the non-associativity, the larger the decrease in the mean effective stress under undrained cyclic conditions.

2.1.2 Cyclic elasto-plastic constitutive model

Oka and Kimoto (2018) proposed an elasto-plastic cyclic model for sandy soils based on the non-linear kinematic hardening rule by combining two yield surfaces, namely, one for the change in the stress ratio and the other for the change in the mean effective stress. In addition, the degradation of the material has been taken into account by the decrease in the plastic modulus, the shrinkage of the overconsolidation boundary surface, and the non-associativity.

2.1.2.1 Total strain rate tensor

The total strain rate tensor, $\dot{\varepsilon}_{ij}$, can be decomposed into the elastic and plastic strain rate tensors, $\dot{\varepsilon}_{ij}^e$ and $\dot{\varepsilon}_{ij}^p$, as follows:

$$\dot{\varepsilon}_{ij} = \dot{\varepsilon}_{ij}^e + \dot{\varepsilon}_{ij}^p \tag{2.1}$$

The elastic strain rate tensor is assumed as

$$\dot{\varepsilon}_{ij}^e = \frac{1}{2G^E}\dot{s}_{ij} + \frac{\kappa}{3(1+e)}\frac{\dot{\sigma}_m'}{\sigma_m'}\delta_{ij} \tag{2.2}$$

where G^E is the elastic shear modulus; s_{ij} is the deviatoric stress tensor ($s_{ij} = \sigma_{ij}' - \sigma_m'\delta_{ij}$); σ_m' is the mean effective stress – the superimposed dot denotes the time differentiation; κ is the swelling index; and e is the void ratio.

Following the experimental and empirical studies by Fahey (1992) and Hardin and Drnevich (1972), it is assumed here that the strain dependency of the elastic shear modulus is given by

$$G^E = \frac{G_0^E}{1+\alpha_e\left(\gamma_{ap}^{P*}/\gamma_r^{E*}\right)^{r_{e1}}}\left(\frac{\sigma'_m}{\sigma'_{m0}}\right)^{r_{e2}}$$

$$\gamma_{ap}^{p*} = \int (de_{ij}^p de_{ij}^p)^{1/2} \tag{2.3}$$

where r_{e1}, r_{e2}, and α_e are material constants; G_0^E is the initial value of G^E; de_{ij}^p is the deviatoric plastic strain increment; γ_{ap}^{P*} is the accumulated plastic deviatoric strain after the stress ratio has reached M_m^*, which will explained in Section 2.1.2.5; and γ_r^{E*} is a referential strain.

The elastic volumetric modulus, K^E, is given in a manner similar to that in Equation 2.3, while the initial elastic volumetric modulus is given by the swelling index κ.

2.1.2.2 Overconsolidation boundary surface

An overconsolidation boundary surface is used to define the boundary between the normally consolidated (NC) region ($f_b \geq 0$) and the overconsolidated (OC) region ($f_b < 0$) in the stress space, which is given by

$$f_b = \bar{\eta}_\xi^* + M_m^* \ln\left(\frac{\sigma'_m}{\sigma'_{mb}}\right) = 0,\ \bar{\eta}_\xi^* = \left\{\left(\eta_{ij}^* - \xi_{ij}\right)\left(\eta_{ij}^* - \xi_{ij}\right)\right\}^{1/2},\ \eta_{ij}^* = s_{ij}/\sigma'_m \tag{2.4}$$

where σ'_m is the mean effective stress; s_{ij} is the deviatoric stress tensor; M_m^* is the stress ratio $\eta^* = (\eta_{ij}\eta_{ij})^{1/2}$ when the maximum volumetric strain (maximum compression) takes place and is called the phase transformation stress ratio; $\eta_{ij(0)}^*$ is the value of η_{ij}^* after the end of consolidation; σ'_{mb} is a parameter depending on the plastic volumetric strain; and ξ_{ij} is a variable that controls the change in the overconsolidation boundary, the initial value of ξ_{ij} is equal to $\eta_{ij(0)}^*$ which is the value of η_{ij}^* after the end of consolidation, and ξ_{ij} could follow an evolutional equation as

$$d\xi_{ij} = -C_d\xi_{ij}d\gamma^{p*},\ \gamma^{p*} = \int d\gamma^{p*} = \int\left(de_{ij}^p de_{ij}^p\right)^{1/2} \tag{2.5}$$

where γ^{p*} is the accumulated plastic shear strain from the initial state; de_{ij}^p is the plastic deviatoric strain increment; and C_d is a material constant that controls the disappearance of anisotropy.

The integrated form of Equation 2.5 is given by

$$\xi_{ij} = \eta^*_{ij(0)} \exp\left(-C_d \int d\gamma^{P*}\right) \tag{2.6}$$

Equation 2.5 indicates that the effect of anisotropy is assumed to fade with the increase in the cyclic shear strain history inside the overconsolidation boundary surface as in the previous model by Oka et al. (1999). In contrast, anisotropy may grow according to the strain history outside of the overconsolidated boundary surface as

$$\begin{aligned} &\xi_{ij} = \eta^*_{ij(0)} \exp\left(-C_d \int d\gamma^{P*}\right) \quad (f_b < 0) \\ &\xi_{ij} = B^*_4(\sigma'_{ij} - \xi_{ij}) \int dz \quad (f_b \geq 0) \\ &dz = (d\varepsilon^p_{ij} d\varepsilon^p_{ij})^{1/2} \end{aligned} \tag{2.7}$$

where B^*_4 is a material constant that controls the change in the overconsolidation boundary surface and $d\varepsilon^P_{ij}$ is the plastic strain increment tensor.

In Equation 2.4, σ'_{mb} is given by the function of the plastic volumetric strain:

$$\sigma'_{mb} = \sigma'_{mbi} \exp\left(\frac{1+e}{\lambda - \kappa} \varepsilon^P_v\right) = OCR^* \sigma'_{m0} \exp\left(\frac{1+e}{\lambda - \kappa} \varepsilon^P_v\right) \tag{2.8}$$

where σ'_{mbi} is the initial value of σ'_{mb}; σ'_{m0} is the initial mean effective stress; e is the void ratio; λ is the compression index; κ is the swelling index; and ε^p_v is the plastic volumetric strain.

In general, the σ'_{mbi} of the isotropically consolidated soil corresponds to the preconsolidation pressure σ'_{m0}. For sandy soils, however, it could be affected by the material anisotropy, the degree of compaction, aging, etc., and σ'_{mbi} does not necessarily correspond toσ'_{m0}. Hence, it is herein assumed that σ'_{mbi} is related to the quasi-overconsolidation ratio $OCR^*(=\sigma'_{mbi} / \sigma'_{m0})$. For soils that exhibit strong structural changes, σ'_{mb} may change due to changes in the plastic shear strain as well as the plastic volumetric strain shown in Figure 2.1. The strain softening, induced by the plastic strain considering the decrease in σ'_{mb}, can be described as follows:

$$\begin{aligned} &\sigma'_{mb} = \left\{\sigma'_{mbf} + \left(\sigma'_{mbi} - \sigma'_{mbf}\right)\exp(-\beta z)\right\} \exp\left(\frac{1+e}{\lambda - \kappa} \varepsilon^p_v\right) \\ &z = \int dz = \int \left(d\varepsilon^p_{ij} d\varepsilon^p_{ij}\right)^{1/2} \end{aligned} \tag{2.9}$$

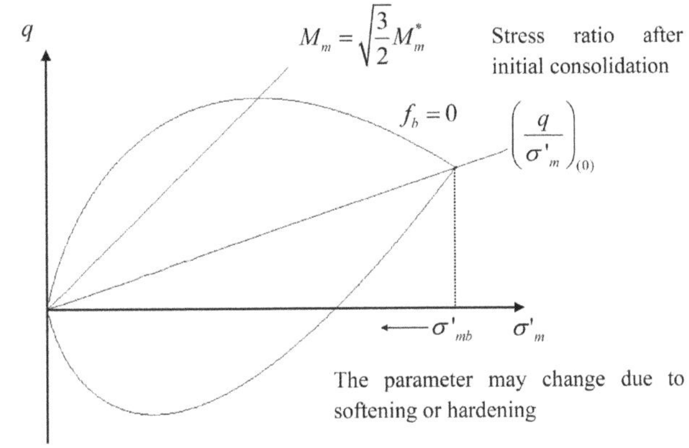

Figure 2.1 Overconsolidation boundary surface and the parameter σ'_{mb}.

where σ'_{mbf} is the final value of σ'_{mbi} and β is a parameter of the reduction rate.

2.1.2.3 *Yield function*

The yield function is assumed to be a non-linear function of the kinematic hardening parameters of the stress ratio and the mean effective stress; it is given by

$$f_y = \overline{\eta}_\chi^* + C_{ns}\tilde{M}^* \left(\ln \frac{\sigma'_{mk}}{\sigma'_{ma}} + \left| \ln \frac{\sigma'_m}{\sigma'_{mk}} - y_m^* \right| \right) = 0 \tag{2.10}$$

$$C_{ns} = \begin{cases} C_{ns} & \text{(After stress ratio has reached } M_m^*) \\ 1.0 & \text{(Before stress ratio has reached } M_m^*) \end{cases}$$

$$\overline{\eta}_\chi^* = \left\{ \left(\eta_{ij}^* - \chi_{ij}^* \right) \left(\eta_{ij}^* - \chi_{ij}^* \right) \right\}^{1/2} \tag{2.11}$$

where σ'_{mk} is a material constant; y_m^* is the kinematic hardening parameter for the change in the mean effective stress; σ'_{ma} is a parameter for the structural change; C_{ns} is a non-associativity parameter; and χ_{ij}^* is the non-linear kinematic hardening parameter for the change in the stress ratio. It is well known that non-associativity plays an important role in the modeling of geomaterials, which comes from the deformation mechanism and depends on the friction of the soil particles. Parameter C_{ns} controls the non-associativity which is less than 1.0; C_{ns} = 1.0 for the associative flow rule and $0 \le C_{ns} < 1.0$ for the non-associative flow rule. In the present model, it is assumed that

$C_{ns} = 1.0$ before the state reaches the phase transformation line and $0 \leq C_{ns} < 1.0$ after the state reaches the phase transformation line. The kinematic hardening parameter, χ^*_{ij}, has a dimension of η^*_{ij} and is called "back stress". The non-linear hardening rule, as an evolutional equation of χ^*_{ij} is given by

$$d\chi^*_{ij} = B^*\left(A^* de^p_{ij} - \chi^*_{ij} d\gamma^{P*}\right) \tag{2.12}$$

where de^p_{ij} is the deviatoric plastic strain increment; $d\gamma^{P*}$ is an accumulated deviatoric plastic strain; and A^* and B^* are material parameters related to the stress ratio at failure, M^*_f, and the non-dimensional initial plastic shear modulus, G^P, as follows:

$$A^* = M^*_f, \quad B^* = \frac{G^P}{M^*_f} \tag{2.13}$$

In addition, B^* follows an evolutional equation as

$$dB^* = C_f\left(B^*_1 - B^*\right) d\gamma^{P*} \tag{2.14}$$

where C_f and B^*_1 are material constants; and B^* could be given by the referential strain, γ^{P*}_r, as

$$B^* = B^*_0 / (1 + \gamma^{P*}_{ap} / \gamma^{P*}_r) \tag{2.15}$$

where γ^{P*}_{ap} is the accumulated plastic deviatoric strain after the stress ratio has reached M^*_m.

The softening due to the internal structural change is described by the same equation as for the overconsolidation surface and is given by

$$\sigma'_{ma} = \frac{\left\{\sigma'_{mbf} + \left(\sigma'_{mbi} - \sigma'_{mbf}\right)\exp(-\beta z)\right\}}{\sigma'_{mbi}} \sigma'_{mai} \tag{2.16}$$

where σ'_{mai} coincides with the initial mean effective stress. The kinematic hardening associated with the change in the mean effective stress is given by

$$y^*_m = y^*_{m1} + y^*_{m2} \tag{2.17}$$

$$dy^*_{m1} = B^*_2\left(A^*_2 d\varepsilon^p_v - y^*_{m1}\left|d\varepsilon^p_v\right|\right) \tag{2.18}$$

$$dy^*_{m2} = B^*_3 d\varepsilon^p_v \tag{2.19}$$

where A^*_2, B^*_2, and B^*_3 are material constants and $d\varepsilon^p_v$ is the volumetric plastic increment.

2.1.2.4 Failure conditions

In general, the stress ratio at the failure state depends on Lode's angle, and several failure criteria have been used, such as the extended von Mises criterion and Mohr–Coulomb's criterion. Mohr–Coulomb's criterion, however, gives smaller stress ratios at failure in many cases.

In many models, the stress ratio at failure is arbitrarily controlled with Lode's angle (e.g., Yasufuku 1990). Of course, other failure criteria such as the von Mises, Mohr–Coulomb, Lade (1975), and Matsuoka and Nakai (1977) criterion can be used in the model. Yasufuku (1990) proposed a failure criterion which depends on Lode's angle:

$$M_f^* = \frac{M_{fc}^* \omega_f^*}{\left[\omega_f^{*2} \cos^2\left\{\frac{3}{2}\left(\theta + \frac{\pi}{6}\right)\right\} + \sin^2\left\{\frac{3}{2}\left(\theta + \frac{\pi}{6}\right)\right\}\right]^{1/2}} \tag{2.20}$$

where M_f^* is the value of η^* at failure (η^* is the second invariant of the stress ratio tensor); θ is Lode's angle, $\theta = -(\pi/6)$ for triaxial compression; and ω_f^* is the parameter that controls the ratio of the stress ratio at triaxial compression and triaxial extension.

$$\omega_f^* = \frac{M_{fe}^*}{M_{fc}^*} \tag{2.21}$$

where M_{fc}^* is the stress ratio at triaxial compression and M_{fe}^* is the stress ratio at triaxial extension. In the case of $\omega_f^* = 1.0$, the criterion becomes the extended von Mises criterion. As an example, the failure surface on the π plane in the case of $\phi = 30°$, $\omega_f^* = 0.714$. The Yasufuku model is relatively simple and applicable to the failure of soils. In addition, the convexity of the failure surface should be satisfied with a wider range of friction angles, as shown in Appendix A2.1. Morio et al. (1994) showed that the Yasufuku criterion can reproduce the failure stress ratio of the Fuji River sand under the multi-axial stress conditions obtained by Yamada and Ishihara (1979).

On the other hand, for the stress ratio M_m^* (value of η^*) at the phase transformation state, which will be described in the following, a similar model is used. The ratio of the stress ratios at triaxial compression and triaxial extension is given by

$$\omega_m^* = \frac{M_{me}^*}{M_{mc}^*} \tag{2.22}$$

where M_{mc}^* and M_{me}^* are the stress ratios at phase transformation under triaxial compression and extension conditions, respectively.

2.1.2.5 Plastic potential function

The plastic potential function is given by

$$g = \overline{\eta}_{\chi}^{*} + \tilde{M}^{*}\left(\ln\frac{\sigma'_{mk}}{\sigma'_{mp}} + \left| \ln\frac{\sigma'_{m}}{\sigma'_{mk}} - y_{m}^{*} \right| \right) = 0 \tag{2.23}$$

where σ'_{mp} is a constant; σ'_{mk} is a material constant; and $\tilde{M}^{*}$ is given by

$$\tilde{M}^{*} = \begin{cases} M_{m}^{*} & f_b \geq 0 \text{ (Normally consolidated region and after stress ratio has reached } M_{m}^{*}) \\ \left(\sigma_{m}^{*}/\sigma'_{mb}\right)^{n2} M_{m}^{*} & f_b < 0 \text{ (Overconsolidated region)} \end{cases} \tag{2.24}$$

where n_2 is a material parameter; σ_{m}^{*} is the value of the mean effective stress at the intersecting point between the surface passing through the present stress point, which is similar to the overconsolidation surface, and the consolidation stress axis shown in Figure 2.2; and $\tilde{M}^{*}$, which controls the direction of the plastic strain increment, depends on the stress state and is smaller than the stress ratio at phase transformation M_{m}^{*}.

$$\tilde{M}^{*} = A_{cm} M_{m}^{*} \text{ after } \tilde{M}^{*} \text{ has reached } M_{m}^{*} \tag{2.25}$$

where A_{cm} is a material constant.

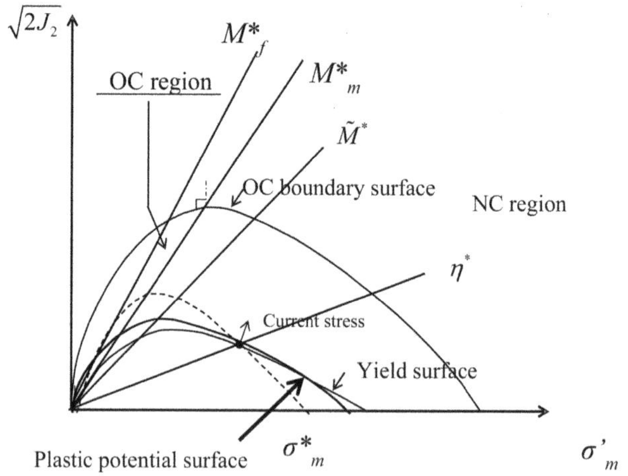

Figure 2.2 Overconsolidation boundary surface, yield function, and plastic potential function for isotropically consolidated state. (After Oka, F. and Kimoto, S. 2018. A cyclic elastoplastic constitutive model and effect of non-associativity on the response of liquefiable sandy soils. *Acta Geotech.* 13(6): 1283–1297.)

2.1.2.6 *Plastic flow rule*

The following generalized flow rule, including the fourth-order isotropic tensor H_{ijkl} (Oka et al. 1999), is used in the present model to more precisely reproduce the dilatancy characteristics as

$$d\varepsilon_{ij}^{p} = H_{ijkl}\frac{\partial g}{\partial \sigma'_{kl}}, \quad H_{ijkl} = a\delta_{ij}\delta_{kl} + b(\delta_{ik}\delta_{jl} + \delta_{il}\delta_{jk}) \tag{2.26}$$

where a and b are material parameters that depend on the stress and strain.

The stress–dilatancy relationship is given by

$$\frac{d\varepsilon_{v}^{p}}{d\gamma^{P*}} = D^{*}\left(\tilde{M}^{*} - \bar{\eta}_{\chi}^{*}\right), \quad D^{*} = \frac{3a}{2b} + 1 \tag{2.27}$$

where D^* is a so-called dilatancy coefficient and controls the plastic volumetric strain associated with shearing. In the case of $a = 0$ in Equation 2.27, the relationship corresponds to the conventional stress dilatancy relation.

The dilatancy coefficient D^* is given by

$$D^{*} = D_{0}^{*}(\tilde{M}^{*}/M_{m}^{*})^{n} \tag{2.28}$$

where D_0^* and n are material constants.

2.1.3 Simulation results

2.1.3.1 *Determination of parameters*

The material parameters included in the present model are as follows: e is the void ratio and e_0 is the initial value of e; λ, κ are the compression and swelling indices, respectively, which are determined by compression and swelling tests; M_f^*, M_m^* are the stress ratios at failure and at the phase transformation, respectively; G_0^E is the initial elastic shear modulus that can be determined by the initial slope of the stress–strain relationship and/or the shear wave velocity. OCR^* is the quasi-overconsolidation ratio that is related to the degree of compaction, which can be determined by the volumetric deformation response (OCR^* is not always 1.0 after compaction); K_0 is the anisotropic consolidation ratio; D_0^*, n, C_d, and n_2 are the parameters for the dilatancy characteristics; $\sigma'_{mbf}/\sigma'_{mbi}$ and β are the parameters for the shear strain-dependent degradation; $B_0^*, B_1^*, A_1^*(=M_f^*)$, and C_f are the parameters for the non-linear kinematic hardening rule for deviatoric plastic strain and B_2^*, A_2^* are those for the volumetric plastic strain; B_3^* corresponds to the plastic volumetric hardening parameter, such as $B_3^* = (1+e)/(\lambda - \kappa)$; C_{ns} is a parameter that controls the non-associativity;

r_{e1}, r_{e2}, and α_e are the parameters for the strain dependency of the elastic shear modulus and are determined by the cyclic deformation test results; $B_0^*, B_1^*, B_2^*, A_1^*(=M_f^*), C_f$, C_d, $\gamma_r^{E^*}$, $\gamma_r^{P^*}$, D_0^*, and n can be determined by referring to the previous study by Oka et al. (1999); B_0^* can be determined by the initial shear modulus of the stress–strain relation taking account of the elastic shear modulus; B_1^* and C_f are determined by the stress–plastic strain relation and the stress path mainly in the cyclic mobility region; C_d and B_4^* control the disappearance of the initial anisotropy and growth, respectively, and can be determined by tests for anisotropically consolidated and isotropically consolidated materials; $\gamma_r^{E^*}$ and $\gamma_r^{P^*}$ are determined by the stress–strain relation – it is known that the value of $\gamma_r^{E^*}$ is larger than that of $\gamma_r^{P^*}$; and D_0^*, n, and n_2 are determined by the stress path under undrained conditions and the volumetric strain under drained conditions, i.e., they control the dilatancy characteristics.

In the present section, $B_2^* = 0.0$ due to the lack of data on the mean effective stress versus the volumetric strain relationship and $B_3^* = (1+e)/(\lambda - \kappa)$ considering the Cam-clay type of hardening. In addition, $r_{e2} = 1.0$, $\alpha_e = 1.0$, and $r_{e1} = 1.0$ due to the insufficient amount of data; C_d and B_4^* are not necessary because the following analyses are for isotropically consolidated sand; A_{cm} exhibits the change in M_m^* after reaching cyclic mobility (A_{cm} = 1.0 in the following analysis due to the lack of data). The above parameters are determined by the data-adjusting method for the stress–strain relation and the stress path of the undrained cyclic shearing tests, monotonic shearing tests, and consolidation tests. In this section, focus is placed on the effect of parameters C_{ns}, $\sigma'_{mbf} / \sigma'_{mbi}$, and β on the cyclic responses, which are newly introduced in the present model in the following section.

2.1.3.2 Comparison with experimental results

In order to show the applicability of the present model, the responses of loose Toyoura silica sand under undrained simple shear conditions by Oka et al. (1999) will be discussed. The simple shear test result of the loose silica sand is shown in Figure 2.3. The test conditions of the simple shear test carried out using the hollow cylinder test device are given as the maximum shear stress ratio $\tau_{xy} / \sigma'_{m0} = 0.22$ (T-A-2), $\tau_{xy} / \sigma'_{m0} = 0.15$ (T-A-4); the initial void ratio e_0 = 0.772; the initial mean effective stress $\sigma'_{m0} = 98$ kPa; and K_0 = 1.0. The fundamental properties of Toyoura silica sand used in the analysis are G_s = 2.66, e_{max} = 0.998, e_{min} = 0.620, D_{50} = 0.20 mm, and U_c = 1.24.

The material parameters used in the simulations are listed in Table 2.1. Figure 2.4 shows the stress paths and stress–strain relations under undrained simple shear tests after isotropic consolidation using the hollow cylinder torsional test apparatus for two different shear stress amplitudes (Case-1 and Case-2). The number of cycles by which the stress paths reach

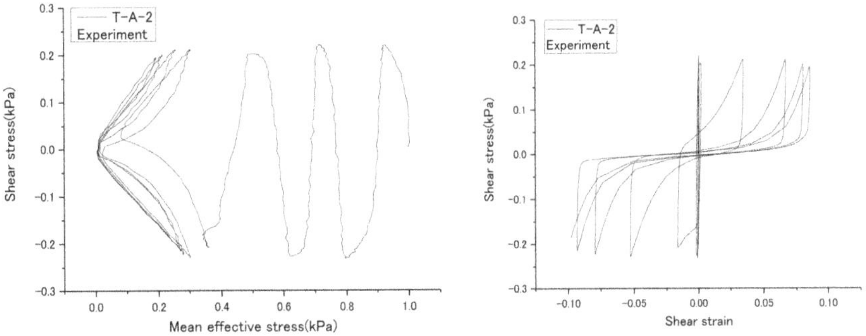

(a) T-A-2 Initial normalized shear stress=0.22

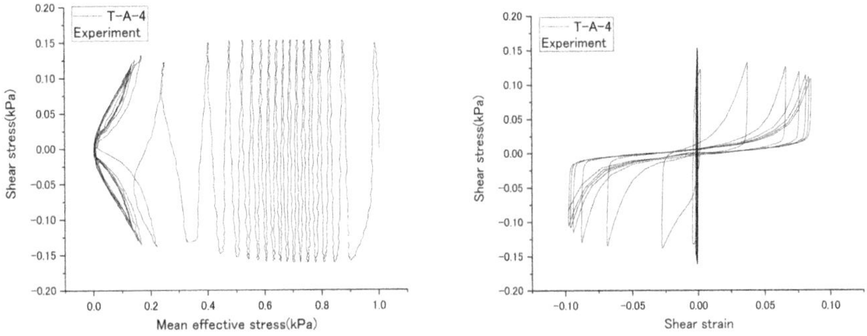

(b) T-A-4 Initial normalized shear stress=0.15

Figure 2.3 Undrained cyclic simple shear test results. (After Oka, F., Yashima, A., Tateishi, A., Taguchi, Y. and Yamashita, S. 1999. A cyclic elasto-plastic constitutive model for sand considering a plastic-strain dependence of the shear modulus. *Geotechnique* 49(5): 661–680.)

cyclic mobility and the strain amplitudes are almost in good agreement with the experimental results. The stress ratio magnitudes and the parameters used in the simulation cases are

$$q / p' = 0.22 \text{ (Case-1, Case-4, Case-5)}, \; q / p' = 0.15 \text{ (Case-2, Case-9, Case-10)}$$

$$K_0 = 1.0 \text{ (Case-1, Case-2, Case-4, Case-5, Case-9, Case-10)}$$

$$n_2 = 0.5 \text{ (Case-1, Case-2, Case-4, Case-5)}, \; n_2 = 1.0 \text{ (Case-9, Case-10)}$$

$$C_{ns} = 0.01 \text{ (Case-1, Case-2, Case-9, Case-10)}, C_{ns} = 0 \text{ (Case-4)}, C_{ns} = 1.0 \text{ (Case-5)}$$

$$\beta = 0 \text{ (Case-1, Case-2, Case-4, Case-5, Case-10)}, \; \beta = 30 \text{ (Case-9)}$$

$$n' = \sigma'_{mbf} / \sigma'_{mbi} = 0.7 \text{ (Case-9, Case-10)}$$

Table 2.1 Material parameters used in the simulations

Parameter	*Value*
Compression index, λ	0.0091
Swelling index, κ	0.00052
Quasi-overconsolidation ratio, OCR^*	1.2
Non-dimensional initial elastic shear modulus, G_0 / σ'_{m0}	1023.6
Stress ratio at failure under simple shear conditions, $M_f^* = A^*$	0.990
Stress ratio at phase transformation under simple shear conditions, M_m^*	0.707
Initial value of the parameter for strain hardening, B_0^*	4089
Final value of the parameter for strain hardening, B_1^*; when Equation 2.15 is adopted, B_1^* is not used	–
Degradation parameter, C_f; when Equation 2.15 is adopted, C_f is not used	–
Parameter for the non-dimensionalization (kPa)	1.0
Strain-dependent parameter for plastic shear modulus (referential strain), γ_r^{P*}	0.002
Strain-dependent parameter for elastic shear modulus (referential strain), γ_r^{E*}	0.012
Strain-dependent parameter for elastic shear modulus, α_e	1.0
Parameter for strain-dependent elastic shear modulus, r_{e1}	1.0
Parameter for effective stress-dependent elastic shear modulus, r_{e2}	1.0
Coefficient of dilatancy, D_0^*	0.85
Coefficient of dilatancy, n	5.1
Parameter for vanishing anisotropy, C_d	500
Parameter for internal structure degradation, β	Variable
Parameter for structural degradation, $\sigma'_{mbf} / \sigma'_{mbi}$	Variable
Parameter for dilatancy, n_2	Variable
Parameters for kinematical strain hardening, A_2^*, B_2^*, B_3^*	0, 0, 204.4
Parameter for growth of overconsolidation boundary surface	0.0
Parameter for dilatancy, A_{cm}	1.0
Parameter for non-associativity, C_{ns}	Variable

Source: Data from Oka, F. and Kimoto, S. 2018. A cyclic elastoplastic constitutive model and effect of non-associativity on the response of liquefiable sandy soils. *Acta Geotechnica*, 13(6): 1283-1297.)

2.1.3.3 Effect of non-associativity parameter

The simulation results under cyclic undrained simple shear test conditions are illustrated in Figure 2.5 with different values for the non-associativity parameter C_{ns}. In Case-5, in which $C_{ns} = 1.0$, i.e., for the case of the associated flow rule, the minimum mean effective stress is rather far from zero. On the other hand, in Case-4, in which $C_{ns} = 0$, the minimum mean effective stress is close to zero at 2 kPa. This indicates that non-associativity leads to a large decrease

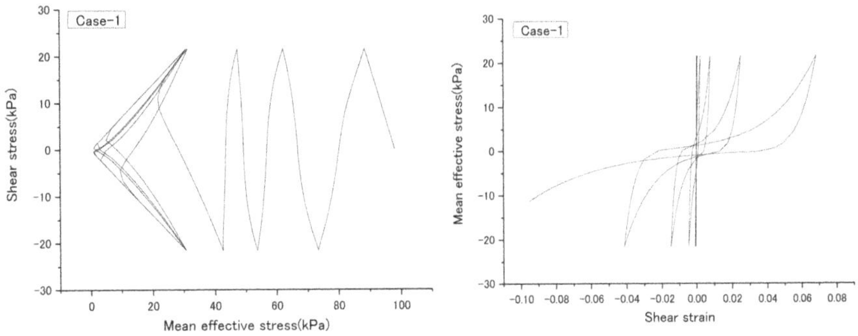

(a) Initial stress ratio=0.22

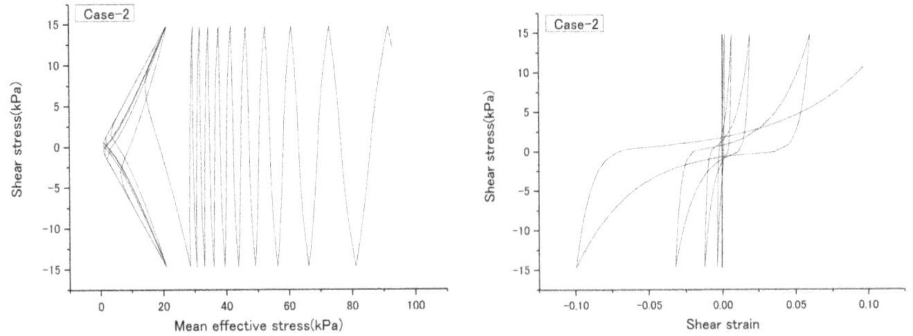

(b) Initial stress ratio=0.15

Figure 2.4 Simulated undrained cyclic simple shear test results. (Case-1 and Case-2) (After Oka, F. and Kimoto, S. 2018. A cyclic elasto-plastic constitutive model and effect of non-associativity on the response of liquefiable sandy soils. *Acta Geotech.* 13(6): 1283–1297.)

in the mean effective stress. As for the stress–strain relationship, the stress–strain curves are relatively stable for the case of C_{ns} = 1.0, while for the case of a small value for C_{ns}, i.e., C_{ns} = 0.01, the stress–strain curve is S-shaped with a large maximum strain that is typical in cyclic liquefaction tests. Oka and Kimoto (2018) showed that the minimum effective stress monotonically increases as the coefficient C_{ns} increases from zero to unity.

Strong non-associativity, which is manifested by the smaller value for C_{ns}, has been seen to lead to a large decrease, i.e., a mean effective stress of almost zero during cyclic deformation under undrained conditions, while the model with the associated flow rule has not. This indicates that the model with strong non-associativity has a high potential for simulating the liquefaction of a sandy ground, which is appropriate for a water-soil fully coupled dynamic analysis. This result is quite salient because it has been observed that the mean effective stress becomes almost zero at the state of full liquefaction.

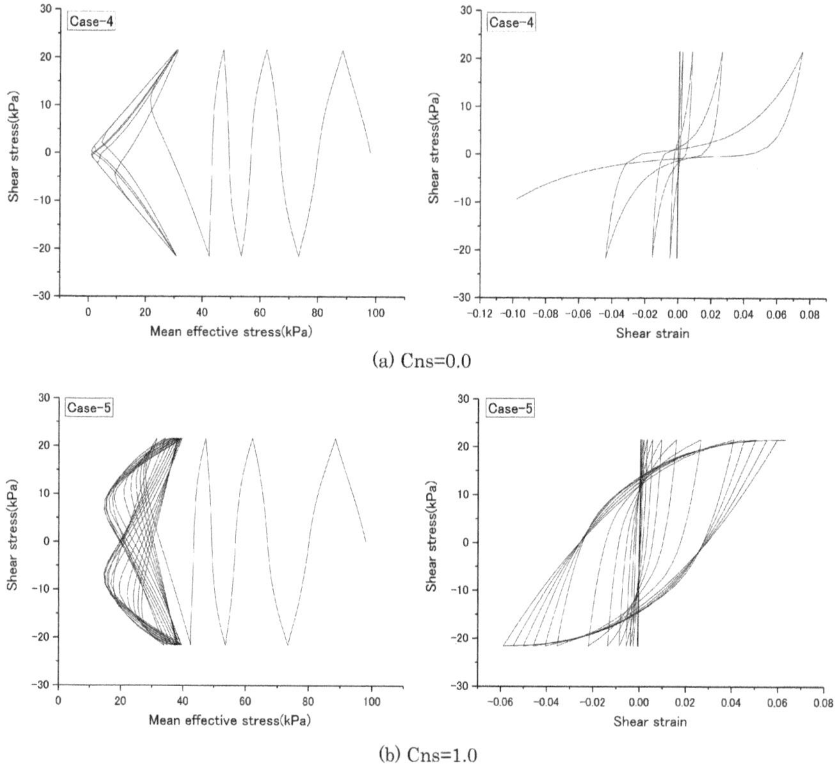

Figure 2.5 Simulated results with C_{ns} = 0.0 (Case-4) and C_{ns} = 1.0 (Case-5). (After Oka, F. and Kimoto, S. 2018. A cyclic elasto-plastic constitutive model and effect of non-associativity on the response of liquefiable sandy soils. *Acta Geotech.* 13(6): 1283–1297.)

2.1.3.4 Effect of degradation parameters

Figure 2.6 shows the effect of the degradation parameters $n' = \sigma'_{mbf} / \sigma'_{mbi}$ and β on the cyclic response; $n' = \sigma'_{mbf} / \sigma'_{mbi}$ is the decreasing ratio of the size of the overconsolidation surface and β is the decreasing rate. They showed that the stress path stops around the mean effective stress of 30 kPa with zero degradation, i.e., $\beta = 0$, while the mean effective stress decreases to almost zero with $\beta = 30$ for Case-9 and Case-10, respectively. The induced stress–strain relations are quite different; the stress–strain curve is S-shaped for the case of $\beta = 30$, while the shape of the stress–strain curve is stable with a small maximum strain for the case of $\beta = 0$. These results indicate that the introduction of the degradation parameters, $n' = \sigma'_{mbf} / \sigma'_{mbi}$ and β, brings about the shrinkage of the overconsolidation surface and the weaker response of the material.

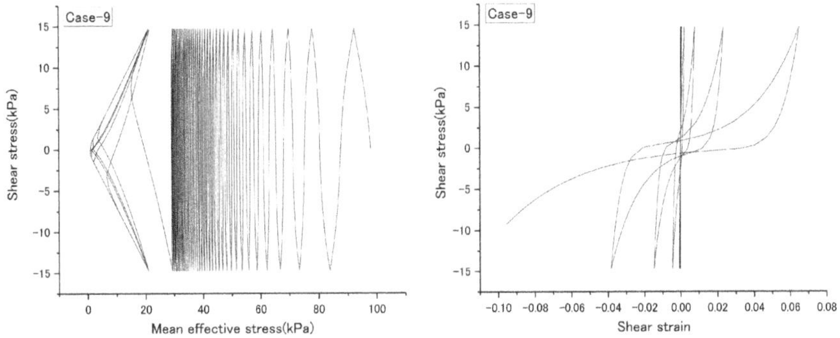

(a) $n' = 0.7$ *and* $\beta = 30.0$

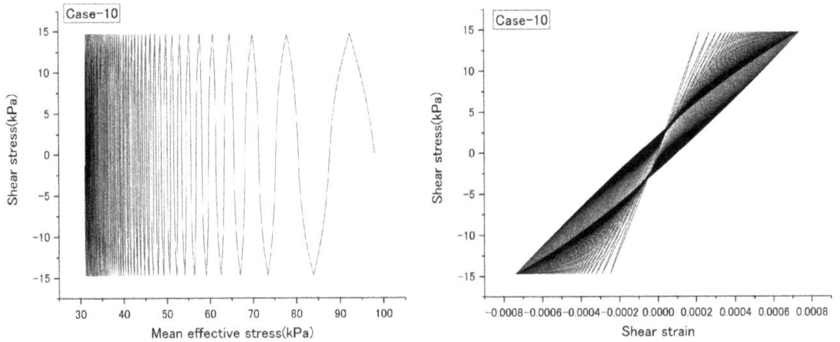

(b) $n' = 0.7$ *and* $\beta = 0.0$

Figure 2.6 Simulated results with and without degradation effects (Case-9 and Case-10). (After Oka, F. and Kimoto, S. 2018. A cyclic elasto-plastic constitutive model and effect of non-associativity on the response of liquefiable sandy soils. *Acta Geotech.* 13(6): 1283–1297.)

2.2 CYCLIC ELASTO-VISCOPLASTIC CONSTITUTIVE MODEL

2.2.1 Introduction

The behavior of soft clay under dynamic loading conditions is of considerable importance in a dynamic analysis of multilayered ground and soil structures. This behavior includes the interaction between the sandy and the clayey layers, and the effect of the clay layer on liquefaction. In order to accurately predict the behavior of the foundation ground and soil structures during earthquakes, ocean wave storms, traffic vibrations, and any other similar phenomena, it is necessary to employ constitutive models that properly reproduce the dynamic behavior of both sandy and clayey soils. The prediction of soft clay behavior is rather complex due to rate-dependent and cyclic plastic behaviors. This becomes even more complicated when the destructuration and microstructural changes of soils are taken into consideration. Several constitutive models have been proposed to describe the rheological behavior of clay under static loading conditions. However, few viscoplastic constitutive models are available for the analysis under dynamic and cyclic loading conditions (e.g., Oka 1992; Modaressi and Laloui 1997; Oka et al. 2004; Maleki and Cambou 2009). Kimoto et al. (2015) proposed a cyclic elasto-viscoplastic model for clayey soils considering non-linear kinematic hardening and strain-dependent degradation.

In this section, the overall feature of the cyclic elasto-viscoplastic constitutive model (Kimoto et al. 2015) is described. In order to evaluate the capability of the model, modeling of soft clay samples has been evaluated under cyclic triaxial test conditions. Comparisons are made between the simulated results and the corresponding experimental data.

2.2.2 Cyclic elasto-viscoplastic constitutive equation

In this section, a cyclic elasto-viscoplastic constitutive model is presented based on the work by Kimoto et al. (2015) in which the non-linear kinematic hardening rules (Armstrong and Frederick 1966; Chaboche and Rousselier 1983) and the concept of structural degradation in the elasto-viscoplastic model (Kimoto and Oka 2005) are incorporated. The model includes the structural degradation of the soil skeleton by the shrinkage of both the static yield surface and the overconsolidation boundary surface with respect to the accumulation of viscoplastic strain. The model is classified as an overstress type of viscoplasticity model (Perzyna 1963) and adopts the generalized flow rule including the associated flow rule. The non-linear kinematic hardening rules for both the deviatoric and volumetric strain components and the viscoplastic strain dependency of elastic shear modulus are used. The model has been applied to the analysis of the consolidation and strain localization problems (Mirjalili 2010; Shahbodagh et al. 2014).

2.2.2.1 *Elastic strain rate tensor*

In the constitutive model, Terzaghi's effective stress σ'_{ij} is used; $\sigma'_{ij} = \sigma_{ij} - p\delta_{ij}$, where σ_{ij} is the total stress tensor, p is the pore water pressure (compression is positive), and δ_{ij} is Kronecker's delta.

The elastic strain rate tensor, $\dot{\varepsilon}^e_{ij}$, is given as

$$\dot{\varepsilon}^e_{ij} = \frac{1}{2G}\dot{s}_{ij} + \frac{\kappa}{3(1+e)}\frac{\dot{\sigma}'_m}{\sigma'_m}\delta_{ij} \tag{2.29}$$

where G is the elastic shear modulus; s_{ij} is the deviatoric stress tensor; $s_{ij} = \sigma'_{ij} - \sigma'_m\delta_{ij}$, σ'_m is the mean effective stress and the superimposed dot denotes the time differentiation; κ is the swelling index; and e is the void ratio. Under the undrained conditions and in the infinitesimal analysis, the initial void ratio e_0 is used in Equation 2.29.

Various empirical formulations have been provided to express the strain dependency of the shear modulus (e.g., Hardin and Drnevich 1972; Fahey 1992). In the formulation, the degradation of the elastic shear modulus from the beginning of loading is assumed to be dependent on the accumulated viscoplastic shear strain:

$$G = \frac{G_0}{\left(1+\alpha\left(\gamma^{vp}\right)^r\right)}\sqrt{\frac{\sigma'_m}{\sigma'_{m0}}} \tag{2.30}$$

where $\gamma^{vp} = \int_0^t \dot{\gamma}^{vp} dt = \int (\dot{e}_{ij}^{vp}\dot{e}_{ij}^{vp})^{1/2} dt$; $\dot{e}_{ij}^{vp}$ is the viscoplastic deviatoric strain rate; and r and α are the strain-dependent parameters, which are determined from laboratory test results. In this section, $r = 0.4$ is experimentally chosen.

2.2.2.2 Overconsolidation boundary surface

In order to define the boundary in the stress space between the normally consolidated region and the overconsolidated region, an overconsolidation boundary surface, f_b, is used:

$$f_b = \bar{\eta}^*_{(0)} + M^*_m \ln\left(\sigma'_m / \sigma'_{mb}\right) = 0 \tag{2.31}$$

where $f_b < 0$ indicates the overconsolidated region and $f_b \geq 0$ indicates the normally consolidated region. The overconsolidation boundary surface is introduced to describe the different dilatancy behavior in the NC and OC regions, and controls the value of the dilatancy coefficient $\tilde{M}^*$ in the viscoplastic potential function and the yield function. The relative stress ratio, $\bar{\eta}^*_{(0)}$, is defined as

$$\bar{\eta}^*_{(0)} = \left\{\left(\eta^*_{ij} - \eta^*_{ij(0)}\right)\left(\eta^*_{ij} - \eta^*_{ij(0)}\right)\right\}^{\frac{1}{2}} \tag{2.32}$$

in which subscript (0) denotes the initial state after consolidation; η^*_{ij} is the stress ratio tensor; and σ'_{mb} controls the size of the OC boundary surface. The overconsolidation boundary surface, in general, may change following Equation 2.32.

The value of $\eta^* = \sqrt{\eta^*_{ij}\eta^*_{ij}}$ is M^*_m at critical state, at which the volumetric strain increment changes from compression to dilation. For clay, we assume

M_m^* is the same as that at the failure state and, in general, the value is a function of Lode's angle.

Any kind of failure criterion can be used, such as the Mohr–Coulomb failure criterion or others by Matsuoka-Nakai (1977), Lade (1975), and Yasufuku (1990). Yasufuku's criterion is given by Equation 2.1.20. Adopting the Mohr–Coulomb criterion, M_m^* is defined by the friction angel ϕ. In conventional cyclic triaxial tests, the stress path follows the compression and the extension. In that case, failure under the triaxial cyclic tests is characterized by the stress ratio at triaxial compression failure state, M_{mc}^*, and the stress ratio at the extension failure state, M_{me}^*. If the material follows the Mohr–Coulomb criterion, M_m^* is related to the internal frictional angle ϕ as

$$M_{mc}^* = \sqrt{\frac{2}{3}} \frac{6\sin\phi}{3-\sin\phi}, \quad M_{me}^* = \sqrt{\frac{2}{3}} \frac{6\sin\phi}{3+\sin\phi} \tag{2.33}$$

Specific failure criteria are necessary for solving the boundary value problem.

Kimoto and Oka (2005) introduced a degradation term with respect to the accumulated viscoplastic strain, which describes unstable behavior such as strain softening during both shearing and compaction processes. The parameter of the overconsolidation boundary surface that controls the strain hardening and softening with respect to the viscoplastic strain is given as

$$\sigma'_{mb} = \sigma'_{ma} \exp\left(\frac{1+e_0}{\lambda-\kappa}\varepsilon_v^{vp}\right) \tag{2.34}$$

where λ is the compression index; ε_v^{vp} is the viscoplastic volumetric strain; and σ'_{ma} is assumed to decrease with an increase in the viscoplastic accumulated strain as

$$\sigma'_{ma} = \sigma'_{maf} + \left(\sigma'_{mai} - \sigma'_{maf}\right)\exp(-\beta z) \tag{2.35}$$

where z is the accumulation of the second invariant of the viscoplastic strain rate given by

$$z = \int_0^t \dot{z}dt; \; \dot{z} = \sqrt{\dot{\varepsilon}_{ij}^{vp}\dot{\varepsilon}_{ij}^{vp}} \tag{2.36}$$

The initial and final values for σ'_{ma} are σ'_{mai} and σ'_{maf}, respectively, in Equation 2.35, and β is a parameter that stands for the changing rate of σ'_{ma}, while the proportion of $n = \sigma'_{maf}/\sigma'_{mai}$ provides the degree of possible collapse of the soil structure at the initial state. The structural parameters n and β control the strain-softening process. Parameter n is related to the decrease from the peak stress to the residual stress in shear tests and β affects the strain-softening process.

2.2.2.3 Static yield function

The static yield function is given by the non-linear kinematic hardening rule for the changes in the stress ratio and the mean effective stress (Kimoto et al. 2015; Oka and Kimoto 2018):

$$f_y = \bar{\eta}_\chi^* + \tilde{M}^* \left(\ln \frac{\sigma'_{mk}}{\sigma'^{(s)}_{my}} + \left| \ln \frac{\sigma'_m}{\sigma'_{mk}} - y_m^* \right| \right) = 0 \tag{2.37}$$

$$\bar{\eta}_\chi^* = \left\{ \left(\eta_{ij}^* - \chi_{ij}^* \right) \left(\eta_{ij}^* - \chi_{ij}^* \right) \right\}^{\frac{1}{2}} \tag{2.38}$$

In Equations 2.37 and 2.38, σ'_{mk} is a material constant; y_m^* is the scalar kinematic hardening parameter; $\tilde{M}^*$ is the dilatancy coefficient; $\sigma'^{(s)}_{my}$ denotes the static hardening parameter; and the kinematic hardening parameter χ_{ij}^* is a so-called back stress which has the same dimensions as the stress ratio η_{ij}^*.

As with the OC boundary surface, the strain softening for the structural degradation is manifested through the changes in $\sigma'^{(s)}_{my}$ with the viscoplastic strain as

$$\sigma'^{(s)}_{my} = \frac{\sigma'_{maf} + \left(\sigma'_{mai} - \sigma'_{maf} \right) \exp(-\beta z)}{\sigma'_{mai}} \sigma'^{(s)}_{myi} \tag{2.39}$$

where $\sigma'^{(s)}_{myi}$ is the initial value of $\sigma'^{(s)}_{my}$.

2.2.2.4 Viscoplastic potential function

The viscoplastic potential function, f_p, is given by a similar function to the static yield function as

$$f_p = \bar{\eta}_\chi^* + \tilde{M}^* \left(\ln \frac{\sigma'_{mk}}{\sigma'_{mp}} + \left| \ln \frac{\sigma'_m}{\sigma'_{mk}} - y_m^* \right| \right) = 0 \tag{2.40}$$

where σ'_{mp} is determined from the current stress and gives the size of the surface. The dilatancy coefficient, $\tilde{M}^*$, is defined separately for the normally consolidated region and the overconsolidated region as

$$\tilde{M}^* = \begin{cases} M_m^* & : \text{NC region} \\ \left(\sigma_m^* / \sigma'_{mb} \right) M_m^* & : \text{OC region} \end{cases} \tag{2.41}$$

$$\sigma_m^* = \sigma_m' \exp\left(\frac{\overline{\eta}_{(0)}^*}{M_m^*}\right) \tag{2.42}$$

where σ_m^* denotes the mean effective stress at the intersection of the surface, which has the same shape as f_b, and σ_m' is the current mean effective stress. Figure 2.7 illustrates the overconsolidation boundary surface, the static yield surface, and the viscoplastic potential surface for the isotropically consolidated soil in both the OC and NC regions.

2.2.2.5 *Viscoplastic flow rule*

Following the overstress type of viscoplastic theory adopted by Perzyna (1963), the viscoplastic strain rate tensor, $\dot{\varepsilon}_{ij}^{vp}$, is defined as

$$\dot{\varepsilon}_{ij}^{vp} = C_{ijkl}\left\langle \Phi\left(f_y\right)\right\rangle \frac{\partial f_p}{\partial \sigma_{kl}'} \tag{2.43}$$

$$\left\langle \Phi\left(f_y\right)\right\rangle = \begin{cases} \Phi\left(f_y\right) & : f_y > 0 \\ 0 & : f_y \le 0 \end{cases} \tag{2.44}$$

$$C_{ijkl} = a\delta_{ij}\delta_{kl} + b\left(\delta_{ik}\delta_{jl} + \delta_{il}\delta_{jk}\right) \tag{2.45}$$

where $\langle\ \rangle$ are Macaulay's brackets; $\Phi(f_y)$ is the rate-sensitive material function; C_{ijkl} is a fourth-order isotropic tensor; and a and b in Equation 2.45 are material constants. A fourth-order isotropic tensor, C_{ijkl}, is introduced in the viscoplastic flow rule so that the model can well describe the dilatancy characteristics of soil. The material function, $\Phi(f_y)$, is determined as

$$\Phi\left(f_y\right) = \sigma_m' \exp\left\{ m'\left(\overline{\eta}_\chi^* + \tilde{M}^*\left(\ln\frac{\sigma_{mk}'}{\sigma_{ma}'} + \left| \ln\frac{\sigma_m'}{\sigma_{mk}'} - y_m^* \right| \right)\right)\right\} \tag{2.46}$$

where m' is the viscoplastic parameter.

2.2.2.6 *Kinematic hardening rules*

The evolutional equation for the non-linear kinematic hardening parameter, χ_{ij}^*, is given by

$$d\chi_{ij}^* = B^*\left(A^* de_{ij}^{vp} - \chi_{ij}^* d\gamma^{vp}\right) \tag{2.47}$$

where A^* and B^* are material parameters; de_{ij}^{vp} is the viscoplastic deviatoric strain increment tensor and $d\gamma^{vp} = \sqrt{de_{ij}^{vp} de_{ij}^{vp}}$; A^* is related to the stress ratio

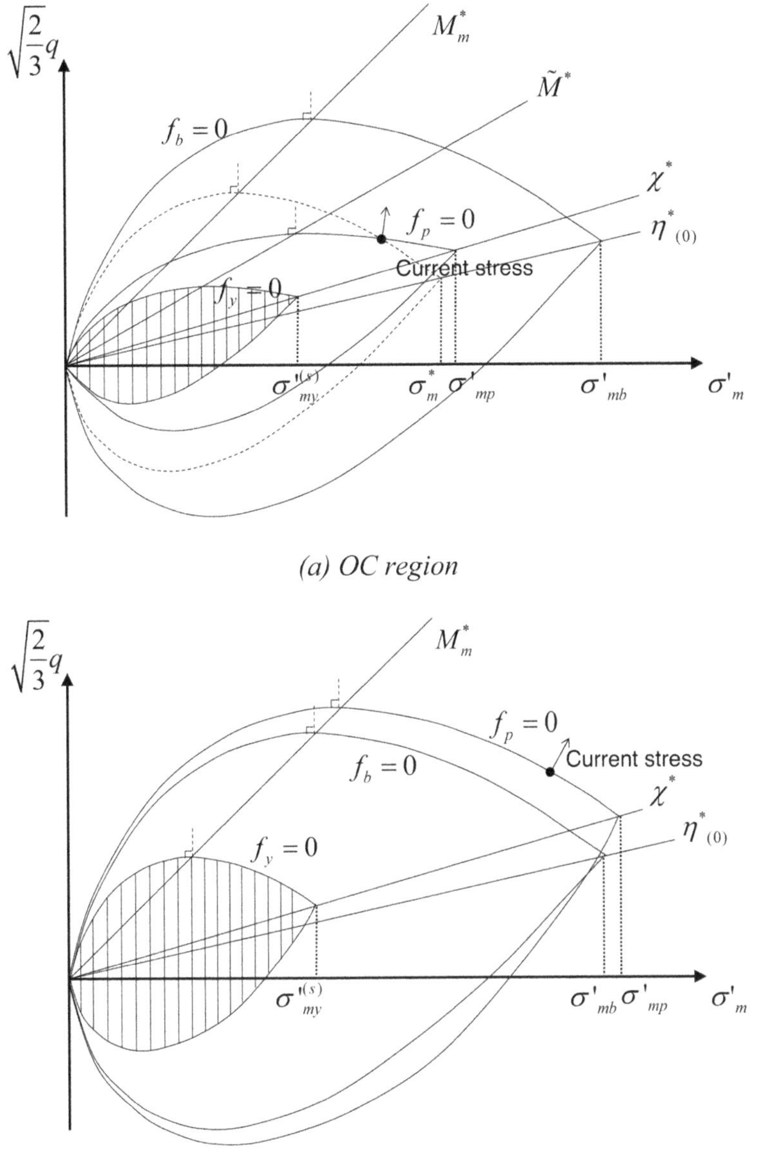

(b) NC region

Figure 2.7 Overconsolidation boundary surface, static yield surface, and viscoplastic potential surface. (After Kimoto, S., Shahbodagh Khan, B., Mirjalili, M. and Oka, F. 2015. A cyclic elasto-viscoplastic constitutive model for clay considering the nonlinear kinematic hardening rules and the structural degradation. *Int. J. Geomech., ASCE*, 15(5): A4014005-1-A4014005-14.)

at failure, namely, $A^* = M_f^*$; and B^* is proposed to be dependent on the viscoplastic shear strain as

$$B^* = \left(B_{\max}^* - B_1^*\right)\exp\left(-C_f \gamma_{(n)}^{vp*}\right) + B_1^* \tag{2.48}$$

In Equation 2.48, B_1^* is the lower boundary of B^*; C_f is the parameter controlling the rate of reduction; $\gamma_{(n)}^{vp*}$ is the accumulated value of the viscoplastic shear strain between two sequential stress reversal points in the previous cycle; and $B_{\max}^*$ is the maximum value of parameter B^* as

$$B_{\max}^* = \begin{cases} B_0^* & : \text{Before reaching failure line} \\ \dfrac{B_0^*}{1 + \gamma_{(n)\max}^{vp*} / \gamma_{(n)r}^{vp*}} & : \text{After reaching failure line} \end{cases} \tag{2.49}$$

where B_0^* is the initial value of B^*; $\gamma_{(n)\max}^{vp*}$ is the maximum value of $\gamma_{(n)}^{vp*}$ in the past cycles; and $\gamma_{(n)r}^{vp*}$ is the viscoplastic referential strain.

In order to include the effect of the volumetric viscoplastic strains, the scalar non-linear kinematic hardening parameter, y_m^*, is adopted (Oka et al. 1999) and is decomposed into two parts, such as linear and non-linear kinematic hardening, as

$$y_m^* = y_{m1}^* + y_{m2}^* \tag{2.50}$$

where y_{m1}^* and y_{m2}^* are the non-linear and linear kinematic hardening parameters, respectively.

The scalar non-linear kinematic hardening parameter, y_{m1}^*, is assumed to follow the evolutional equation as

$$dy_{m1}^* = B_2^*\left(A_2^* d\varepsilon_v^{vp} - y_{m1}^* \left|d\varepsilon_v^{vp}\right|\right) \tag{2.51}$$

where A_2^* and B_2^* are material parameters, and $d\varepsilon_v^{vp}$ is the increment of the viscoplastic volumetric strain. The values of A_2^* and B_2^* are determined by the data-adjusting method from the laboratory test data.

For the isotropic hardening, we can use a linear kinematic hardening parameter, y_{m2}^*, as

$$dy_{m2}^* = B_3^* d\varepsilon_v^{vp} \tag{2.52}$$

where B_3^* is a hardening parameter.

2.2.2.7 *Total strain rate tensor*

Finally, by combining Equations 2.37–2.52, the viscoplastic deviatoric strain rate, $\dot{e}_{ij}^{vp}$, and the viscoplastic volumetric strain rate, $\dot{\varepsilon}_{kk}^{vp}$ (or $\dot{\varepsilon}_{v}^{vp}$), are given by

$$\dot{e}_{ij}^{vp} = C_1 \exp\left\{ m' \left(\bar{\eta}_\chi^* + \tilde{M}^* \left(\ln\frac{\sigma'_{mk}}{\sigma'_{ma}} + \left| \ln\frac{\sigma'_m}{\sigma'_{mk}} - y_m^* \right| \right) \right) \right\} \frac{\eta_{ij}^* - \chi_{ij}^*}{\bar{\eta}_\chi^*} \tag{2.53}$$

$$\dot{\varepsilon}_{kk}^{vp} = C_2 \exp\left\{ m' \left(\bar{\eta}_\chi^* + \tilde{M}^* \left(\ln\frac{\sigma'_{mk}}{\sigma'_{ma}} + \left| \ln\frac{\sigma'_m}{\sigma'_{mk}} - y_m^* \right| \right) \right) \right\}$$
$$\left\{ \tilde{M}^* \frac{\ln\dfrac{\sigma'_m}{\sigma'_{mk}} - y_m^*}{\left| \ln\dfrac{\sigma'_m}{\sigma'_{mk}} - y_m^* \right|} - \frac{\eta_{mn}^*\left(\eta_{mn}^* - \chi_{mn}^*\right)}{\bar{\eta}_\chi^*} \right\} \tag{2.54}$$

where $C_1 = 2b$ and $C_2 = 3a + 2b$ are the viscoplastic parameters for the deviatoric and the volumetric strain components, respectively.

Finally, the total strain rate tensor, $\dot{\varepsilon}_{ij}$, is assumed to be divided into two parts, namely,

$$\dot{\varepsilon}_{ij} = \dot{\varepsilon}_{ij}^{e} + \dot{\varepsilon}_{ij}^{vp} \tag{2.55}$$

where $\dot{\varepsilon}_{ij}^{e}$ denotes the elastic strain rate tensor and $\dot{\varepsilon}_{ij}^{vp}$ is the viscoplastic strain rate tensor.

2.2.2.8 *Determination of material parameters*

The material parameters can be determined by conventional tests, physical property tests, and triaxial compression tests. The initial void ratio can be calculated with the specific gravity of the soil particles, the water content, and the bulk density based on the results of tests conducted on undisturbed samples. The compression index and the swelling index can be calculated as the slope of the *e–ln p* relation during isotropic consolidation and the swelling tests, respectively. The initial elastic shear modulus is calculated from the slope of the stress–strain relation curve at the earlier stage of loading. The stress ratios at failure are calculated from the internal friction angle using Equation 2.33. The viscoplastic parameter, m', is determined from undrained triaxial compression tests conducted at different strain rates (Kimoto and Oka 2005). By having m', the other viscoplastic parameters, C_1 and C_2, are obtained from Equations 2.53 and 2.54 in the monotonic triaxial stress state. In the case of adequate laboratory test data from the long-term consolidation test, the viscoplastic parameters, C_1 and C_2, are determined by the data-adjusting method from the existing triaxial test results. The rest of the parameters are conventionally determined by the data-adjusting method through the comparison of the simulated results with the experimental values. However, often some empirical relations can be used to determine the parameters as the tentative one.

2.2.2.9 Simulation results of cyclic triaxial compression tests

In order to verify the performance of the proposed model, the simulation of soft clayey soil behavior has been carried out by integrating the constitutive equations under undrained triaxial conditions. Verification of the model is accomplished by a comparison of the simulated results with the experimental data through the stress–strain relations and the stress paths. The soft clay was used for validation, which was sampled at Nakanoshima in Osaka city, Japan, namely, Nakanoshima clay.

The specimens were obtained through in situ borehole drilling with an undisturbed tube sampling and block sampling. For Nakanoshima clay, two individual specimens are considered: N-1 for the cyclic triaxial test and N-2 for the monotonic triaxial test. The cyclic triaxial test on the N-1 sample has been carried out at the cyclic stress ratio($CSR = (\sigma_1 - \sigma_3)/2\sigma'_{m0} = q/2\sigma'_{m0}$) equal to 0.30. The monotonic triaxial test on the N-2 sample has been conducted as a strain-controlled testing method with a strain rate of 0.05%/min.

Following the described procedure, the material parameters of Nakanoshima clay are listed in Table 2.2. The Nakanoshima clay samples used for the cyclic triaxial test and the monotonic test are slightly different. Therefore, different sets of parameters are obtained for each sample. Comparing the initial void ratio of the two samples, the N-1 sample has a smaller void ratio that exhibits stiffer behavior compared with the sample for monotonic test N-2. Accordingly, the values of the hardening parameter and the structural parameters for the monotonic sample have been determined providing more softening behavior.

The simulated and the experimental results of the stress–strain relations and the stress paths for N-1 sample under cyclic triaxial loading conditions are presented in Figure 2.8. The stress–strain relations of the simulated results demonstrate good agreement with the experimental data, in terms of the strain levels in both the compression and extension sides and the number of cycles. Efforts were made to adjust the parameters such as the structural parameters, the hardening parameters, and the strain-dependent parameter, so that the best possible agreement could be achieved between the simulation and the experimental results. The simulated result of the stress paths, on the other hand, does not show such good agreement with the experimental data. This might be attributed to the measurement technique of the pore water pressure during the experiments, which was measured at the base of the specimen, while in the simulation, the constitutive equation is integrated neglecting the size of the specimen as usual. The

Table 2.2 Material parameters of Nakanoshima clay

		Nakanoshima clay	
Sample no.		*N-1*	*N-2*
Test		*Cyclic test*	*Monotonic test*
Parameters			
Initial void ratio	e_0	1.373	1.573
Compression index	Λ	0.2173	
Swelling index	K	0.0344	
Initial elastic shear modulus (kPa)	G_0	22670	
Stress ratio at compression	M^*_{mc}	1.143	
Stress ratio at extension	M^*_{me}	1.061	
Viscoplastic parameter	m'	22.7	
Viscoplastic parameter (1/s)	C_1	1.00×10^{-5}	
Viscoplastic parameter (1/s)	C_2	3.30×10^{6}	
Structural parameter	$n = \sigma'_{maf} / \sigma'_{mai}$	0.325	
Structural parameter	β	3.7	5.7
Hardening parameter	B^*_0	105	
Hardening parameter	B^*_1	1.0	
Hardening parameter	C_f	5	75
Reference value of viscoplastic strain (%)	$\gamma^{vp*}_{(n)r}$	3.5	
Strain-dependent parameter	α	10	1
Scalar hardening parameter	A^*_2	5.1	
Scalar hardening parameter	B^*_2	2.6	
Scalar hardening parameter	B^*_3	0.0	
Overconsolidation ratio	OCR	1.0	

Source: Data from Kimoto, S., Shahbodagh Khan, B., Mirjalili, M. and Oka, F. 2015. A cyclic elasto-viscoplastic constitutive model for clay considering the nonlinear kinematic hardening rules and the structural degradation. *Int. J. Geomech., ASCE*, 15(5):A4014005-1-A4014005-14.

Note: $\sigma'_{mai} = \sigma'_{m0} \times OCR$

results of the N-2 sample under monotonic triaxial conditions are illustrated in Figure 2.9, in which the symbols show the experimental results and the solid lines represent the simulated results. The simulated results for the stress–strain relation and the stress path provide good agreement with the experimental data. Kimoto et al. (2015) successfully applied the model to the simulation of cyclic behavior under isotropic compression.

2.3 TRANSVERSELY ANISOTROPIC AND PSEUDO-ANISOTROPIC VISCOPLASTIC MODELS

It is well known that geomaterials exhibit anisotropic mechanical behavior. The anisotropy of geomaterials can be divided into two categories: an initial structural anisotropy and the stress and strain-induced anisotropy. The

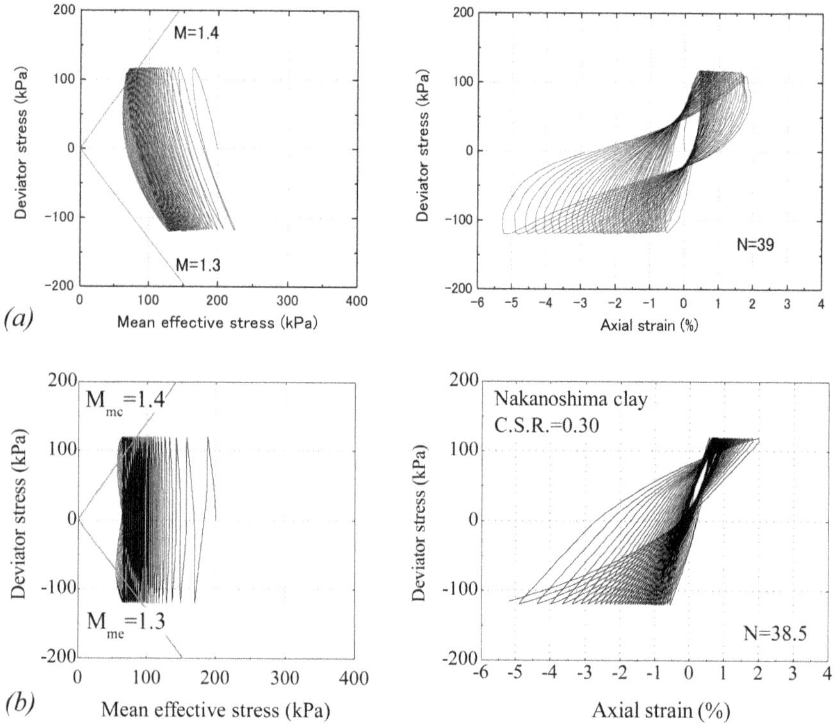

Figure 2.8 Stress paths and stress–strain relations for N-1 sample under cyclic triaxial conditions: (a) experimental results and (b) simulated results. (After Kimoto, S., Shahbodagh Khan, B., Mirjalili, M. and Oka, F. 2015. A cyclic elasto-viscoplastic constitutive model for clay considering the nonlinear kinematic hardening rules and the structural degradation. *Int. J. Geomech., ASCE*, 15(5): A4014005-1-A4014005-14.)

initial structural isotropy includes an inherent anisotropy due to the anisotropic shape of soil particles and the initial anisotropy due to deposition under gravitational conditions. In addition, the structure of soil depends on the current stress condition. In this section, we deal with the current stress-induced pseudo-anisotropy and the initial anisotropy developed during the deposition process.

2.3.1 Transformed stress tensor

Using a structural tensor, Boehler and Sawczuk (1977) explained the directional strength of solids. Following the work by Boehler and Sawczuk (1977), we introduce a structural tensor, A_{ij}, made up of dyadic or tensor products from the unit normal vector that describes the characteristic plane of the material, i.e.,

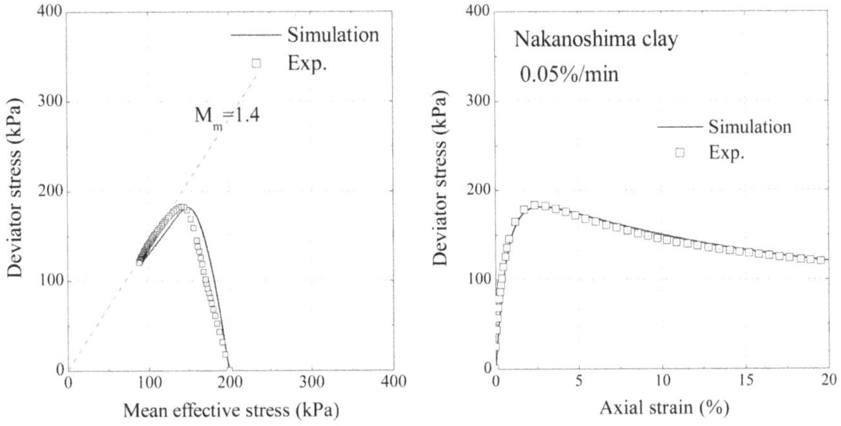

Figure 2.9 Stress paths and stress–strain relations for N-2 sample under monotonic triaxial conditions. (After Kimoto, S., Shahbodagh Khan, B., Mirjalili, M. and Oka, F. 2015. A cyclic elasto-viscoplastic constitutive model for clay considering the nonlinear kinematic hardening rules and the structural degradation. *Int. J. Geomech., ASCE*, 15(5): A4014005-1-A4014005-14.)

$$A_{ij} = m_i m_j \tag{2.56}$$

where m_i is the stress-dependent unit normal to the characteristic plane in the general stress space.

It is assumed that a stress history-dependent structural tensor, A^*_{ij}, given by the linear functional of A_{ij} with respect to the plastic strain measure can be expressed by

$$A^*_{ij} = \int_0^z K(z-z')A_{ij}(z')dz' \tag{2.57}$$

$$dz = (d\varepsilon^p_{ij} d\varepsilon^p_{ij})^{1/2} \tag{2.58}$$

where the kernel function, $K(z)$, should satisfy the principle of fading memory, and $d\varepsilon^p_{ij}$ is the inelastic strain increment tensor such as plastic and viscoplastic one.

Next, a transformed stress tensor is derived as a function of the real stress and the structural history tensor, A^*_{ij}, in the following form:

$$\sigma^A_{ij} = F_{ij}(\sigma_{kl}, A^*_{kl}) \tag{2.59}$$

where σ_{ij} is the Cauchy stress tensor and σ^A_{ij} is the transformed stress tensor.

When function F_{ij} is the isotropic tensor function and Equation 2.59 is form-invariant with respect to σ_{ij} and A^*_{ij}, the transformed stress tensor can be expressed with the representation theorem (Wang 1970, 1971) as

$$\begin{aligned}\sigma^A_{ij} &= \phi_0\delta_{ij} + \phi_1 A^*_{ij} + \phi_2\sigma_{ij} + \phi_3\sigma_{ik}\sigma_{kj} + \phi_4(\sigma_{ik}A^*_{kj} + A^*_{ik}\sigma_{kj}) \\ &\quad + \phi_5(A^*_{ik}\sigma_{kl}\sigma_{lj} + \sigma_{ik}\sigma_{kl}A^*_{lj})\end{aligned} \tag{2.60}$$

where ϕ_i; $i = 0-5$ is a scalar function of the fundamental joint invariants of σ_{ij} and A^*_{ij}. Herein, only the first-order term will be considered, i.e.,

$$\sigma^A_{ij} = \alpha(A^*_{mn}\sigma_{mn})A^*_{ij} + \beta\sigma_{ij} + \gamma(A^*_{ik}\sigma_{kj} + \sigma_{ik}A^*_{kj}) + \zeta\delta_{ij} \tag{2.61}$$

where ζ is a function of the joint invariants ($\zeta = \zeta(A^*_{mn}\sigma_{mn}, \sigma_{mm})$), and α, β, γ are material constants.

It is worth noting that the transformed tensor, σ^A_{ij}, is a symmetric tensor. The second-order terms of A^*_{ij} are excluded in Equation 2.60 because of the following relation:

$$A^*_{ik}A^*_{kj} = A^*_{ij} \tag{2.62}$$

If the kernel function $K(z)$ is the delta function:

$$A^*_{ij} = A_{ij} \tag{2.63}$$

A special case, in which $K(z)$ is the delta function, will be considered in the following; A^*_{ij} is a structural tensor that describes the current stress-induced pseudo-anisotropy.

2.3.2 Transversely isotropic model of clay

It is known that the soil deposit, which was vertically sedimented under gravitational conditions, is transversely isotropic. A clear definition is given by Malvern (1969, p. 410). Materials are transversely isotropic if they have a single axis of rotational symmetry and if, moreover, every plane containing this axis is a plane of reflection symmetry.

Adachi et al. (1991, 1995) reported the anisotropic behavior of Eastern Osaka clay. They conducted drained triaxial tests and undrained torsional shear tests on sensitive Eastern Osaka clay samples trimmed at different angles to the sedimentation plane. It was found that the undrained shear strength and deformation in the lightly overconsolidated region depend on the angle between the bedding plane and the major principal stress plane.

In order to describe the anisotropic behavior of Eastern Osaka clay, we proceed in a similar manner as in Section 2.3.1 following the work by

Oka (1993a). The transformed stress tensor is constructed using the structural tensor M_{ij}. We assume that the Eastern Osaka clay is transversely isotropic soil because of the deposition process under gravitational conditions. In the case of transversely isotropic material with respect to the rotation around the x_2 axis, the characteristic tensor, A^*_{ij}, is given by the structural tensor, M_{ij}, which describes the effect of anisotropy. The structural tensor, M_{ij}, is defined as a tensor product of the unit vector component, n_i, which is normal to the sedimentation plane.

$$M_{ij} = n_i n_j, \quad (i,j = 1,2,3) \tag{2.64}$$

As shown in Figure 2.10, when the sedimentation plane coincides with the X-Z plane, namely, angle θ = 0°, the components of the structural tensor can be expressed by the (X, Y, Z) coordinate which corresponds to the (x_1, x_2, x_3) coordinate as

$$M_{ij} = \begin{bmatrix} 0 \\ 1 \\ 0 \end{bmatrix} \otimes \begin{bmatrix} 0 \\ 1 \\ 0 \end{bmatrix} = \begin{bmatrix} 0 & 0 & 0 \\ 0 & 1 & 0 \\ 0 & 0 & 0 \end{bmatrix} \tag{2.65}$$

If $\theta \neq 0$, as shown in Figure 2.10, the structural tensor, M_{ij}, must be expressed in the (x_1, x_2, x_3) coordinate which corresponds to the principal stress direction as

$$\begin{aligned} \hat{M}_{ij} &= [R][M][R]^T \\ &= \begin{bmatrix} \cos\theta & -\sin\theta & 0 \\ \sin\theta & \cos\theta & 0 \\ 0 & 0 & 1 \end{bmatrix} \begin{bmatrix} 0 & 0 & 0 \\ 0 & 1 & 0 \\ 0 & 0 & 0 \end{bmatrix} \begin{bmatrix} \cos\theta & \sin\theta & 0 \\ -\sin\theta & \cos\theta & 0 \\ 0 & 0 & 1 \end{bmatrix} \\ &= \begin{bmatrix} \sin^2\theta & -\sin\theta\cos\theta & 0 \\ -\sin\theta\cos\theta & \cos^2\theta & 0 \\ 0 & 0 & 0 \end{bmatrix} \end{aligned} \tag{2.66}$$

in which $[R]$ is the matrix of the orthogonal tensor components that expresses the rotation $-\theta$ around the Z axis as

$$[R] = \begin{bmatrix} \cos(-\theta) & \sin(-\theta) & 0 \\ -\sin(-\theta) & \cos(-\theta) & 0 \\ 0 & 0 & 1 \end{bmatrix} = \begin{bmatrix} \cos\theta & -\sin\theta & 0 \\ \sin\theta & \cos\theta & 0 \\ 0 & 0 & 1 \end{bmatrix} \tag{2.67}$$

The transformed stress tensor needs to satisfy the objectivity, and it is assumed to be given by a linear form of the stress tensor. Furthermore,

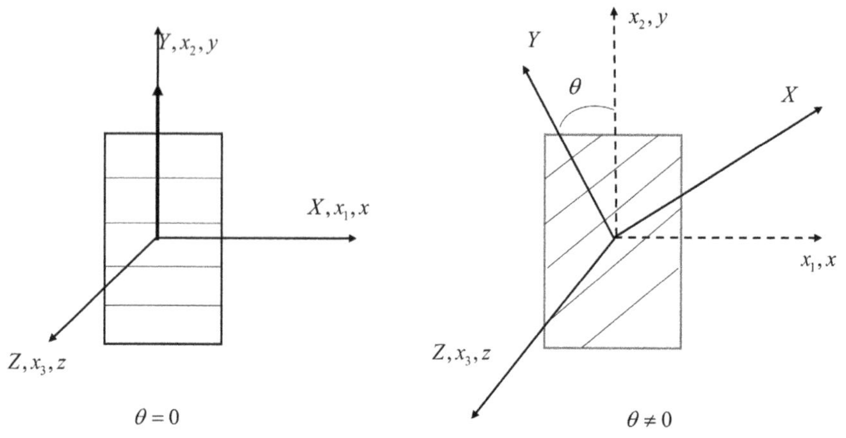

Figure 2.10 Sedimentation plane and coordinate.

it can be expressed by a linear form of the structural tensor, $\hat{M}_{ij}$, since $\hat{M}_{ik}\hat{M}_{kj} = \hat{M}_{ij}$ holds. Following Boehler's theory (1977), a simplified form of the transformed stress tensor, σ_{ij}^M, is given as

$$\sigma_{ij}^M = (\alpha + \gamma - 2\beta)(\hat{M}_{mn}\sigma_{mn})\hat{M}_{ij} + \gamma\sigma_{ij} + (\beta - \gamma)(\hat{M}_{ik}\sigma_{kj} + \sigma_{ik}\hat{M}_{kj}) \tag{2.68}$$

The above transformed tensor is introduced into the viscoplastic part of the model presented in section 5.5 of Oka and Kimoto (2012), although the elastic part of the model is isotropic.

Adachi et al. (1995) performed undrained triaxial tests with different angles between the sedimentation plane and the principal stress plane shown in Figure 2.11. Figures 2.12 and 2.13 show the simulated and the experimental results for the undrained stress–strain relations and stress paths during the undrained triaxial compression tests of Eastern Osaka clay. The material parameters used in the analysis are listed in Table 2.3. It is seen that the transversely anisotropic model can capture the anisotropic behavior of natural clay.

2.3.3 Elastic anisotropic model

In general, a fully anisotropic elastic model includes 81 independent elastic parameters. Since the soil deposit has been formed under the gravitational field, the behavior is transversely isotropic. Materials are transversely isotropic if they have a single axis of rotational symmetry and if, moreover, every plane containing this axis is a plane of reflection symmetry. In the case of transversely isotropic elastic material, five independent parameters

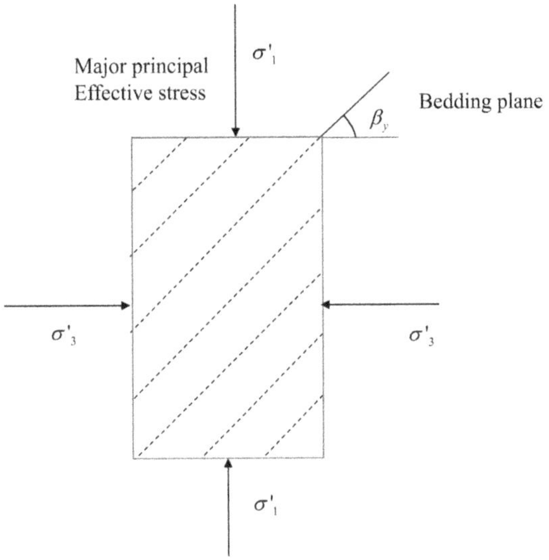

Figure 2.11 Bedding plane and the principal stress direction.

are required for the formulation. Graham and Houlsby (1983) proposed a convenient parameter that expresses the ratio between the horizontal and the vertical stiffness for the modified elastic matrix.

The generalized Hooke's law for the transversely isotropic elastic body is given by

$$\{\sigma\} = [D]\{\varepsilon\} \tag{2.69}$$

where $[D]$ is the elastic stiffness matrix which is symmetric as

$$\begin{Bmatrix} \varepsilon_{xx} \\ \varepsilon_{yy} \\ \varepsilon_{zz} \\ \varepsilon_{xy} \\ \varepsilon_{yz} \\ \varepsilon_{zx} \end{Bmatrix} = \begin{bmatrix} 1/E_x & -\nu_{yx}/E_y & -\nu_{xz}/E_x & 0 & 0 & 0 \\ -\nu_{yx}/E_y & 1/E_y & -\nu_{yx}/E_y & 0 & 0 & 0 \\ -\nu_{xz}/E_x & -\nu_{yx}/E_y & 1/E_x & 0 & 0 & 0 \\ 0 & 0 & 0 & 1/2G_{yx} & 0 & 0 \\ 0 & 0 & 0 & 0 & 1/2G_{yx} & 0 \\ 0 & 0 & 0 & 0 & 0 & (1+\nu_{xz})/E_x \end{bmatrix} \begin{Bmatrix} \sigma_{xx} \\ \sigma_{yy} \\ \sigma_{zz} \\ \sigma_{xy} \\ \sigma_{yz} \\ \sigma_{zx} \end{Bmatrix} \tag{2.70}$$

in which the y axis is perpendicular to the symmetry plane shown in Figure 2.10 and five independent constants are included.

If the angle θ in Figure 2.10 is not zero, the axis for the matrix $[D]$ has to be rotated by the matrix $[Q]$ as

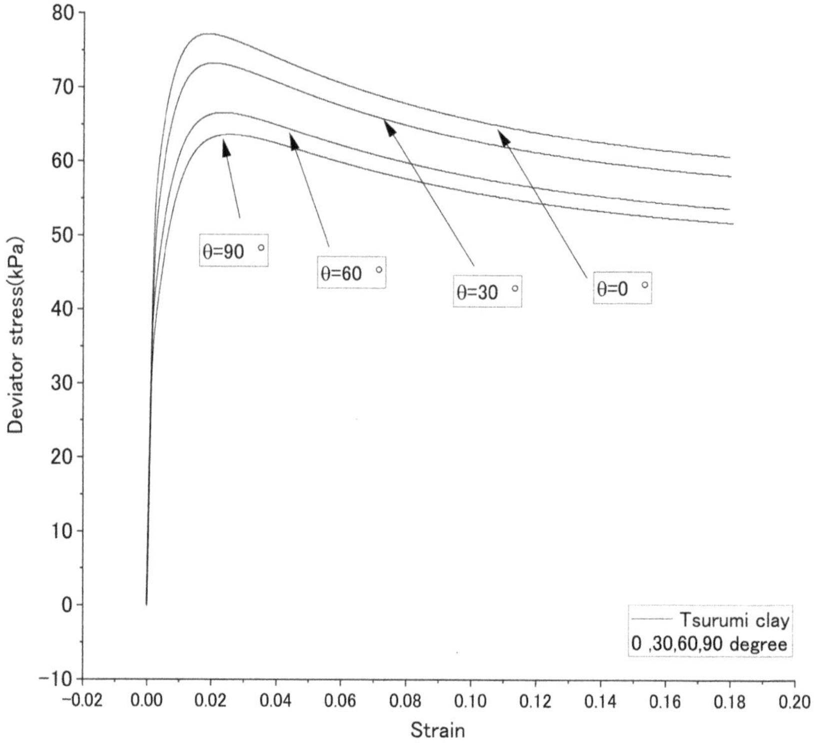

Figure 2.12a Simulated deviator stress–axial strain relations.

$$\left[\hat{D}\right] = [Q]^T [D][Q] \tag{2.71}$$

$$[Q] = \begin{bmatrix} l_1^2 & l_2^2 & l_3^2 & l_1 l_2 & l_2 l_3 & l_3 l_1 \\ m_1^2 & m_2^2 & m_3^2 & m_1 m_2 & m_2 m_3 & m_3 m_1 \\ n_1^2 & n_2^2 & n_3^2 & n_1 n_2 & n_2 n_3 & n_3 n_1 \\ 2l_1 m_1 & 2l_2 m_2 & 2l_3 m_3 & l_1 m_2 + m_1 l_2 & l_2 m_3 + m_2 l_3 & l_3 m_1 + m_3 l_1 \\ 2m_1 n_1 & 2m_2 n_2 & 2m_3 n_3 & m_1 n_2 + m_2 n_1 & m_2 n_3 + m_3 n_2 & m_3 n_1 + m_1 n_3 \\ 2n_1 l_1 & 2n_2 l_2 & 2n_3 l_3 & n_1 l_2 + n_2 l_1 & n_2 l_3 + n_3 l_2 & n_3 l_1 + n_1 l_3 \end{bmatrix} \tag{2.72}$$

where l_i, m_i, n_i $(l_i = (e_i^{'} \bullet e_x), m_i = (e_i^{'} \bullet e_y), n_i = (e_i^{'} \bullet e_z))$ are the inner product of the unit base vector of two different orthogonal coordinates,$X_i^{'}, x_i$:

$$\begin{bmatrix} l_1 & l_2 & l_3 \\ m_1 & m_2 & m_3 \\ n_1 & n_2 & n_3 \end{bmatrix} = \begin{bmatrix} \cos\theta & \sin\theta & 0 \\ -\sin\theta & \cos\theta & 0 \\ 0 & 0 & 1 \end{bmatrix} \tag{2.73}$$

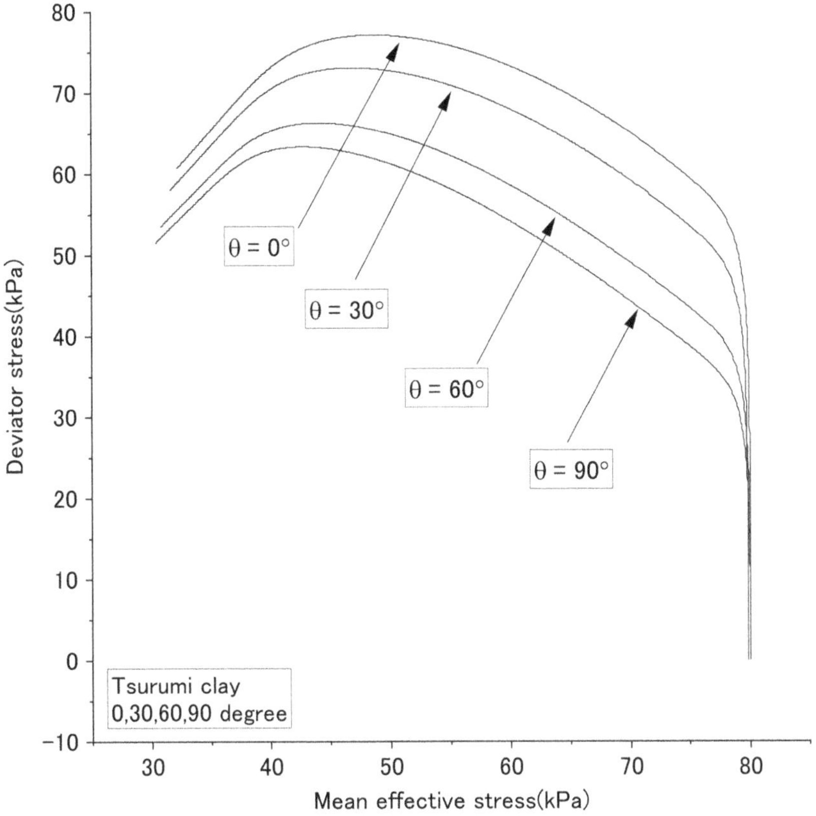

Figure 2.12b Simulated stress paths of triaxial tests for Tsurumi clay.

Finally, Hooke's law for the inclined transversely isotropic elastic body is given by

$$\{\sigma\} = \left[\hat{D}\right]\{\varepsilon\} \tag{2.74}$$

2.3.4 Current stress-induced pseudo-anisotropic failure conditions

It has been pointed out that the behavior of granular materials is characterized by the characteristic planes associated with the sedimentation histories of stress and strain, and current stress. One of the most well-known theories, in which a special plane is used, is the double shearing model (Spencer 1964). Others are the particle maximum mobilized plane theory (Murayama 1964), the mobilized plane theory, and the spatial mobilized

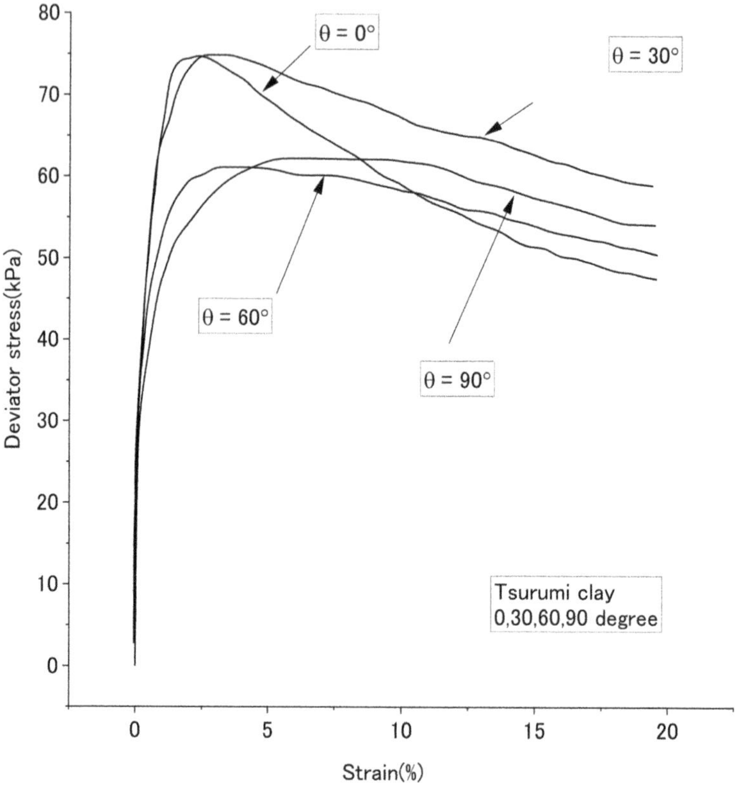

Figure 2.13a Experimental deviator stress–axial strain relations.

plane (SMP) theory (Matsuoka and Nakai 1977). Oka (1993a, 1993b) proposed a pseudo-anisotropic model using the spatial mobilized plane by Matsuoka and Nakai (1977) as a special plane. Nakai and Mihara (1984) developed a constitutive model for soil based on a kind of transformed stress tensor. The fabric tensor can be used to formulate the anisotropic constitutive model for soils (e.g., Oda and Nakayama 1988; Tobita 1988).

Let us specify the structural tensor, A_{ij}. The unit vector which is normal to the spatial mobilized plane, advocated by Matsuoka and Nakai (1977) and Nakai (2012), will be introduced. If this unit characteristic normal vector is denoted by n_i in the principal stress space, the characteristics tensor, A_{ij}, is given by the dyadic of the unit normal vector, n_i, as

$$A_{ij} = n_i n_j \tag{2.75}$$

The unit normal vector to SMP can be written in the principal stress space (Matsuoka and Nakai (1977) as

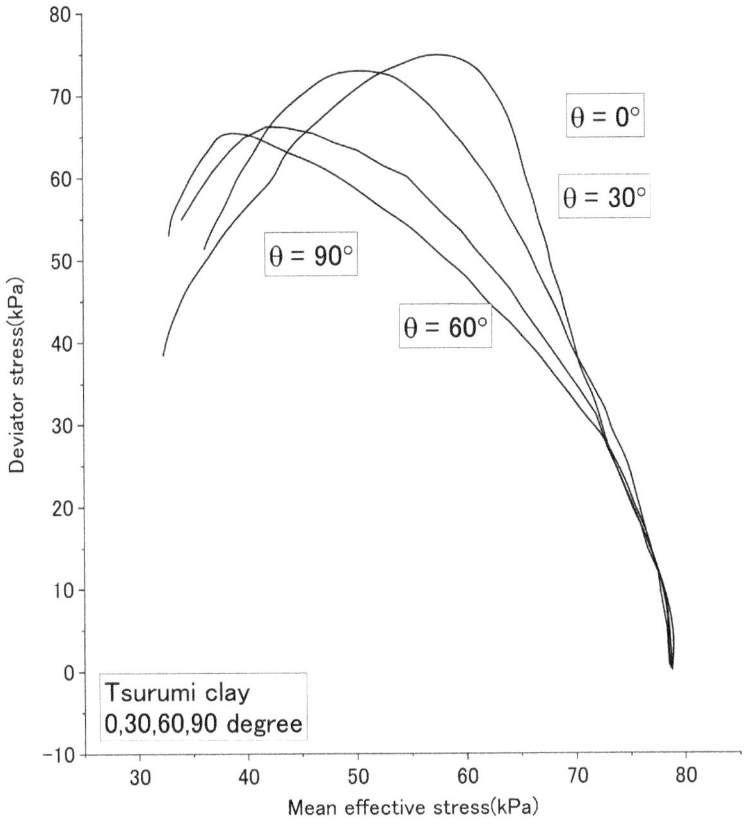

Figure 2.13b Experimental stress paths of triaxial tests for Tsurumi clay.

$$n_i = \sqrt{I_3 / I_2 \sigma_i} \tag{2.76}$$

where I_2 ($=\sigma_1\sigma_2 + \sigma_2\sigma_3 + \sigma_3\sigma_1$) and I_3 ($=\sigma_1\sigma_2\sigma_3$) are the second and third invariants of the stress tensor, respectively, and σ_i ($i = 1, 2, 3$) is the principal stress component.

We generalize the extended von Mises yield or failure criterion by using the transformed stress tensor instead of the Cauchy stress tensor. If α, β, and ζ are zero and $\gamma = 0.5$, the characteristic tensor, A_{ij}, is given by the principal stress as

$$\left[A_{ij}\right] = (I_3 / I_2) \begin{bmatrix} 1/\sigma_1 & 1/\sqrt{\sigma_1\sigma_2} & 1/\sqrt{\sigma_1\sigma_3} \\ 1/\sqrt{\sigma_2\sigma_1} & 1/\sigma_2 & 1/\sqrt{\sigma_2\sigma_3} \\ 1/\sqrt{\sigma_3\sigma_1} & 1/\sqrt{\sigma_3\sigma_2} & 1/\sigma_3 \end{bmatrix} \tag{2.77}$$

Table 2.3 Material parameters and loading conditions for Tsurumi Eastern Osaka clay

Elastic shear modulus (kPa)	G	8000
Compression index	λ	0.355
Swelling index	K	0.0477
Initial void ratio	e_0	1.806
Initial mean effective stress (kPa)	σ'_{m0}	80
Stress ratio at failure	M_f^*	1.22
Stress ratio at maximum compression	M_m^*	1.22
Viscoplastic material constant	m'	14.8
Viscoplastic material constant (1/s)	C_1	5×10^{-10}
Viscoplastic material constant (1/s)	C_2	5×10^{-10}
Hardening parameter	A_2^*	0
Hardening parameter	B_2^*	0
Hardening parameter	$B_3^* = (1+e_0)/(\lambda-\kappa)$	9.131
Material constant for anisotropy	α	1.0
Material constant for anisotropy	β	1.07
Material constant for anisotropy	γ	1.31
Strain rate (%/min)	$\dot{\varepsilon}_{11}$	6×10^{-3}

Then, if $\alpha = \beta = \zeta = 0$, $\gamma = 0.5$ in Equation 2.61, the transformed stress tensor, σ_{ij}^A, under the principal stress conditions is expressed in matrix form as

$$\left[\sigma_{ij}^A\right] = (I_3/I_2)\begin{bmatrix} 1 & \frac{1}{2}\left(\sqrt{\frac{\sigma_1}{\sigma_2}}+\sqrt{\frac{\sigma_2}{\sigma_1}}\right) & \frac{1}{2}\left(\sqrt{\frac{\sigma_1}{\sigma_3}}+\sqrt{\frac{\sigma_3}{\sigma_1}}\right) \\ \frac{1}{2}\left(\sqrt{\frac{\sigma_2}{\sigma_1}}+\sqrt{\frac{\sigma_1}{\sigma_2}}\right) & 1 & \frac{1}{2}\left(\sqrt{\frac{\sigma_2}{\sigma_3}}+\sqrt{\frac{\sigma_3}{\sigma_2}}\right) \\ \frac{1}{2}\left(\sqrt{\frac{\sigma_3}{\sigma_1}}+\sqrt{\frac{\sigma_1}{\sigma_3}}\right) & \frac{1}{2}\left(\sqrt{\frac{\sigma_3}{\sigma_2}}+\sqrt{\frac{\sigma_2}{\sigma_3}}\right) & 1 \end{bmatrix} \tag{2.78}$$

The first and second invariants of the transformed stress tensor are

$$J_1^A = \sigma_{kk}^A = 3(I_3/I_2) \tag{2.79}$$

$$\begin{aligned} J_2^A = s_{kl}^A s_{kl}^A/2 = \big[&(\sigma_1/\sigma_2 + \sigma_2/\sigma_1 + \sigma_2/\sigma_3 + \sigma_3/\sigma_2 \\ &+ \sigma_1/\sigma_3 + \sigma_3/\sigma_1)/2 + 3\big](I_3/I_2)^2 \end{aligned} \tag{2.80}$$

where s_{kl}^A is the deviatoric part of the transformed tensor:

$$s_{ij}^A = \sigma_{ij}^A - \sigma_{mm}^A \delta_{ij}/3 \tag{2.81}$$

Using the three invariants, the extended von Mises yield or failure condition can be generalized as

$$\sqrt{2J_2^A} / J_1^A = \frac{1}{3}\sqrt{\frac{1}{2}(I_1 I_2 / I_3 + 3)} \tag{2.82}$$

where I_1, I_2, and I_3 are the stress invariants.

Matsuoka and Nakai's failure or yield condition (1974) is known to be expressed by

$$I_1 I_2 / I_3 - 3 = \text{constant} \tag{2.83}$$

From Equations 2.82 and 2.83, it is seen that Matsuoka and Nakai's criterion is equivalent to the generalized extended von Mises criterion. When α, γ, and ζ are zero and $\beta = 1.0$, Equation 2.82 can be reduced to the original extended von Mises yield or failure criterion. Figure 2.14 shows an illustration of the shape of the failure surface for arbitrary parameters α, β, and γ and $\zeta = 0$; current stress-dependent condition (CSD: $\alpha = -1.0$, $\beta = 1.0$, $\gamma = 0.5$), Matsuoka and Nakai ($\alpha = 0$, $\beta = 0$, $\gamma = 0.5$), and von Mises ($\alpha = 0$, $\beta = 1.0$, $\gamma = 0$). It is now evident that the proposed failure or yield conditions can describe the various types of conditions.

2.3.5 Elasto-viscoplastic constitutive equation for clay based on the transformed stress tensor

The viscoplastic part of the elasto-viscoplastic model presented by Kimoto and Oka (2005) and in section 5.7 of Oka and Kimoto (2012) is generalized by introducing the transformed tensor instead of the Cauchy stress tensor. Using the transformed stress tensor, σ_{ij}^A, the flow rule, such as Equation 2.43, becomes

$$\dot{\varepsilon}_{ij}^{vp} = C_{ijkl} < \Phi(f_y) > \frac{\partial f_p}{\partial \sigma_{kl}^A} \tag{2.84}$$

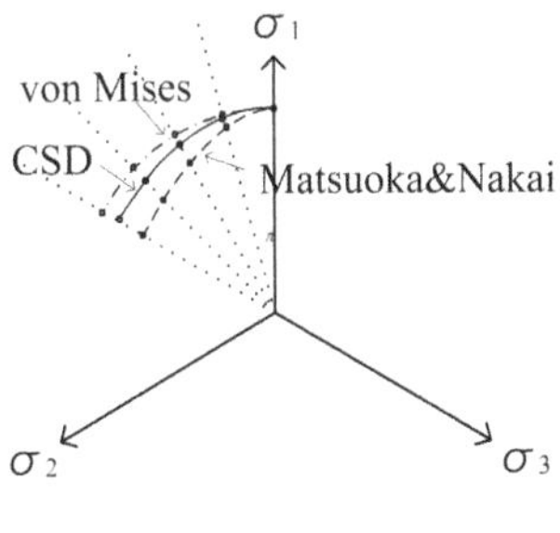

Figure 2.14 Shape of failure surfaces on the π plane.

The elastic component of the model is formulated by the Cauchy stress. Using the generalized viscoplastic model, we have analyzed the undrained compression and extension tests with constant strain rates. Figures 2.15 and 2.16 show the stress–strain relations and the stress paths. The material parameters used in the analysis are listed in Table 2.4. The responses are qualitatively similar to the undrained compression test results for clay (e.g., Adachi et al. 1995). Namely, the maximum stress and the stress ratio in the compression test are larger than those in the extension test.

2.3.6 Non-coaxiality and deviatoric flow rule

From Equation 2.78, the generalized model has non-coaxial terms. Hence, the model is applicable for the prediction of viscoplastic instability phenomena (e.g., Nemat-Nasser 1983). Next, we consider the non-coincidence of the direction of the viscoplastic strain rate tensor and the stress tensor on the π plane (octahedral plane), which is called the "deviatoric flow rule". For clay, Yong and Mckyes (1971), Lade and Musante (1978), and Nakai and Matsuoka (1986) found that the direction of the strain increment vector is not coincident with the direction of the stress on the octahedral plane, except for cases of triaxial compression and extension tests. This trend can be called the break down of the similarity between the directions of the strain rate tensor and the stress tensor. This feature is explained through the use of the generalized viscoplastic model.

Let us imagine an ordinary true triaxial test where only three axial stresses are applied to the specimen and the three axial strains are measured (e.g., Yong and Mckyes 1971). In this type of test, since the measurement of the shear strain is difficult, in practice the shear strains are not measured. The shear strain may occur, however, during the straining process. Namely, only the axial strains are measured even when shear strains occur. Hence, only the three axial strains are discussed herein. From Equation 2.84, the viscoplastic strain components are written as

$$\dot{\varepsilon}_{ij}^{vp} = C_1 < \Phi'(F) > \frac{S_{ij}^A}{\sqrt{2J_2^A}} + \chi\delta_{ij} \tag{2.85}$$

$$\langle \Phi'(F) \rangle = \langle \Phi(F) \rangle / \sigma_m^A \tag{2.86}$$

$$\chi = \frac{1}{3} C_2 \langle \Phi'(F) \rangle \left(M^{*A} - \sqrt{2J_2^A} / \sigma_m^A \right) \tag{2.87}$$

The axial viscoplastic strain rates are expressed as

$$\dot{\varepsilon}_1^{vp} = \dot{\varepsilon}_{11}^{vp} = C_1 \Phi'(F) \frac{S_{11}^A}{\sqrt{2J_2^A}} + \chi \tag{2.88}$$

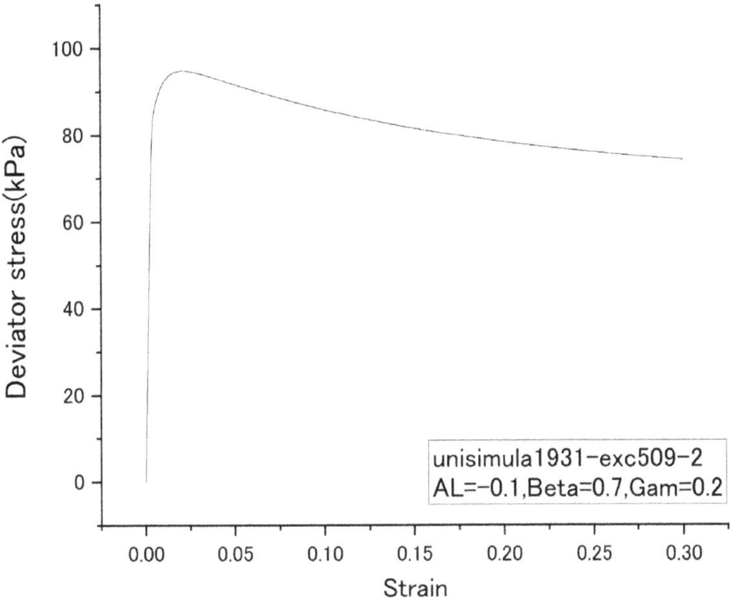

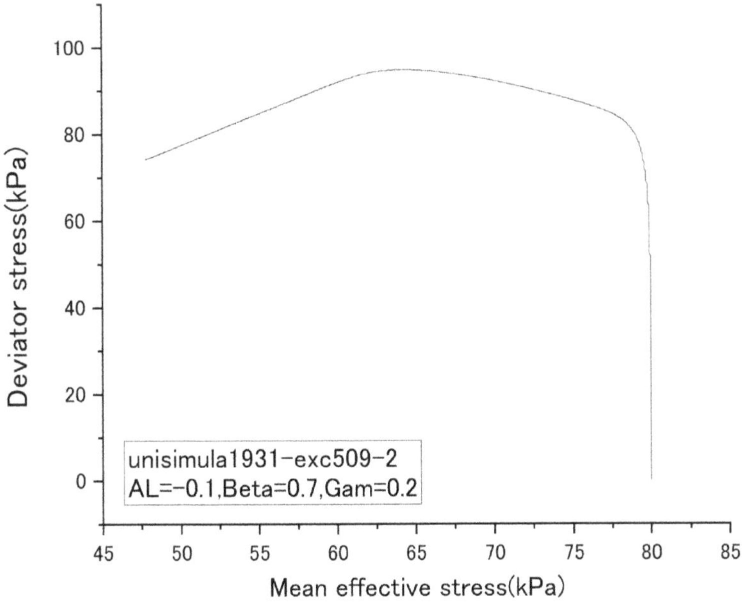

Figure 2.15 Stress–strain relation and stress path under undrained compression test. (AL = α, Beta = β, Gam = γ)

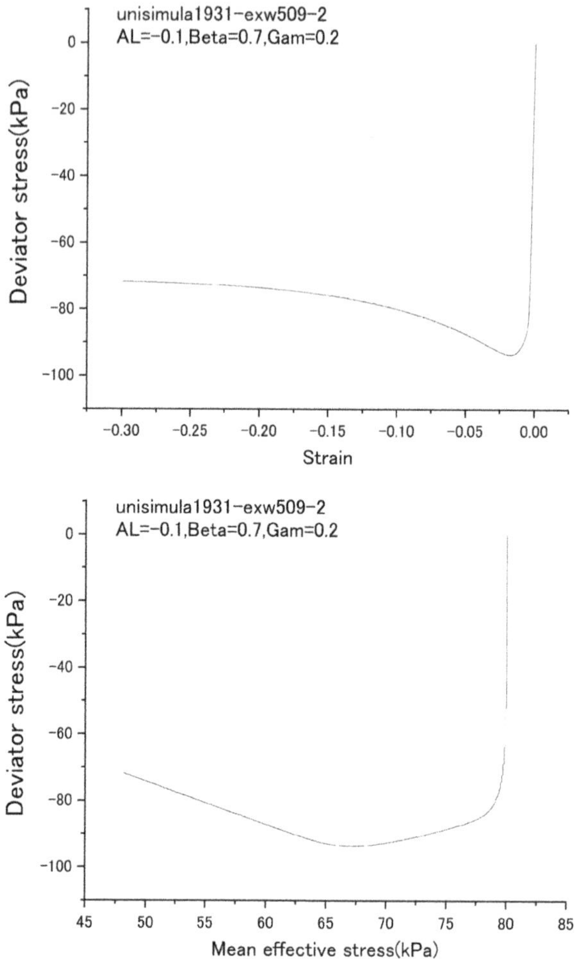

Figure 2.16 Stress–strain relation and stress path under undrained extension test. (AL = α, Beta = β, Gam = γ)

$$\dot{\varepsilon}_2^{vp} = \dot{\varepsilon}_{22}^{vp} = C_1\Phi'(F)\frac{S_{22}^A}{\sqrt{2J_2^A}} + \chi \tag{2.89}$$

$$\dot{\varepsilon}_3^{vp} = \dot{\varepsilon}_{33}^{vp} = C_1\Phi'(F)\frac{S_{33}^A}{\sqrt{2J_2^A}} + \chi \tag{2.90}$$

where S_{ij}^A is the deviatoric transformed stress tensor defined as

$$\begin{aligned} S_{ij}^A &= \sigma_{ij}^A - \sigma_m^A\delta_{ij} \\ &= \alpha(A_{mn}\sigma_{mn})A_{ij}^D + \beta S_{ij} + \gamma(A_{ik}^D\sigma_{kj} + \sigma_{ik}A_{kj}^D) \end{aligned} \tag{2.91}$$

Table 2.4 Material parameters and loading conditions for current stress-dependent model

Elastic shear modulus (kPa)	G	8000
Compression index	λ	0.355
Swelling index	κ	0.0477
Initial void ratio	e_0	1.806
Initial mean effective stress (kPa)	σ'_{m0}	80
Stress ratio at failure	M_f^*	1.22
Stress ratio at maximum compression	M_m^*	1.22
Viscoplastic material constant	m'	14.8
Viscoplastic material constant (1/s)	C_1	5×10^{-10}
Viscoplastic material constant (1/s)	C_2	5×10^{-10}
Hardening parameter	A_2^*	0
Hardening parameter	B_2^*	0
Hardening parameter	$B_3^* = (1+e_0)/(\lambda-\kappa)$	9.131
Material constant for anisotropy	α	−1.0
Material constant for anisotropy	β	0.7
Material constant for anisotropy	γ	0.2
Strain rate (%/min)	$\dot{\varepsilon}_{11}$	6×10^{-3}

where S_{ij} is the deviatoric stress tensor and superscript D denotes the deviatoric part.

Then, A_{ij}^D is the deviatoric part of A_{ij}:

$$A_{ij} = \frac{I_3}{I_2}\sqrt{\frac{1}{\sigma_i \sigma_j}} \tag{2.92}$$

$$A_{mm} = A_{11} + A_{22} + A_{33} = \left(\frac{I_3}{I_2}\right)\left(\frac{1}{\sigma_1} + \frac{1}{\sigma_2} + \frac{1}{\sigma_3}\right) = 1 \tag{2.93}$$

$$\begin{aligned} A_{11}^D &= A_{11} - A_{mm}/3 \\ &= \left(\frac{I_3}{I_2}\right)\left(\frac{1}{\sigma_1} - \frac{1}{3}\frac{I_3}{I_2}\right) = \frac{I_3}{I_2}\frac{1}{\sigma_1} - \frac{1}{3} \end{aligned} \tag{2.94}$$

Similarly, we obtain

$$A_{22}^D = \frac{I_3}{I_2}\frac{1}{\sigma_2} - \frac{1}{3}, \quad A_{33}^D = \frac{I_3}{I_2}\frac{1}{\sigma_3} - \frac{1}{3} \tag{2.95}$$

From Equations 2.85–2.95, the axial strain increments become

$$d\varepsilon_1^{vp} = C_1\Phi'(F)\frac{dt}{\sqrt{2J_2^A}}\left\{\left[\alpha\left(3\frac{I_3}{I_2}\right)\left(\frac{I_3}{I_2}\frac{1}{\sigma_1}-\frac{1}{3}\right)\right]+\frac{1}{3}\beta(2\sigma_1-\sigma_2-\sigma_3)\right\}+\chi$$

$$d\varepsilon_2^{vp} = C_1\Phi'(F)\frac{dt}{\sqrt{2J_2^A}}\left\{\left[\alpha\left(3\frac{I_3}{I_2}\right)\left(\frac{I_3}{I_2}\frac{1}{\sigma_2}-\frac{1}{3}\right)\right]+\frac{1}{3}\beta(2\sigma_2-\sigma_1-\sigma_3)\right\}+\chi \quad (2.96)$$

$$d\varepsilon_3^{vp} = C_1\Phi'(F)\frac{dt}{\sqrt{2J_2^A}}\left\{\left[\alpha\left(3\frac{I_3}{I_2}\right)\left(\frac{I_3}{I_2}\frac{1}{\sigma_3}-\frac{1}{3}\right)\right]+\frac{1}{3}\beta(2\sigma_3-\sigma_1-\sigma_2)\right\}+\chi$$

where dt is a time increment.

Let us define the direction of the stress and the strain rates by θ and θ', respectively (see Figure 2.17); $\tan\theta$ and $\tan\theta'$ are expressed by

$$\tan\theta = \frac{\sqrt{3}(\sigma_2-\sigma_3)}{(\sigma_1-\sigma_2)+(\sigma_1-\sigma_3)} \quad (\sigma_1 \geq \sigma_2 \geq \sigma_3) \quad (2.97)$$

$$\tan\theta' = \frac{\sqrt{3}(d\varepsilon_2-d\varepsilon_3)}{(d\varepsilon_1-d\varepsilon_2)+(d\varepsilon_1-d\varepsilon_3)} \quad (2.98)$$

Substituting Equation 2.96 into Equation 2.98, $\tan\theta'$ is given by

$$\tan\theta' = \sqrt{3}\left[3\alpha\left(\frac{I_3}{I_2}\right)^2\left(\frac{\sigma_3-\sigma_2}{\sigma_2\sigma_3}\right)+\beta(\sigma_2-\sigma_3)\right]$$
$$\times\left\{3\alpha\left(\frac{I_3}{I_2}\right)^2\left(\frac{\sigma_2-\sigma_1}{\sigma_1\sigma_2}+\frac{\sigma_3-\sigma_1}{\sigma_1\sigma_3}\right)+\beta\left[(\sigma_1-\sigma_2)+(\sigma_1-\sigma_3)\right]\right\}^{-1} \quad (2.99)$$

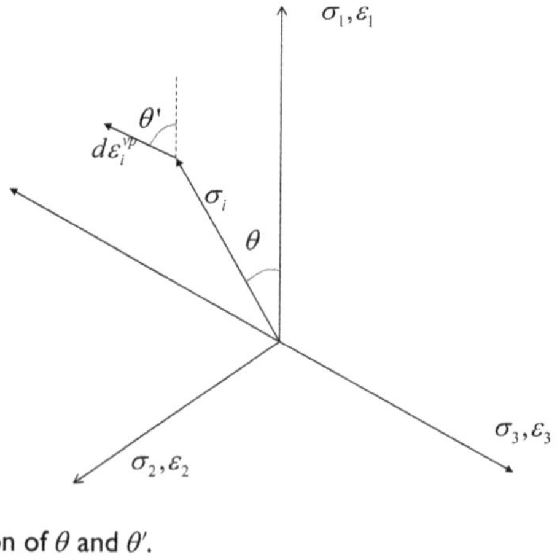

Figure 2.17 Definition of θ and θ'.

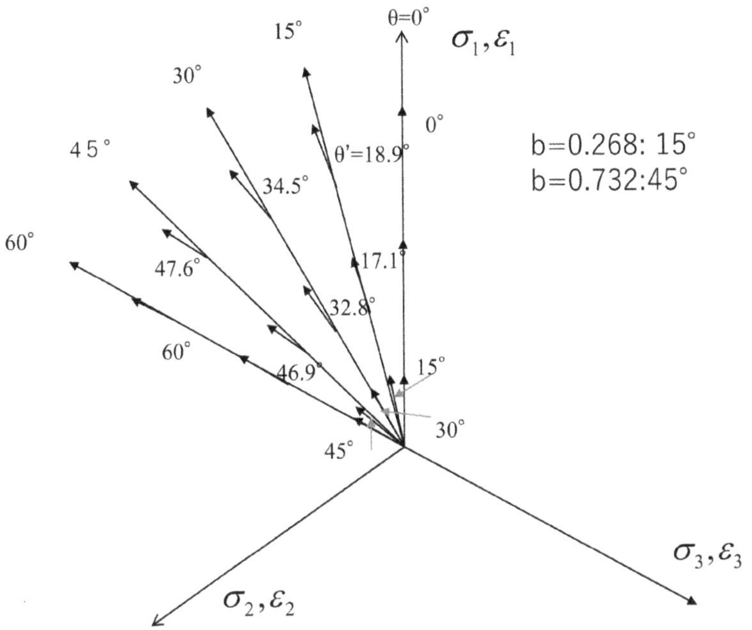

Figure 2.18 Directions of viscoplastic strain rates and stresses on the π plane.

From Equation 2.99, $\theta = \theta'$ in cases for both conventional triaxial compression $\sigma_2 = \sigma_3$ and triaxial extension $\sigma_2 = \sigma_1$. Evidently, $\theta = \theta'$ when $\alpha = 0$. For the sake of illustrating the sign of parameter α, the stress state is arbitrarily assumed where b value ($b = (\sigma_2 - \sigma_3)/(\sigma_1 - \sigma_3)$) is 0.5, e.g., $\sigma_1 = 5$ kPa, $\sigma_2 = 3$ kPa, and $\sigma_3 = 1$ kPa. In addition, it is assumed that $\beta = 1.0$. In this particular case, $\theta = 30°$ and $\theta' = 34.5°$ when $\beta = 1.0$, $\alpha = -1.0$. For θ' to be larger than θ, the sign of the parameter α should be negative in this case. Figure 2.18 shows the direction of the strain rate vector with various b values. From Figure 2.18, it is evident that the numerically evaluated deviatoric flow rule yields similar results as the experimental one (Yong and Mckyes 1971; Lade and Musante 1978; Matsuoka and Nakai 1986).

2.3.7 Constitutive equations for large strain

In the large deformation analysis, we need a corotational stress rate such as the Jaumann rate of Cauchy's stress. Regarding the objective stress rate, we can use the following corotational stress rate as

$$\hat{\sigma}_{ij} = \dot{\sigma}_{ij} - \omega_{ik}\sigma_{kj} + \sigma_{ik}\omega_{kj} \tag{2.100}$$

$$\omega_{ij} = W_{ij} - W_{ij}^P \tag{2.101}$$

where W_{ij} is the continuum spin tensor and W_{ij}^P is the inelastic (plastic) spin tensor (Loret 1983).

Thus, the elastic stretching can be given by

$$D_{ij}^e = E_{ijkl}\hat{\sigma}_{kl} \tag{2.102}$$

and the viscoplastic stretching can be given by

$$D_{ij}^{vp} = C_{ijkl}\langle\Phi(F)\rangle\partial f / \partial\sigma_{kl}^A \tag{2.103}$$

Hence, the total stretching is given by

$$D_{ij} = D_{ij}^e + D_{ij}^{vp} \tag{2.104}$$

Following the work by Dafalias (1983, 1985), the plastic spin is given as a constitutive equation. Recalling the representation theorem by Wang (1970, 1971), we can assume that the plastic spin is expressed by

$$W_{ij}^P = C_3\langle\Phi(F)\rangle(\sigma_{ik}A_{kj} - A_{ik}\sigma_{kj}) \tag{2.105}$$

where C_3 is a material constant.

2.4 CONSTITUTIVE MODELS FOR UNSATURATED SOILS

This section deals with constitutive models for unsaturated soils. The models include the effect of suction in a similar way to that taken by Oka et al. (2008) for partially saturated sand. As shown in Section 1.2, the skeleton stress is used for unsaturated soil instead of Terzaghi's effective stress. Consideration of the effect of suction on the constitutive model has been successfully applied to unsaturated sandy soil behaviors (e.g., Oka et al. 2006, 2008, 2019; Kato et al. 2009, 2014; Shahbodagh 2011; Sadeghi et al. 2014; Akaki et al. 2016; Lee 2012).

In order to take account of the suction effect on the unsaturated constitutive models, i.e., the elasto-plastic and elasto-viscoplastic models in Sections 2.1 and 2.2 can be modified using the overconsolidated boundary surface, the static yield function, and the kinematic hardening parameters with the suction effect. In addition, the suction–saturation relationship is used as a part of the constitutive models.

2.4.1 Overconsolidation boundary surface

The overconsolidation boundary surface, which separates the overconsolidated region ($f_b < 0$) from the normally consolidated region ($f_b \geq 0$), is given by

$$f_b = \overline{\eta}^*_{(0)} + M^*_m \ln \frac{\sigma'_m}{\sigma'_{mb}} = 0 \tag{2.106}$$

$$\overline{\eta}^*_{(0)} = \left\{ \left(\eta^*_{ij} - \eta^*_{ij(0)} \right) \left(\eta^*_{ij} - \eta^*_{ij(0)} \right) \right\}^{\frac{1}{2}} \tag{2.107}$$

where $\eta^*_{ij} (= S_{ij}/\sigma'_m)$ is the stress ratio tensor; S_{ij} is the deviatoric skeleton stress tensor; σ'_m is the mean skeleton stress; $\eta^*_{ij(0)}$ is the initial value of the stress ratio tensor before deformation, e.g., the value of the stress ratio at the end of consolidation; M^*_m is the value of $\eta^* = \sqrt{\eta^*_{ij}\eta^*_{ij}}$ across which dilatancy changes from negative to positive; and σ'_{mb} is a hardening parameter.

In general, the overconsolidation boundary surface may change following the evolutional equation (Equation 2.8) mentioned in Section 2.1.2.2.

The effect of suction is introduced in the hardening-softening rules of σ'_{mb} as

$$\sigma'_{mb} = \sigma'_{mau} \exp\left(\frac{1+e_0}{\lambda - \kappa} \varepsilon^{vp}_v \right) \tag{2.108}$$

$$\sigma'_{mau} = \sigma'_{ma} \left[1 + S_I \exp\left\{ -s_d \left(\frac{p^C_i}{p^C} - 1 \right) \right\} \right] \tag{2.109}$$

where σ'_{ma} is assumed to decrease with the increasing plastic strain; e_0 is the initial void ratio; λ is the compression index; κ is the swelling index; ε^{vp}_v is the viscoplastic or plastic volumetric strain; p^C is the present suction value; S_I is a material parameter that denotes the increase in yield stress when suction increases from zero to the reference value p^C_i; and s_d is a parameter that controls the rate of increase or decrease in σ'_{mb} with suction. The change in σ'_{ma} is given by

$$\sigma'_{ma} = \left\{ \sigma'_{maf} + \left(\sigma'_{mai} - \sigma'_{maf} \right) \exp(-\beta z) \right\} \tag{2.110}$$

$$z = \int dz = \int (\dot{\varepsilon}^{vp}_{ij} \dot{\varepsilon}^{vp}_{ij})^{1/2} dt \tag{2.111}$$

where σ'_{mai} and σ'_{maf} are the initial and final values for σ'_{ma}, respectively; β is a material parameter that controls the rate of structural degradation; and z is an accumulation of the second invariant of the viscoplastic strain rate tensor, $\dot{\varepsilon}^{vp}_{ij}$, or the plastic strain rate tensor, $\dot{\varepsilon}^{p}_{ij}$.

In Equation 2.109, S_I is the strength ratio of unsaturated soil when the value of suction P^C equals P^C_i, and s_d controls the decreasing ratio of strength with the decreasing suction. The term p^C_i can be set to be the initial

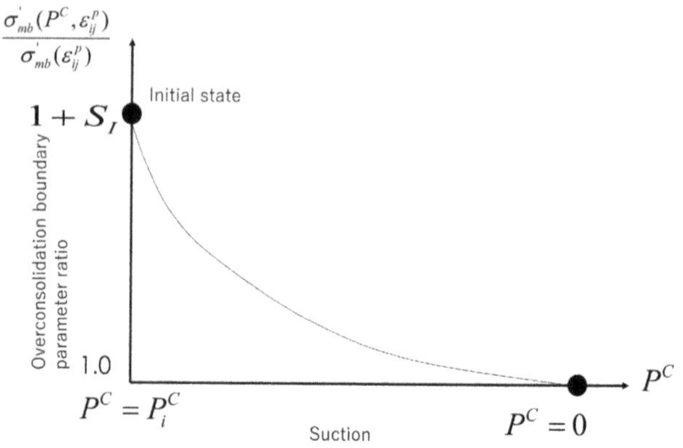

Figure 2.19 Degradation due to the reduction in suction.

value of suction. At the initial state, when $p^C = p_i^C$, the strength ratio of the unsaturated soil to the saturated soil is $1 + S_I$, and it decreases with a drop in suction, as shown in Figure 2.19.

2.4.2 Yield function

The yield function is obtained by considering the non-linear kinematic hardening rule for changes in the stress ratio, the mean effective stress, and the viscoplastic volumetric strain:

$$f_y = \bar{\eta}_\chi^* + \tilde{M}^* \left(\ln \frac{\sigma'_{mk}}{\sigma'^{(s)}_{my}} + \left| \ln \frac{\sigma'_m}{\sigma'_{mk}} - y_m^* \right| \right) = 0 \tag{2.112}$$

$$\bar{\eta}_\chi^* = \left\{ \left(\eta_{ij}^* - \chi_{ij}^* \right) \left(\eta_{ij}^* - \chi_{ij}^* \right) \right\}^{\frac{1}{2}} \tag{2.113}$$

where σ'_{mk} is a material constant; y_m^* is the scalar kinematic hardening parameter; $\tilde{M}^*$ is the dilatancy coefficient; $\sigma'^{(s)}_{my}$ denotes the static hardening parameter that controls the size of the static yield surface; and χ_{ij}^* is the kinematic hardening variable, the so-called back stress, which has the same dimensions as the stress ratio, η_{ij}^*.

Incorporating strain softening for the structural degradation, the hardening rule for $\sigma'^{(s)}_{my}$ can be expressed as

$$\sigma'^{(s)}_{my} = \frac{\sigma'_{mau}}{\sigma'_{mai}} \sigma'^{(s)}_{myi} \tag{2.114}$$

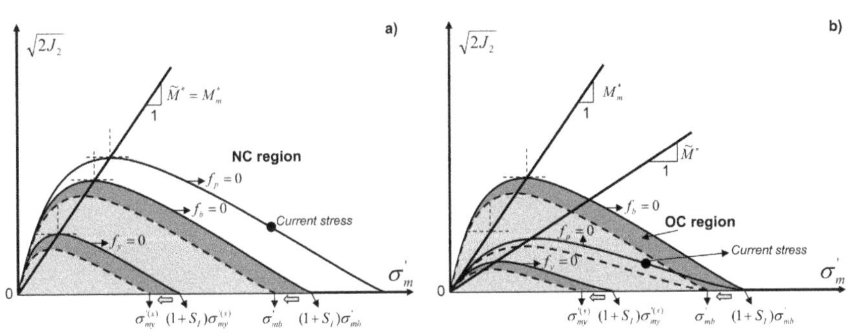

Figure 2.20 Shrinkage of the OC boundary surface, static yield function, and potential function in (a) the NC region and (b) the OC region. (After Oka, F., Shahbodagh, B. and Kimoto, S. 2019. A computational model for dynamic strain localization in unsaturated elasto-viscoplastic soils. *Int. J. Numer. Anal. Meth. Geomech.* 43(1): 138–165.)

Figure 2.20 shows the shrinkage of the OC boundary surface, the yield function, and the viscoplastic potential function in both the NC and OC regions. It is worth noting that the yield function is taken as a static yield function for the elasto-viscoplastic model.

2.4.3 Kinematic hardening rule

For the elasto-plastic model, we can use the plastic strain and the plastic strain rate instead of the viscoplastic strain and the viscoplastic strain rate.

The evolutional equation for the non-linear kinematic hardening parameter, $\overset{*}{\chi}_{ij}$, is

$$d\overset{*}{\chi}_{ij} = B^* \left(A^* de_{ij}^{vp} - \overset{*}{\chi}_{ij} d\gamma^{vp} \right) \tag{2.115}$$

where A^* and B^* are material parameters; de_{ij}^{vp} is the viscoplastic deviatoric strain increment tensor; $d\gamma^{vp} = \sqrt{de_{ij}^{vp} de_{ij}^{vp}}$ is the second invariant of the viscoplastic deviatoric strain increment tensor; A^* is related to the stress ratio at failure, namely, $A^* = M_f^*$; and B^* is proposed to be dependent on the viscoplastic deviatoric strain as

$$B^* = \left[\left(B_{\max}^* - B_1^* \right) \exp\left(-C_f \gamma_{(n)}^{vp*} \right) + B_1^* \right] \left(1 + S_{IB} \exp\left\{ -s_{db} \left(\frac{P_i^C}{P^C} - 1 \right) \right\} \right) \tag{2.116}$$

where S_{IB} is the strength ratio of unsaturated soil when the value of suction P^C equals P_i^C; s_{db} controls the decreasing ratio of strength with the decreasing suction; B_1^* is the lower boundary of B^*; C_f is the parameter controlling the rate of reduction; $\gamma_{(n)}^{vp*}$ is the accumulated value of the viscoplastic shear strain between two sequential stress reversal points in the previous cycle; and $B^*_{\max}$ is the maximum value of parameter B^*, which is defined following the proposed method by Oka et al. (1999), namely,

$$B^*_{\max} = \begin{cases} B^*_0 & \text{: Before reaching failure line} \\ \dfrac{B^*_0}{1+\gamma_{(n)\max}^{vp*}/\gamma_{(n)r}^{vp*}} & \text{: After reaching failure line} \end{cases} \tag{2.117}$$

where B^*_0 is the initial value of B^*; $\gamma_{(n)\max}^{vp*}$ is the maximum value of $\gamma_{(n)}^{vp*}$ in the past cycles; and $\gamma_{(n)r}^{vp*}$ is the viscoplastic referential strain.

In order to improve the predicted results under cyclic loading conditions, the scalar non-linear kinematic hardening parameter, y^*_m, is adopted and decomposed into two parts, namely, linear and non-linear kinematic hardening as follows:

$$y^*_m = y^*_{m1} + y^*_{m2} \tag{2.118}$$

where y^*_{m1} and y^*_{m2} are the non-linear and the linear kinematic hardening parameters.

The scalar non-linear kinematic hardening parameter, y^*_{m1}, is introduced as

$$dy^*_{m1} = B^*_2\left(A^*_2 d\varepsilon_v^{vp} - y^*_{m1}\left|d\varepsilon_v^{vp}\right|\right) \tag{2.119}$$

where A^*_2 and B^*_2 are the material parameters, and $d\varepsilon_v^{vp}$ is the viscoplastic volumetric strain increment. The values for A^*_2 and B^*_2 are determined by the data-adjusting method from laboratory test data. Figure 2.21 shows the relation between the scalar non-linear hardening, y^*_{m1}, and the volumetric strain, ε_v^{vp}. In addition, the linear kinematic hardening rule can be used for y^*_{m2} as follows:

$$dy^*_{m2} = B^*_3 d\varepsilon_v^{vp} \tag{2.120}$$

where B^*_3 is a hardening parameter.

For the above volumetric hardening, the same suction dependence can be used:

$$dy^*_{m1} = B^*_2\left(A^*_2 d\varepsilon_v^{vp} - y^*_{m1}\left|d\varepsilon_v^{vp}\right|\right)\left(1 + S_{IB}\exp\left\{-s_{db}\left(\frac{P_i^C}{P^C} - 1\right)\right\}\right) \tag{2.121}$$

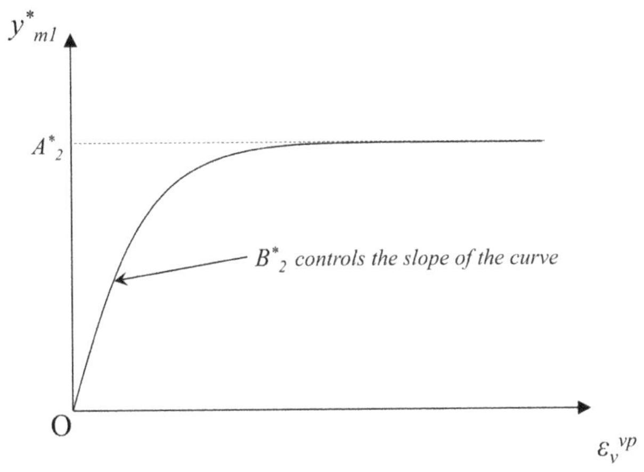

Figure 2.21 Non-linear kinematic hardening function y^*_{ml} and viscoplastic volumetric strain ε_v^{vp}.

$$dy^*_{m2} = B^*_3 d\varepsilon_v^{vp}\left(1 + S_{IB}\exp\left\{-s_{db}\left(\frac{P_i^C}{P^C} - 1\right)\right\}\right) \tag{2.122}$$

2.4.4 Compression parameter

The compression index for unsaturated soil is assumed to be given by

$$\lambda = \lambda_0 + \lambda_s \exp\left(-s_a\left(\frac{P_i^c}{P^c} - 1\right)\right) \tag{2.123}$$

where λ_0 is the compression index for saturated soil and s_a is the parameter of the suction dependency of the compression index.

2.4.5 Hydraulic constitutive equations of unsaturated soil

The interaction between water and air induced by the surface tension can be described by the soil-water characteristic curve. In this section, the van Genuchten model (1980) is adopted:

$$S_{re} = \left(1 + (\alpha P^C)^{n'}\right)^{-m} \tag{2.124}$$

where α, m, and n' are the material parameters; the relation $m = 1 - 1/n'$ holds; $P^C(=P^G - P^W)$ is the suction; and S_{re} is the effective saturation, i.e.,

$$S_{re} = \frac{S_r - S_{r\min}}{S_{r\max} - S_{r\min}} \tag{2.125}$$

where $S_{r\max}$ and $S_{r\min}$ are the maximum and minimum saturation values, respectively.

For the permeability of unsaturated soil, the following equation is used in order to avoid numerical instability (Garcia et al. 2010):

$$k^W = k_s^W S_{re}{}^a \left(1-(1-S_{re}^{1/m})^{n'}\right),\ k^G = k_s^G (1-S_{re})^b \left(1-(S_{re}^{1/m})^{n'}\right) \tag{2.126}$$

where k_s^W and k_s^G depend on the void ratio e in the following form:

$$k_s^W = k_{s0}^W \exp[(e-e_0)/C_k],\ k_s^G = k_{s0}^G \exp[(e-e_0)/C_k] \tag{2.127}$$

where k_{s0}^W and k_{s0}^G are the initial values at $e = e_0$, and C_k is the material parameter.

For the case that the soil-water characteristic curve depends on the void ratio change, Gallipoli et al. (2003, 2008) proposed a soil-water characteristic curve considering the change in the void ratio in which the parameter α in van Genuchten's equation (Equation 2.124) is given by

$$\alpha = \phi e^{\psi} \tag{2.128}$$

where ϕ and ψ are material parameters.

2.4.6 Simulation of cyclic drained tests for unsaturated sandy soil

Figures 2.22–2.24 show the deviator stress–axial strain and the volumetric strain–axial strain relations of the experimental and simulation results for unsaturated sandy soil under cyclic suction-controlled triaxial conditions. The material parameters were determined so that the simulated results provide the closest trend to the experimental results, under three different strain rates, namely, 0.01%/min, 0.1%/min, and 0.75%/min with a constant cyclic stress ratio of 0.2. The number of cycles is 100. The material parameters are listed in Table 2.5. The maximum strain level in the case of $\dot{\varepsilon} = 0.1\%/\text{min}$ is higher than that in the case of $\dot{\varepsilon} = 0.01\%/\text{min}$, although the higher strain level is expected to be observed for the smaller strain rate. This might be due to the experimental problem in the tests. In the cases of $\dot{\varepsilon} = 0.01\%/\text{min}$ and $\dot{\varepsilon} = 0.75\%/\text{min}$, the simulation results demonstrate good agreement with the experimental results.

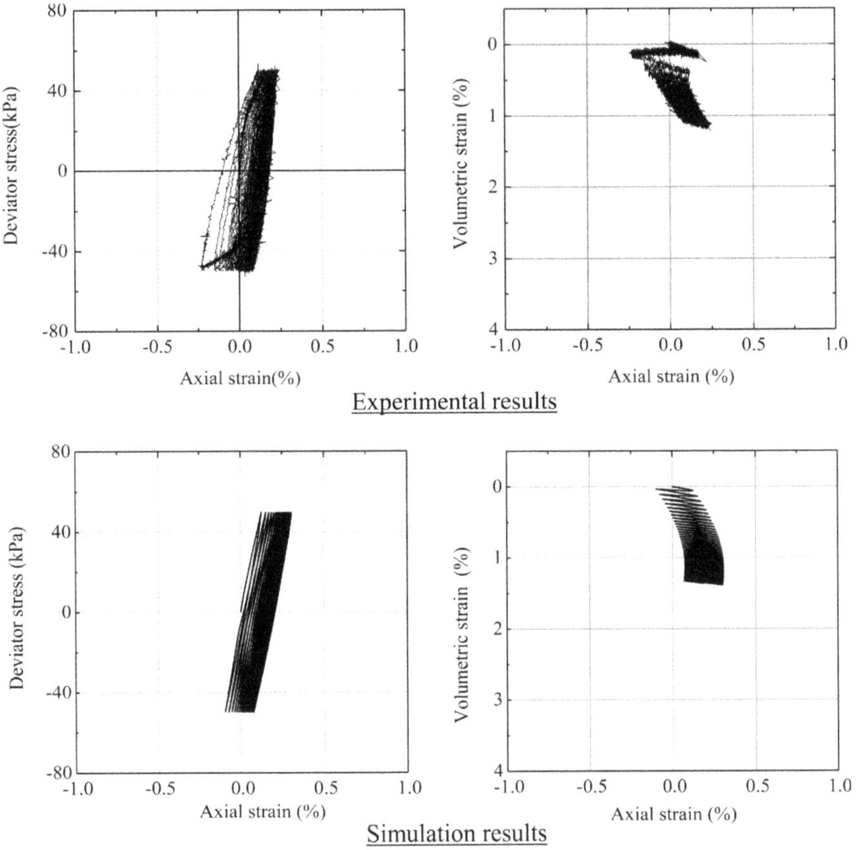

Figure 2.22 Experimental and simulated results for the undrained sandy soil under cyclic suction-controlled triaxial conditions (strain rate = 0.01%/min, suction = 50 kPa).

2.4.7 Simulation of fully undrained triaxial tests for unsaturated sandy soil

2.4.7.1 Test results

Using a special testing method (Oka et al. 2010; Kimoto et al. 2011), Kimoto et al. (2017) carried out fully undrained tests on unsaturated sandy soil. In the fully undrained test, no pore air or pore water is allowed to flow out in the shearing process. A pressure gauge and a diaphragm valve are controlled by air pressure from outside the cell in order to accurately measure the pore air pressure. The volume changes were measured by taking photographs through the acrylic cell from the orthogonal two directions at intervals of 1% axial strain.

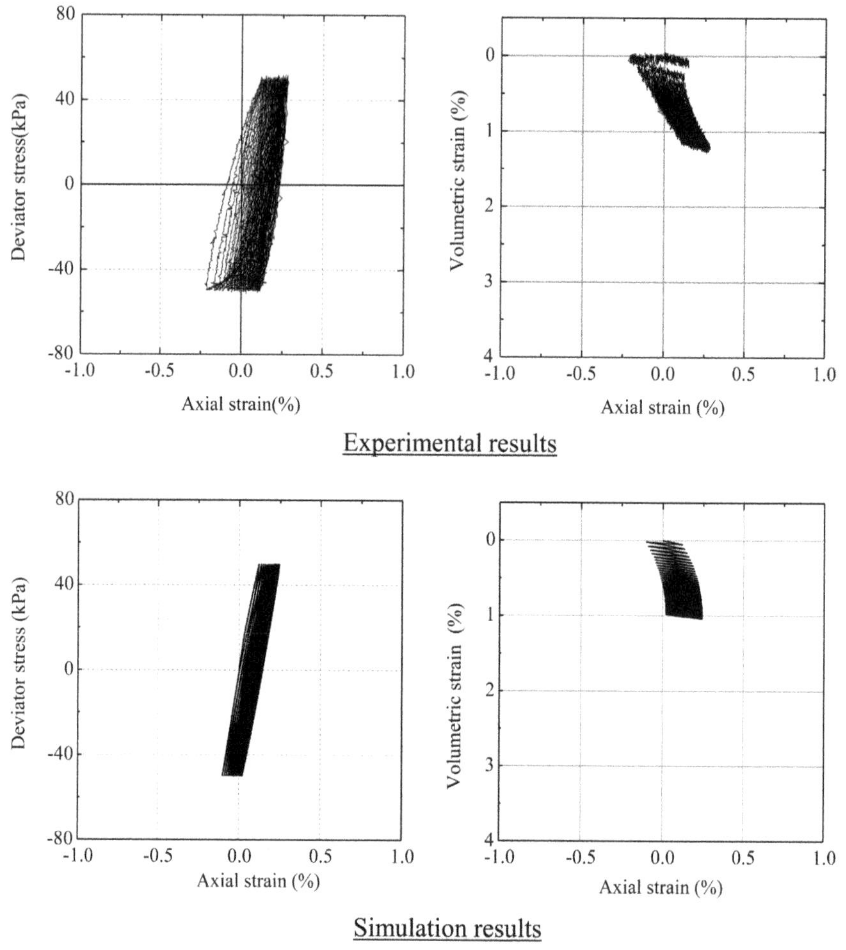

Figure 2.23 **Experimental and simulated results for the undrained sandy soil under cyclic suction-controlled triaxial conditions (strain rate = 0.1%/min, suction = 50 kPa).**

The sandy soil was obtained from a river embankment of the Kizu River of the Yodo River system in Kyoto prefecture and passed through a 2-mm mesh sieve. The soil is composed of 74% sand, 17% silt, and 9.2% clay. The soil was well mixed with water to make up the optimum water content and then compacted in six or seven layers. The specimen was compacted to achieve an 85% degree of compaction which corresponds to a dry density of around 1.61 g/cm^3.

The test conditions and specimen data before loading are given in Table 2.6. The values of the initial mean skeleton stress are set to be about 104 kPa for the cases U1-10, U1-50, and U1-80. The suction values are set to be 10, 50, and 80 kPa. The axial strain rate of 0.1%/min was applied

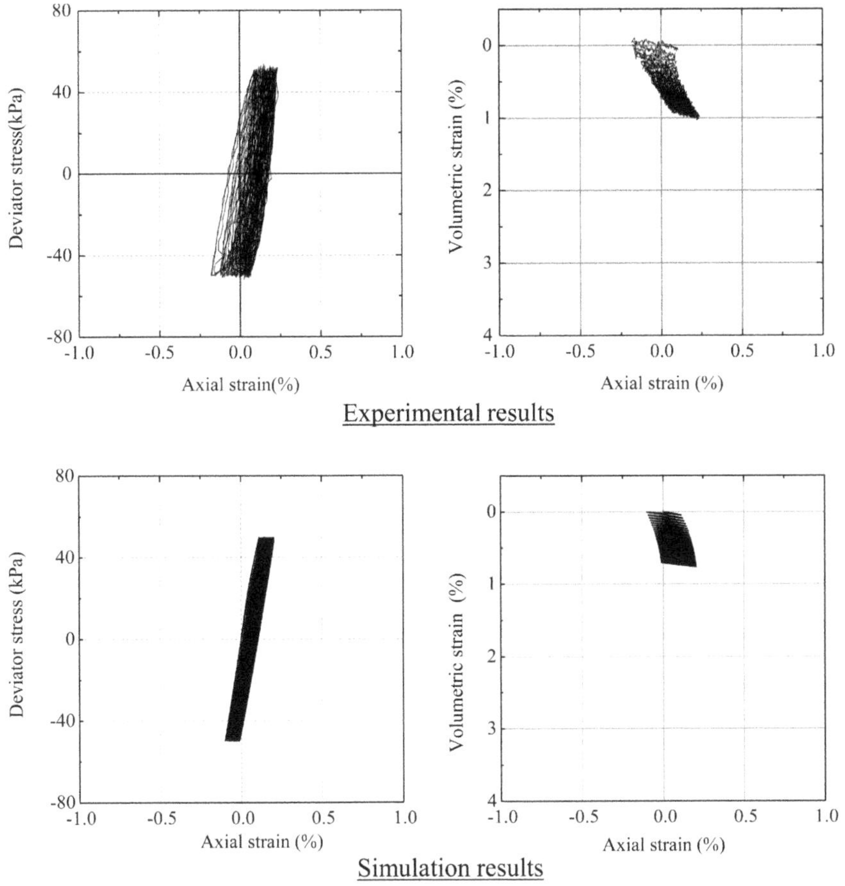

Figure 2.24 Experimental and simulated results for the undrained sandy soil under cyclic suction-controlled triaxial conditions (strain rate=0.75%/min, suction=50kPa).

under fully undrained conditions for both water and air. Figure 2.25 shows the test results with suction levels of 10, 50, and 80 kPa under the initial mean skeleton stress of 104 kPa. The mean skeleton stress, $\sigma'_m = \sigma_m - P^F$; $P^F = S_r u_w + (1 - S_r) u_a$, is used as a stress variable in the stress paths.

2.4.7.2 Simulation results

The constitutive model used in the analysis is the elasto-viscoplastic model described in Section 2.4. The water retention curve of Equation 2.124 with Equation 2.128, and the compression index of Equation 2.123 were used. Tables 2.7 and 2.8 show the material parameters used in the simulations. Since the water content, w, is constant under fully undrained conditions,

Table 2.5 Material parameters and initial conditions for Yodogawa sandy soil

Initial elastic shear modulus (kPa)	G_0	20000
Compression index	λ	0.061
Swelling index	κ	0.004
Initial void ratio	e_0	0.65
Initial mean effective stress (kPa)	σ'_{m0}	127.0
Stress ratio at failure	M_f^*	1.18
Stress ratio at maximum compression	M_m^*	0.799
Viscoplastic material constant	m'	40.0
Viscoplastic material constant (1/s)	C_1	5×10^{-5}
Viscoplastic material constant (1/s)	C_2	5×10^{-5}
Structural parameter	$n = \sigma'_{maf} / \sigma'_{mai}$	0.53
Structural parameter	β	5.0
Hardening parameter	B_0^*	200
Hardening parameter	B_1^*	15.0
Hardening parameter	C_f	5.0
Strain-dependent modulus parameter	α	1.0
Hardening parameter	A_2^*	9.0
Hardening parameter	B_2^*	3.6
Initial suction (kPa)	P_i^C	50.0
Suction parameter	S_I	0.2
Suction parameter	s_d	0.6

Table 2.6 Test conditions for fully undrained triaxial tests

	UI-10	*UI-50*	*UI-80*
Cell pressure (kPa)	300	281.5	270
Pore air pressure (kPa)	200	200	200
Pore water pressure (kPa)	190	150	120
Suction (kPa)	10	50	80
Void ratio e	0.614	0.607	0.595
Saturation *Sr* (%)	47.17	43.25	42.46
Initial mean skeleton stress (kPa)	104	103	104

$S_r - P^C$ relations can be obtained for each value of w using the relations of $e = wG_s / S_r$. The parameters ϕ and ψ are determined by the curve fitting of soil-water characteristic curves during the shearing for each suction level. Figure 2.26a–d shows the simulation results for the initial mean skeleton stress of 104 kPa. The model used in the present study describes the air–water–soil coupled mechanical behavior such as an increase in the pore pressure, change in the suction, volumetric compression, and the suction effect on the strength.

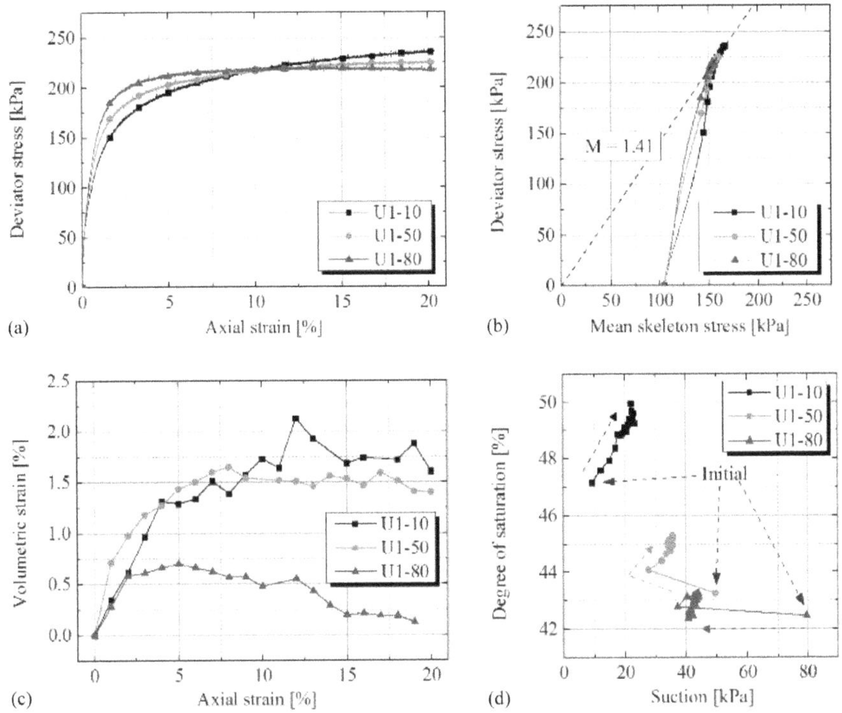

Figure 2.25 Fully undrained triaxial test results (initial skeleton stress = 104 kPa).

Table 2.7 Material parameters for elasto-viscoplastic model

Case	*U1-10*	*U1-50*	*U1-80*
Void ratio e_0	0.614	0.607	0.595
Compression index λ_0		0.044	
Compression index parameter λ_s, s_a		0.050, 0.050	
Reference suction P_i^C (kPa)		80.0	
Swelling index k		0.009	
Initial shear modulus G_0 (kPa)		15800	
Over consolidation ratio OCR^*		1.20	
Viscoplastic parameter m'		50.0	
Viscoplastic parameters C_{01}, C_{02} (1/s)		$5.0 \times 10^{-10}, 3.0 \times 10^{-10}$	
Microstructure n, b		0.50, 1.00	
Stress ratio at critical state M_m, M_f		1.410	
Suction parameters S_I, s_d		1.20, 0.90	
Hardening parameter B_0^*		1.50	
Hardening parameters A_2^*, B_2^*		5.0, 10.0	

Table 2.8 Material parameters for water retention curve

Case	*UI-10*	*UI-50*	*UI-80*
n'		1.50	
α		3.00	
S_{rmax} (%)		77.6	
S_{rmin} (%)		40.0	
ϕ	6.50×10^6	1.10×10^6	5.00
ψ	30.0	12.0	1.0

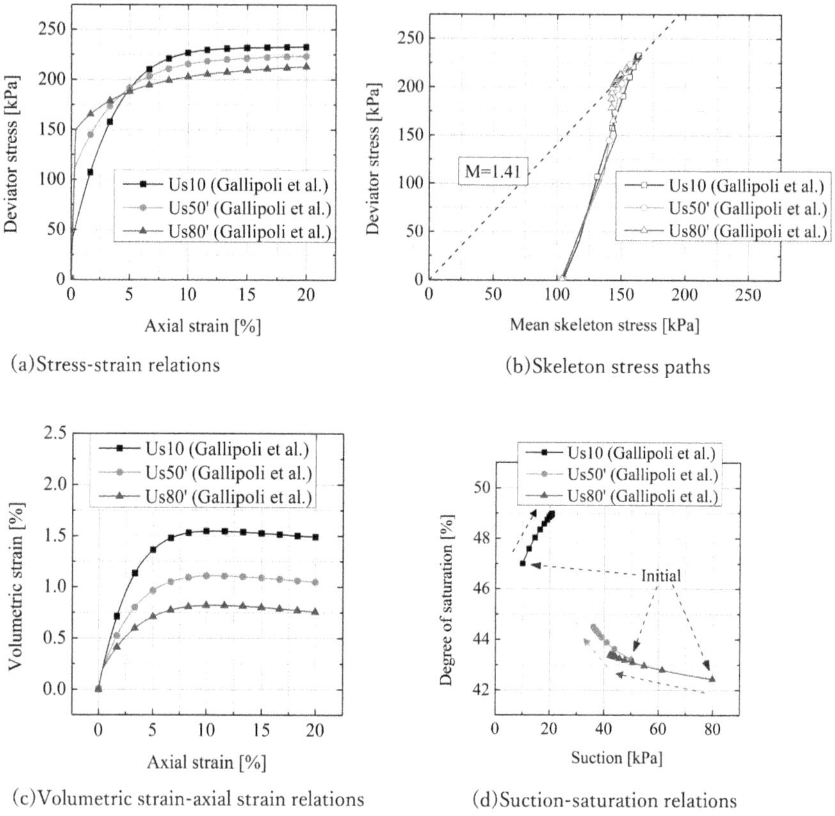

Figure 2.26 Simulation results for fully undrained triaxial tests (initial skeleton stress = 104 kPa).

2.5 NON-COAXIAL CONSTITUTIVE MODELS

2.5.1 Introduction

The coaxial concept has been used in plasticity theory, in which the principal stress direction coincides with the direction of the strain rate. On the other hand, many researchers have argued that this concept does not hold for geomaterials (e.g., Roscoe et al. 1967; Roscoe 1970; Drescher and de Jong 1972; Rudnicki and Rice 1975; Arthur et al. 1980). In general, the direction of the strain rate does not coincide with the principal stress direction, i.e., non-coaxial behavior has been observed in the experiment (e.g., Tsutsumi and Hashiguchi 2005).

Although anisotropic models reproduce the non-coaxiality, herein we consider the isotropic models that can describe the non-coaxiality. As mentioned in Section 2.3.4, the current stress-induced pseudo-anisotropic model (Oka 1993a,b) describes a kind of non-coaxiality such as the deviatoric flow rule, in which the transformed stress tensor is adopted in the formulation. In addition, the hypo-plastic model (Kolymbas 1991) and the t_{ij}-model (Nakai and Mihara 1984) could reproduce the non-coaxiality.

Another type of non-coaxial model, the vertex non-coaxial model, has been developed by Rudnicki and Rice (1975). They proposed a non-coaxial plasticity theory for fissured rocks considering the vertex yield surface. They assumed that the additional strain rate which is tangential to the yield surface and the strain increments are purely deviatoric. Yu (2006) historically reviewed non-coaxial plasticity models with tangential strain rate components to the yield surface.

On the other hand, De Jong (1959, 1971) proposed a planar model called the "double sliding model" for granular materials using the stress characteristics of the partial differential equations composed of the equations of equilibrium and Coulomb's yield condition. Then, following Jong's idea, Spencer (1964) constructed a rigid plastic plane model, called the "double shearing model" for non-dilatant granular materials. Spencer assumed that the stress and velocity characteristics coincide. Mehrabadi and Cowin (1978) constructed a dilatant double shearing model for dilatant granular materials adopting the points by Butterfield and Harkness (1972). Butterfield and Harkness (1972) assumed the kinematical proposition "As successive material points along a slip line are considered, any change in velocity relative to the slip line field that occurs between one point and the next is in a direction at to the conjugate slip line". Then, Anand (1983) generalized the Mehrabadi–Cowin model, relaxing the hypothesis of the coincidence of the stress and velocity characteristics to derive a model that is applicable to the pre-failure behavior. Vermeer (1981) and Teunisen and Vermmer (1988) formulated a double shearing model following the work by de Josselin de Jong's model. Yang and Yu (2006a,b) incorporated the non-coaxiality to the critical state model by Yu (1998) and found that the non-coaxial term softens the stress–strain response. Yatomi et al. (1989) showed that the non-coaxial model can more easily predict the strain localization by numerical analysis. Hashiguchi (1998) proposed a non-coaxial plasticity model called tangential plasticity.

In this section, the double shearing model by Mehrabadi and Cowin (1978) and Mehrabadi (1979), and the model derived by Anand (1983) are briefly presented based on their work. Then, we discuss the relation between the Rudnicki and Rice (1975) theory and the double sliding theory.

2.5.2 Double shearing model

2.5.2.1 Derivation of velocity equations

Let us consider the two-dimensional stress field and assume that the tension is taken as positive.

Considering the maximum and minimum principal stress, T_I, T_{II} (see Figure 2.27), p and q are defined as

$$p = -\frac{1}{2}(T_I + T_{II}) \tag{2.129}$$

$$q = \frac{1}{2}(T_I - T_{II}) \tag{2.130}$$

The stress components are given by

$$\begin{aligned} T_{xx} &= -p + q\cos 2\psi \\ T_{yy} &= -p - q\cos 2\psi \\ T_{xy} &= q\sin 2\psi \end{aligned} \tag{2.131}$$

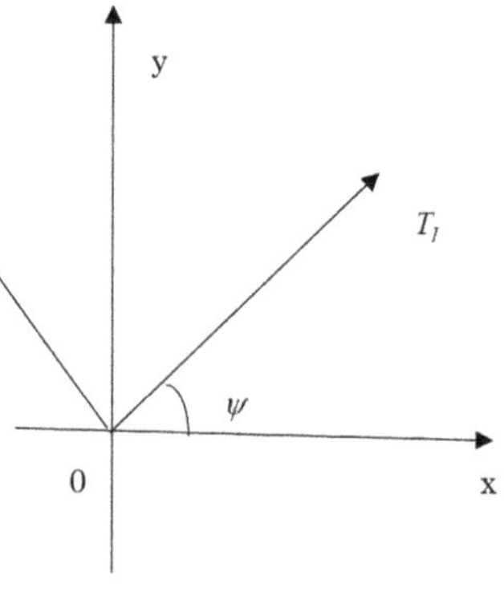

Figure 2.27 Principal stress direction.

where ψ is the angle between the maximum principal stress direction and the x axis.

$$p = -\frac{1}{2}(T_{xx} + T_{yy}), \quad q = \frac{1}{2}\left\{(T_{xx} - T_{yy})^2 + 4T_{xy}^2\right\} \tag{2.132}$$

The Mohr–Coulomb failure criterion denotes that the shear stress on the slip plane satisfies $\tau_n = c + \sigma_n \tan\phi$, where τ_n is the shear stress, c is the cohesion, σ_n is the normal stress, and ϕ is the internal friction angle.

The equilibrium equations in two-dimensional form are given as

$$\begin{aligned} &\frac{\partial T_{xx}}{\partial x} + \frac{\partial T_{xy}}{\partial y} + X = 0, \\ &\frac{\partial T_{yx}}{\partial x} + \frac{\partial T_{yy}}{\partial y} + Y = 0 \end{aligned} \tag{2.133}$$

Substituting Equation 2.131 into Equation 2.133 with $q = p\sin\phi + c\cos\phi$ yields

$$[A]\begin{pmatrix} \partial q / \partial x \\ \partial q / \partial y \\ \partial \psi / \partial x \\ \partial \psi / \partial y \end{pmatrix} = \begin{pmatrix} -X \\ -Y \\ dq / dx \\ d\psi / dx \end{pmatrix} \tag{2.134}$$

where

$$[A] = \begin{bmatrix} \cos 2\psi - \operatorname{cosec}\phi & \sin 2\psi & -2q\sin 2\psi & 2q\cos 2\psi \\ \sin 2\psi & -(\cos 2\psi + \operatorname{cosec}\phi) & 2q\cos 2\psi & 2q\sin 2\psi \\ 1 & dy / dx & 0 & 0 \\ 0 & 0 & 1 & dy / dx \end{bmatrix}$$

Discontinuous solutions are obtained if Det([A]) = 0. Hence, the following characteristic curves are given as

$$\begin{aligned} &\frac{dy}{dx} = \tan\left(\psi - \frac{\pi}{4} - \frac{\phi}{2}\right); \text{ along } \alpha \text{ line,} \\ &\frac{dy}{dx} = \tan\left(\psi + \frac{\pi}{4} + \frac{\phi}{2}\right); \text{ along } \beta \text{ line} \end{aligned} \tag{2.135}$$

In the slip line theory, by integrating the ordinary differential equation (Equation 2.134) along the characteristics equation (Equation 2.135), Kötter's equation is obtained.

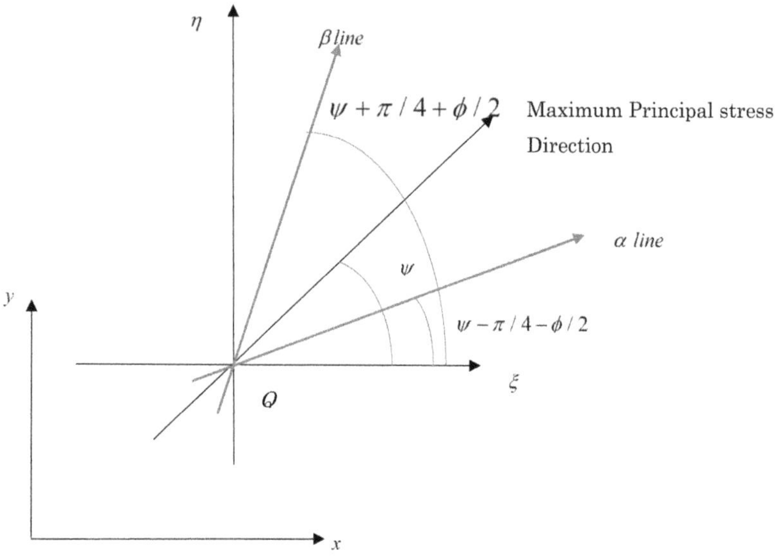

Figure 2.28 Principal stress direction and slip lines.

As shown in Figure 2.28, the origin of the (ξ,η) axes coincides with the position of the particle and the (ξ,η) axes are parallel to the (x,y) axes; (v'_ξ, v'_η) denotes the velocity components relative to the (ξ,η) axes; (v'_α, v'_β) are the velocity components in the directions of the characteristic curve α, β. Considering the transformation between the oblique and Cartesian coordinate systems, we have

$$\begin{pmatrix} v'_\xi \\ v'_\xi \end{pmatrix} = \begin{bmatrix} \cos(\psi - \pi/4 - \phi/2) & \cos(\psi + \pi/4 + \phi/2) \\ \sin(\psi - \pi/4 - \phi/2) & \sin(\psi + \pi/4 + \phi/2) \end{bmatrix} \begin{pmatrix} v'_\alpha \\ v'_\beta \end{pmatrix}$$

$$v'_\xi = v'_\alpha \cos\left(\psi - \frac{\pi}{4} - \frac{\phi}{2}\right) + v'_\beta \cos\left(\psi + \frac{\pi}{4} + \frac{\phi}{2}\right) \tag{2.136}$$

$$v'_\eta = v'_\alpha \sin\left(\psi - \frac{\pi}{4} - \frac{\phi}{2}\right) + v'_\beta \sin\left(\psi + \frac{\pi}{4} + \frac{\phi}{2}\right) \tag{2.137}$$

where ϕ is the internal friction angle.

Following Butterfield and Harkness's hypothesis, namely, $(\partial v'_\xi / \partial s_\beta, \partial v'_\xi / \partial s_\beta)$ at point Q is a vector lying in the α direction and $(\partial v'_\xi / \partial s_\alpha, \partial v'_\eta / \partial s_\alpha)$ at Q is in the β direction, the following relations are assumed along the slip lines shown in Figure 2.29:

$$\frac{\partial v'_\eta}{\partial s_\beta} \Big/ \frac{\partial v'_\xi}{\partial s_\beta} = \tan\left(\psi - \frac{\pi}{4} - \frac{\phi}{2} + \nu\right) \tag{2.138}$$

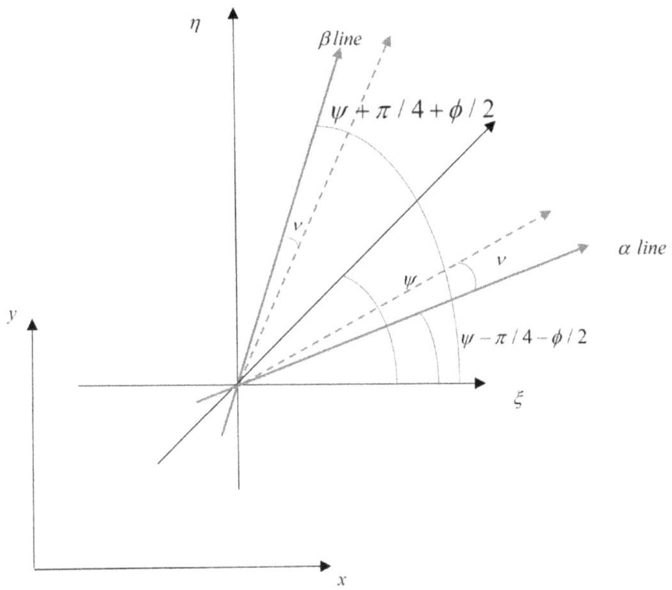

Figure 2.29 Slip lines and the direction of dilatancy.

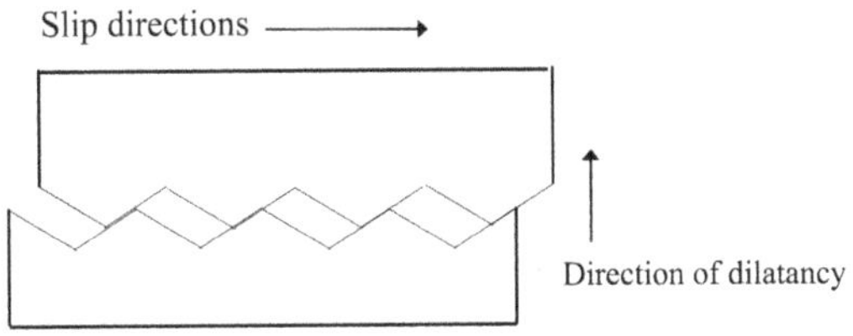

Figure 2.30 Direction of dilatancy.

$$\frac{\partial v'_\eta}{\partial s_\alpha} \bigg/ \frac{\partial v'_\xi}{\partial s_\alpha} = \tan\left(\psi + \frac{\pi}{4} + \frac{\phi}{2} - \nu\right) \tag{2.139}$$

where s_α, s_β are length parameters along the α, β characteristic curves.

The direction of the slip line is illustrated in Figure 2.29 and the direction of dilatancy is shown in Figure 2.30 as a saw shape.

Differentiating Equations 2.136 and 2.137 with respect to s_α, s_β and substituting the results into Equations 2.138 and 2.139, we have the following relations at point Q ($v'_\alpha = v'_\beta = 0$).

$$\begin{aligned} &\frac{\partial v'_\beta}{\partial s_\beta}\cos(\phi - \nu) - \frac{\partial v'_\alpha}{\partial s_\beta}\sin\nu = 0 \\ &\frac{\partial v'_\alpha}{\partial s_\alpha}\cos(\phi - \nu) - \frac{\partial v'_\beta}{\partial s_\alpha}\sin\nu = 0 \end{aligned} \tag{2.140}$$

In the case of $\nu = 0$ (non-dilatant case), Equation 2.140 corresponds to Spencer's postulate (1964) as $\partial v'_\beta / \partial s_\beta = 0, \partial v'_\alpha / \partial s_\alpha = 0$. Namely, at point Q, $\partial(v'_\xi, v'_\eta) / \partial s_\alpha$ is a vector that is directed in the β line direction. Hence, $\partial v'_\alpha / \partial s_\alpha = 0$.

The turning rates of the α, β lines, equivalently, the (ξ,η) coordinate system relative to the material is denoted by $\dot{\psi}$ and relative to the spatial (x,y) coordinate system by $\partial\psi / \partial t$.

Let V_x and V_y be the components of V relative to the (x,y) system at point Q and $\tilde{r}$ the position vector. From the outer-product relation, $(v_x, v_y) = (v'_\xi, v'_\eta) + (V_x, V_y) + \Omega \otimes (\xi,\eta)$; $\mathbf{\Omega}$: angular velocity vector.

Hence, we have Equations 2.141 and 2.142:

$$\begin{aligned} v'_\xi &= v_x - V_x + \dot{\psi}\eta, \\ v'_\eta &= v_y - V_y - \dot{\psi}\xi \end{aligned} \tag{2.141}$$

$$\begin{aligned} v'_\alpha \cos\phi &= v_\alpha \cos\phi - (V_x - \dot{\psi}\eta)\sin\left(\psi + \frac{\pi}{4} + \frac{\phi}{2}\right) + (V_y + \dot{\psi}\xi)\cos\left(\psi + \frac{\pi}{4} + \frac{\phi}{2}\right) \\ v'_\beta \cos\phi &= v_\beta \cos\phi + (V_x - \dot{\psi}\eta)\sin\left(\psi - \frac{\pi}{4} - \frac{\phi}{2}\right) - (V_y + \dot{\psi}\xi)\cos\left(\psi - \frac{\pi}{4} - \frac{\phi}{2}\right) \end{aligned} \tag{2.142}$$

where

$$\begin{aligned} \frac{\partial\xi}{\partial s_\alpha} &= \cos(\psi - \pi/4 - \phi/2), \quad \frac{\partial\eta}{\partial s_\alpha} = \sin(\psi - \pi/4 - \phi/2) \\ \frac{\partial\xi}{\partial s_\beta} &= \cos(\psi + \pi/4 + \phi/2), \quad \frac{\partial\eta}{\partial s_\beta} = \sin(\psi + \pi/4 + \phi/2) \end{aligned} \tag{2.143}$$

Using Equation 2.143 and substituting Equation 2.142 into Equation 2.140, the following relations are obtained at point Q:

$$\begin{aligned} &\cos(\phi - \nu)\left[\frac{\partial}{\partial s_\alpha}(v_\alpha \cos\phi) - (v_\beta - v_\alpha \sin\phi)\left(\frac{\partial\psi}{\partial s_\alpha} + \frac{1}{2}\frac{\partial\phi}{\partial s_\alpha}\right) - \dot{\psi}\sin\phi\right] \\ &\quad - \sin\nu\left[\frac{\partial}{\partial s_\alpha}(v_\beta \cos\phi) + (v_\alpha - v_\beta \sin\phi)\left(\frac{\partial\psi}{\partial s_\alpha} - \frac{1}{2}\frac{\partial\phi}{\partial s_\alpha}\right) - \dot{\psi}\right] = 0 \\ &\cos(\phi - \nu)\left[\frac{\partial}{\partial s_\beta}(v_\beta \cos\phi) + (v_\alpha - v_\beta \sin\phi)\left(\frac{\partial\psi}{\partial s_\beta} - \frac{1}{2}\frac{\partial\phi}{\partial s_\beta}\right) + \dot{\psi}\sin\phi\right] \\ &\quad - \sin\nu\left[\frac{\partial}{\partial s_\beta}(v_\alpha \cos\phi) - (v_\beta - v_\alpha \sin\phi)\left(\frac{\partial\psi}{\partial s_\alpha} + \frac{1}{2}\frac{\partial\phi}{\partial s_\alpha}\right) + \dot{\psi}\right] = 0 \end{aligned} \tag{2.144}$$

The following relations hold:

$$\begin{aligned} v_\alpha \cos\phi &= v_x \sin(\psi + \pi/4 + \phi/2) - v_y \cos(\psi + \pi/4 + \phi/2) \\ v_\beta \cos\phi &= -v_x \sin(\psi - \pi/4 - \phi/2) + v_y \cos(\psi - \pi/4 - \phi/2) \end{aligned} \tag{2.145}$$

$$\begin{aligned} \frac{\partial}{\partial s_\alpha} &= \cos(\psi - \pi/4 - \phi/2)\frac{\partial}{\partial x} + \sin(\psi - \pi/4 - \phi/2)\frac{\partial}{\partial y} \\ \frac{\partial}{\partial s_\beta} &= \cos(\psi + \pi/4 + \phi/2)\frac{\partial}{\partial x} + \sin(\psi + \pi/4 + \phi/2)\frac{\partial}{\partial y} \end{aligned} \tag{2.146}$$

Using the above relations, we have

$$\sin\nu\left[(D_{11} - D_{22})\cos 2\psi + 2D_{12}\sin 2\psi\right] - (D_{11} + D_{22})\cos(\phi - \nu) = 0 \tag{2.147}$$

$$\cos\nu\left[(D_{11} - D_{22})\sin 2\psi - 2D_{12}\cos 2\psi\right] - 2(W_{12} + \dot{\psi})\sin(\phi - \nu) = 0 \tag{2.148}$$

where the following relations are used:

$$D_{11} = \frac{\partial v_x}{\partial x},\ D_{22} = \frac{\partial v_y}{\partial y},\ D_{12} = \frac{1}{2}\left(\frac{\partial v_x}{\partial y} + \frac{\partial v_y}{\partial x}\right),\ W_{12} = \frac{1}{2}\left(\frac{\partial v_x}{\partial y} - \frac{\partial v_y}{\partial x}\right) \tag{2.149}$$

Then, using Equations 2.147, 2.148, and 2.10, we obtain Equations 2.151 and 2.152:

$$\tan\delta = \frac{2D_{12}}{D_{11} - D_{22}} \tag{2.150}$$

$$(D_{11} + D_{22})\cos(\phi - \nu) = \left[(D_{11} - D_{22})^2 + 4D_{12}^2\right]^{1/2} \sin\nu \cos 2(\psi - \delta) \tag{2.151}$$

$$2(W_{12} + \dot{\psi})\sin(\phi - \nu) = \left[(D_{11} - D_{22})^2 + 4D_{12}^2\right]^{1/2} \cos\nu \sin 2(\psi - \delta) \tag{2.152}$$

2.5.2.2 *Double shearing constitutive theory*

Using Cauchy's stress tensor, T_{ij}, in two-dimensional form, p and τ are defined:

$$\begin{aligned} p &= -\frac{1}{2}(T_I + T_{II}) = -\frac{1}{2}(T_{11} + T_{22}) \\ \tau &= \frac{1}{2}(T_I - T_{II}) = \left\{\frac{1}{4}(T_{11} - T_{22})^2 + T_{12}^2\right\}^{1/2} \end{aligned} \tag{2.153}$$

The angle ψ between the maximum principal stress and the x coordinate is written as

$$\tan 2\psi = \frac{2T_{12}}{T_{11} - T_{22}} \tag{2.154}$$

Hence,

$$4\tau^2\dot{\psi} = (T_{11} - T_{22})\dot{T}_{12} - T_{12}(\dot{T}_{11} - \dot{T}_{22}) \tag{2.155}$$

Using the Jaumann stress rate as a corotational stress rate:

$$\begin{aligned} \hat{\mathbf{T}} &= \dot{\mathbf{T}} + \mathbf{TW} - \mathbf{WT} \\ \hat{T}_{ij} &= \dot{T}_{ij} + T_{ik}W_{kj} - W_{ik}T_{kj} \end{aligned} \tag{2.156}$$

Equation 2.155 becomes

$$4\tau^2(W_{12} + \dot{\psi}) = (T_{11} - T_{22})\hat{T}_{12} - T_{12}(\hat{T}_{11} - \hat{T}_{22}) \tag{2.157}$$

Therefore, Equations 2.147 and 2.148 become,

$$(T_{11} - T_{22})(D_{11} - D_{22}) + 4T_{12}D_{12} = \frac{2\tau\cos(\phi - \nu)}{\sin\nu}(D_{11} + D_{22}) \tag{2.158}$$

$$(T_{11} - T_{22})\hat{T}_{12} - T_{12}(\hat{T}_{11} - \hat{T}_{22}) = \frac{2\tau\cos\nu}{\sin(\phi - \nu)}\left[T_{12}(D_{11} - D_{22}) - D_{12}(T_{11} - T_{22})\right] \tag{2.159}$$

Rewriting the above equations, we have

$$T_{ij}D_{ij} + \left(p - \frac{\tau\cos(\phi - \nu)}{\sin\nu}\right)D_{kk} = 0 \tag{2.160}$$

$$T_{ik}\hat{T}_{kj} - \hat{T}_{ik}T_{kj} = \frac{2\tau\cos\nu}{\sin(\phi - \nu)}(D_{ik}T_{kj} - T_{ik}D_{kj}) \tag{2.161}$$

Equations 2.160 and 2.161 are denoted in direct notations as

$$tr\mathbf{TD} + \left(p - \frac{\tau\cos(\phi - \nu)}{\sin\nu}\right)tr\mathbf{D} = 0, \quad \mathbf{T}\hat{\mathbf{T}} - \hat{\mathbf{T}}\mathbf{T} = \frac{2\tau\cos\nu}{\sin(\phi - \nu)}(\mathbf{DT} - \mathbf{TD})$$

It is worth noting that the constitutive theory includes stress rates because Equation 2.152 contains $\dot{\psi}$. Equations 2.160 and 2.161 constitute a constitutive theory with Coulomb's yield condition.

2.5.3 Anand's model

Asaro (1979) proposed a constitutive model for ductile single crystals undergoing symmetric double slip. Nemat-Nasser et al. (1981) and Anand (1983) derived a similar model for granular material. They assumed that the slip line directions are always symmetric about the maximum principal stress directions.

In the following, we will explain the extended double shearing theory based on the paper by Anand (1983). The characteristic directions, i.e., slip directions $\mathbf{s}^{(\alpha)}$ and its normal $\mathbf{n}^{(\alpha)}$ are given by

$$\begin{aligned}
\mathbf{s}^{(1)} &= \cos\left(\psi - \frac{\pi}{4} - \frac{\xi}{2}\right)\mathbf{e}_1 + \sin\left(\psi - \frac{\pi}{4} - \frac{\xi}{2}\right)\mathbf{e}_2 \\
\mathbf{n}^{(1)} &= \cos\left(\psi + \frac{\pi}{4} - \frac{\xi}{2}\right)\mathbf{e}_1 + \sin\left(\psi + \frac{\pi}{4} - \frac{\xi}{2}\right)\mathbf{e}_2 \\
\mathbf{s}^{(2)} &= \cos\left(\psi + \frac{\pi}{4} + \frac{\xi}{2}\right)\mathbf{e}_1 + \sin\left(\psi + \frac{\pi}{4} + \frac{\xi}{2}\right)\mathbf{e}_2 \\
\mathbf{n}^{(2)} &= \cos\left(\psi - \frac{\pi}{4} + \frac{\xi}{2}\right)\mathbf{e}_1 + \sin\left(\psi - \frac{\pi}{4} + \frac{\xi}{2}\right)\mathbf{e}_2
\end{aligned} \tag{2.162}$$

where ψ is the angle that the maximum principal stress direction $\hat{\mathbf{e}}_1$ makes with the $\mathbf{e}_1$ axis of the fixed rectangular Cartesian coordinate system; $(\mathbf{s}^{(\alpha)}, \mathbf{n}^{(\alpha)})$ are local direction pairs and the directions $\mathbf{s}^{(\alpha)}$ is the slip direction; and ξ is a material constant determined by experiments or micro-structural consideration.

If $\xi = \phi$ (internal friction angle), the theory corresponds to that by Mehrabadi and Cowin (1978).

The component of traction in the slip direction is denoted as

$$\begin{aligned}
\tau^{(\alpha)} &= \mathbf{s}^{(\alpha)} \cdot \mathbf{T}\mathbf{n}^{(\alpha)}, \\
\sigma^{(\alpha)} &= -\mathbf{n}^{(\alpha)} \cdot \mathbf{T}\mathbf{n}^{(\alpha)} \\
(\alpha &= 1,2)
\end{aligned} \tag{2.163}$$

where $\mathbf{T}$ is the Cauchy stress tensor.

The material obeys the yield condition as

$$\tau\cos\xi \leq c + \tan\phi(p - \tau\sin\xi),\ c \geq 0, \text{ and } 0 \leq \phi < \pi/2.$$

If $\xi = 0$, it corresponds to Coulomb's law: $\tau \leq c + p\tan\phi$; and c and ϕ are the cohesion and the angle of internal friction.

Considering the plastic constitutive model of crystals, the stretching and spin are assumed to be

$$\mathbf{D}=\sum_{\alpha=1}^{2}\dot{\gamma}^{(\alpha)}\mathbf{A}^{(\alpha)}:\text{stretching},\quad \sum_{\alpha=1}^{2}\dot{\gamma}^{(\alpha)}\mathbf{B}^{(\alpha)}:\text{spin} \tag{2.164}$$

where

$$\begin{aligned}
\mathbf{A}^{(\alpha)} &= sym(\mathbf{s}^{(\alpha)}\otimes\mathbf{n}^{(\alpha)})+\tan\nu(\mathbf{n}^{(\alpha)}\otimes\mathbf{n}^{(\alpha)})\\
\mathbf{B}^{(\alpha)} &= skw(\mathbf{s}^{(\alpha)}\otimes\mathbf{n}^{(\alpha)})\\
\tan\nu &= \dot{\delta}^{(\alpha)}/\dot{\gamma}^{(\alpha)}
\end{aligned} \tag{2.165}$$

$\dot{\delta}^{(\alpha)}$: extension rates, $\dot{\gamma}^{(\alpha)}$: simple shearing rates.

In the above, the direction pairs are supposed not to alter relative to the fixed coordinate frame. Since the direction pairs rotate with respect to the fixed frame, the total spin is given by adding the rotation of the pairs $\mathbf{W}^{*}$:

$$\mathbf{W}=\sum_{\alpha=1}^{2}\dot{\gamma}^{(\alpha)}\mathbf{B}^{(\alpha)}+\mathbf{W}^{*} \tag{2.166}$$

$$\mathbf{W}^{*}=\dot{\psi}\mathbf{E},\quad \mathbf{E}=\begin{pmatrix}0 & -1\\ 1 & 0\end{pmatrix} \tag{2.167}$$

From Equations 2.162 and 2.165, $\mathbf{A}^{(\alpha)},\mathbf{B}^{(\alpha)}$ are

$$\begin{aligned}
\mathbf{A}^{(1)} &= \frac{1}{\cos\nu}\{\cos[2\psi-(\xi-\nu)]\mathbf{I}_1+\sin[2\psi-(\xi-\nu)]\mathbf{I}_2-\sin\nu\mathbf{I}_3\}\\
\mathbf{A}^{(2)} &= \frac{1}{\cos\nu}\{\cos[2\psi+(\xi-\nu)]\mathbf{I}_1+\sin[2\psi+(\xi-\nu)]\mathbf{I}_2-\sin\nu\mathbf{I}_3\}\\
\mathbf{B}^{(1)} &= -\tfrac{1}{2}\mathbf{E},\quad \mathbf{B}^{(2)}=\tfrac{1}{2}\mathbf{E}
\end{aligned} \tag{2.168}$$

where

$$\mathbf{I}_1=1/2\begin{pmatrix}1 & 0\\ 0 & -1\end{pmatrix},\quad \mathbf{I}_2=1/2\begin{pmatrix}0 & 1\\ 1 & 0\end{pmatrix},\quad \mathbf{I}_3=1/2\begin{pmatrix}-1 & 0\\ 0 & -1\end{pmatrix} \tag{2.169}$$

The above matrices $\mathbf{I}_1$, $\mathbf{I}_2$, $\mathbf{I}_3$ are orthogonal to each other.

The following relations are set for later use as

$$\begin{aligned}
&\sin 2\psi = T'_{12}/\tau,\ \cos 2\psi=(T'_{11}-T'_{22})/(2\tau)\\
&\dot{\psi}=\left[(T'_{11}-T'_{22})\dot{T}'_{12}-T'_{12}(\dot{T}'_{11}-\dot{T}'_{22})\right]/(4\tau^2)
\end{aligned} \tag{2.170}$$

where T'_{ij} is the deviatoric component of the stress tensor.

From Equations 2.164 and 2.166–2.168, we obtain

$$\mathbf{W} = \left[\dot{\psi} - (\dot{\gamma}^{(1)} - \dot{\gamma}^{(2)})/2\right]\mathbf{E} \tag{2.171}$$

$$(\dot{\gamma}^{(1)} - \dot{\gamma}^{(2)}) = 2(W_{12} + \dot{\psi}) \tag{2.172}$$

From Equations 2.170–2.172 and the definition of the Jaumann stress rate:

$$(\dot{\gamma}^{(1)} - \dot{\gamma}^{(2)}) = \left[(\cos 2\psi)2\hat{T}'_{12} - (\sin 2\psi)(\hat{T}'_{11} - \hat{T}'_{22})\right]/(2\tau) \tag{2.173}$$

where $\dot{\gamma}$ is defined as

$$\dot{\gamma} \equiv [(\dot{\gamma}^{(1)} + \dot{\gamma}^{(2)})\cos(\xi - \nu)/\cos\nu \geq 0 \tag{2.174}$$

The deviatoric stretching is decomposed for two slip directions as

$$\mathbf{D}' = \bar{\mathbf{D}}^{(1)} + \bar{\mathbf{D}}^{(2)} \tag{2.175}$$

where

$$\bar{\mathbf{D}}^{(1)} = \left\{\frac{\dot{\gamma}^{(1)}}{\cos\nu}\right\}\bar{\mathbf{N}}^{(1)},\ \bar{\mathbf{D}}^{(2)} = \left\{\frac{\dot{\gamma}^{(2)}}{\cos\nu}\right\}\bar{\mathbf{N}}^{(2)} \tag{2.176}$$

$$\begin{aligned}\bar{\mathbf{N}}^{(1)} &= \{\cos[2\psi - (\xi - \nu)]\mathbf{I}_1 + \sin[2\psi - (\xi - \nu)]\mathbf{I}_2\}\\ \bar{\mathbf{N}}^{(2)} &= \{\cos[2\psi + (\xi - \nu)]\mathbf{I}_1 + \sin[2\psi + (\xi - \nu)]\mathbf{I}_2\}\end{aligned} \tag{2.177}$$

On the other hand, the deviatoric stretching $\mathbf{D}'$ may be decomposed as

$$\mathbf{D}' = \mathbf{D}^{(1)} + \mathbf{D}^{(2)} \tag{2.178}$$

where

$$\begin{aligned}&\mathrm{D}^{(1)} = \dot{\gamma}\mathrm{N}^{(1)},\\ &\mathrm{D}^{(2)} = -\langle\dot{\gamma}\rangle\left\{\frac{\sin(\xi - \mu)}{\tau\cos\nu}\right\}(\hat{\mathrm{T}}'\cdot\mathrm{N}^{(2)})\mathrm{N}^{(2)}\\ &\langle\dot{\gamma}\rangle = \begin{Bmatrix}\dot{\gamma} & (\dot{\gamma} > 0)\\ 0 & (\dot{\gamma} = 0)\end{Bmatrix}\end{aligned} \tag{2.179}$$

where · dot denotes the inner product of two matrices and second-order tensors.

$$\mathbf{N}^{(1)} = \{\cos 2\psi \mathbf{I}_1 + \sin 2\psi \mathbf{I}_2\}$$
$$\mathbf{N}^{(2)} = \{\sin 2\psi \mathbf{I}_1 - \cos 2\psi \mathbf{I}_2\} \tag{2.180}$$

The volumetric component $\mathbf{D}^{(3)}$ is given using the dilatancy parameter β:

$$\mathbf{D}^{(3)} = -\beta\dot{\gamma}\mathbf{I}_3$$
$$\beta = \sin\nu / \cos(\xi - \nu), \ -1 < \beta < 1 \tag{2.181}$$

Hence, total stretching is obtained:

$$\mathrm{D} = \dot{\gamma}(\mathrm{N}^{(1)} - \beta \mathrm{I}_3) - \langle\dot{\gamma}\rangle \left\{ \frac{\sin(\xi - \mu)}{\tau\cos\nu} \right\} (\hat{\mathrm{T}}' \cdot \mathrm{N}^{(2)}) \mathrm{N}^{(2)} \tag{2.182}$$

Rewriting Equation 2.182, we have

$$\mathbf{D} = \dot{\gamma}\mathbf{P} + \langle\dot{\gamma}\rangle \frac{1}{2h_1}\left[\hat{\mathbf{T}}' - (\hat{\mathbf{T}}' \cdot \mathbf{T}')\frac{\mathbf{T}'}{2\tau^2}\right] \tag{2.183}$$

$$\mathbf{P} = \left(\frac{\mathbf{T}'}{2} + \frac{1}{2}\beta\mathbf{I}\right) \tag{2.184}$$

$$\beta = \frac{\sin\nu}{\cos(\xi - \nu)} \tag{2.185}$$

$$h_1 = -\left[\frac{\tau\cos\nu}{\sin(\xi - \nu)}\right] \tag{2.186}$$

The above equation is equivalent to the vertex model by Rudnicki and Rice (1975), i.e., the yield surface has a vertex.

It is worth noting that the Jaumann stress rate, which is invariant with respect to the rigid body motion, emerges in a natural way. This may be a reason why the Jaumann stress rate is used for the constitutive model for geomaterials. The stress power is

$$\mathbf{T} \cdot \mathbf{D} = \mathbf{T} \cdot \mathbf{D}' - p(tr\mathbf{D}) = \dot{\gamma}(\tau - \beta p) \tag{2.187}$$

Then,

$$\mathbf{T} \cdot \mathbf{D}' = \dot{\gamma}\tau \tag{2.188}$$

Using this relation and Equations 2.183 and 2.188, we have the following relations in direct notation as

$$trTD' = \frac{\tau}{\beta} trD \tag{2.189}$$

$$T\hat{T} - \hat{T}T = \frac{2\tau\cos\nu}{\sin(\xi - \nu)}(DT - TD) \tag{2.190}$$

If $\xi = \phi$, Equations 2.189 and 2.190 are equal to Equations 2.160 and 2.161 by Mehrabadi and Cowin (1978), namely, Anand's model which includes the double shearing model by Mehrabadi and Cowin (1978). In order to derive a constitutive model, the hardening parameters have to be given. The second term of Equation 2.183 corresponds to the non-coaxial vertex term by Rudnicki and Rice (1975).

Since $\left[\hat{T}' - (\hat{T}' \cdot T')(T'/2\tau^2)\right] \cdot T' = 0$, it is seen that the second term of Equation 2.183 does not contribute to work. This second term is a component that is tangential to the yield surface, as shown in Figure 2.31, and is called the "non-coaxial term".

For geomaterials, the experimental results of sands show that the direction of strain rates does not correspond to that of stress (Roscoe 1970; Drescher and de Josselin de Jong 1972; Arthur et al. 1980; Gutierrez et al. 1991). Since the usual coaxial model cannot explain the experimental results, the

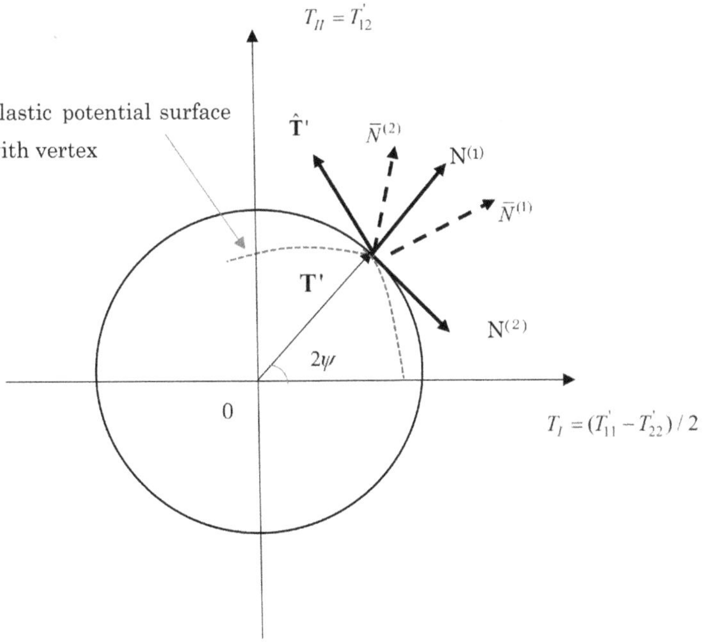

Figure 2.31 Several vectors in the deviatoric stress space. (After Anand, L. 1983. Plane deformations of ideal granular materials. *J. Mech. Phys. Solids* 31(2): 105–122.)

non-coaxial model is preferable for geomaterials such as soils. Rudnicki and Rice (1975) proposed a non-coaxial model by arguing that the co-axial model overestimates the stiffness of the response to the stress increments. They introduced a non-coaxial term as the second term of Equation 2.5.55 considering a vertex yield surface. Yatomi et al. (1989a, 1989b) proposed a non-coaxial Cam-clay model incorporating the non-coaxial term by Rudnicki and Rice (1975) and performed a finite element simulation of the strain localization problem. Yang and Yu (2006) discussed the effect of the non-coaxial term on the simple shear problem.

It is worth noting that, in general, the non-coaxial term should include the term given by $\hat{\mathbf{T}}' - (\hat{\mathbf{T}}' \cdot \mathbf{N})\mathbf{N}$; $\mathbf{N}$ denotes the unit normal to the yield surface on the deviatoric plane for the non-circular yield surface (Tsutsumi and Hashiguchi 2005), although Rudnick and Rice (1975) only consider the circular yield surface on the deviatoric plane as $\mathbf{N} = \mathbf{T}' / \sqrt{\mathbf{T}' : \mathbf{T}'}$.

2.5.4 Total strain theory and the non-coaxial term

Stören and Rice (1975) showed that the non-coaxial term of Equation 2.183 is an extension of total strain theory.

The coefficient h_1 of Equation 2.183 is related to the secant modulus.

The stress–strain relation of the total strain theory is expressed as

$$e_{ij}^p = \frac{3}{2}\left(\frac{1}{E_s} - \frac{1}{E}\right)T'_{ij} \tag{2.191}$$

where e_{ij}^p is the plastic deviatoric strain; T'_{ij} is the deviatoric stress; E is the elastic modulus; and $E_s = \bar{T} / \bar{e}$ is the secant modulus.

$$\bar{T} = \sqrt{\frac{3}{2} T'_{ij} T'_{ij}}, \quad \bar{e} = \sqrt{\frac{3}{2} e_{ij} e_{ij}} \tag{2.192}$$

where $\bar{e}$ is the equivalent strain and $\bar{T}$ is the equivalent deviatoric stress.

The tangential modulus is expressed as $E_t = \dot{\bar{T}} / \dot{\bar{e}}$, shown in Figure 2.32.

Differentiating Equation 2.191 and considering the relation $\dot{\bar{e}} / \bar{e} = (E_s / E_t)(\dot{\bar{T}} / \bar{T})$, we have

$$\begin{aligned} \dot{e}_{ij}^p &= \frac{3}{2}\left(\frac{1}{E_s} - \frac{1}{E}\right)\dot{T}'_{ij} + \frac{3}{2}T'_{ij}\left[-\frac{1}{E_s}\left(\frac{\dot{\bar{T}}}{\bar{e}} - \frac{\bar{T}\dot{\bar{e}}}{\bar{e}^2}\right)\right] \\ &= \frac{3}{2}\left(\frac{1}{E_s} - \frac{1}{E}\right)\dot{T}'_{ij} + \frac{3}{2}\left(\frac{1}{E_t} - \frac{1}{E_s}\right)\frac{\dot{\bar{T}}}{\bar{T}}T'_{ij} \\ &= \frac{3}{2}\left(\frac{1}{E_t} - \frac{1}{E}\right)\frac{\dot{\bar{T}}}{\bar{T}}T'_{ij} + \frac{3}{2}\left(\frac{1}{E_s} - \frac{1}{E}\right)\left(\dot{T}'_{ij} - \frac{\dot{\bar{T}}}{\bar{T}}T'_{ij}\right) \end{aligned} \tag{2.193}$$

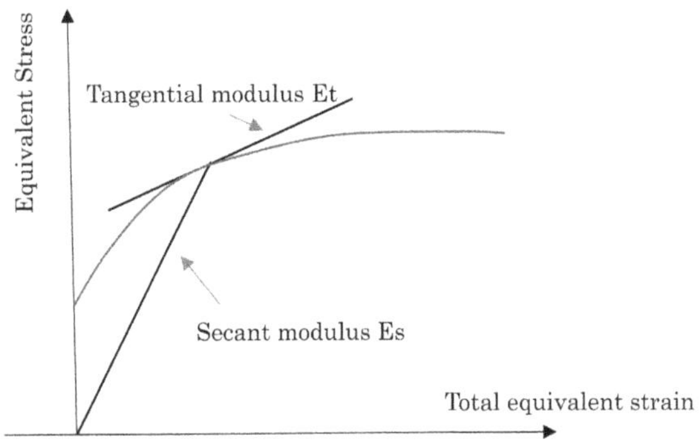

Figure 2.32 Secant modulus and tangential modulus.

The first and second term on the right-hand side of Equation 2.193 correspond to the flow theory and the non-coaxial term, respectively.

2.6 GRADIENT-DEPENDENT ELASTIC MODEL FOR GRANULAR MATERIALS AND STRAIN LOCALIZATION SOLUTION

Gradient-dependent constitutive models have been proposed and studied as a solution to the post-strain localization problem (e.g., Aifantis 1984, 1987; Vardoulakis and Aifantis 1991; Mühlhaus and Aifantis 1991; Zbib and Aifantis 1989; Oka et al. 1992). Strain localization has been studied as a bifurcation problem based on the works of Thomas (1961) and Hill (1962). Within the framework of the bifurcation theory, the conditions for the initiation of shear bands have been derived from the viewpoint of the loss of ellipticity for quasi-static problems. These approaches are not, however, appropriate for predicting the thickness of a shear band or simulating post-localization behavior. To overcome these shortcomings, Aifantis (1984) and Coleman and Hodgdon (1985) advocated a new approach for dealing with the localization problem, which uses the gradient-dependent constitutive model.

The feature of this method is that a characteristic length scale, such as the particle size of the soil, appears explicitly in the model. Although the effectiveness of the gradient model has been studied for predicting the thickness of a shear band and the continuation of the numerical simulation in the post-localization regime, the physical meaning of the higher-order strain gradient term is also important. Mühlhaus and Oka (1996) clarified that the higher-gradient term naturally comes from the fact that the material is discrete and has an inherent length scale.

In this section, we will discuss the gradient-dependent elastic model by Oka (1995). In the first part of the section, we focus on the derivation of a gradient-dependent constitutive model from a micro-mechanical consideration of particulate materials. In the second part, the analytical solutions of strain gradient-dependent models are derived to illustrate the characteristics of the material model with gradient terms. The results show that gradient elastic models have a localized solution and a periodic solution.

2.6.1 First gradient-dependent elastic model

Although the gradient-dependent constitutive model has been applied to granular materials and clay (e.g., Oka et al. 1991, 1992), its physical nature has not yet been sufficiently clarified. The nature of the gradient term should, however, be discussed within a micro-structural context. Herein, we focus on the physical nature of the strain-gradient term for soils. Let us derive a gradient-dependent constitutive model from a particulate discrete spring model.

For a non-linear spring model, the Fermi–Pasta–Ulam oscillator (1965) came to fame in connection with the discovery of solitons in the numerical simulations of the vibration of the non-linear lattice in Figure 2.33. They used the following non-linear spring:

$$F = \kappa(r_i + \alpha r_i^2) \tag{2.194}$$

where F is the force; y_i is the displacement of the ith mass point (or particle); $r_i = y_i - y_{i-1}$ is the relative displacement between two particles; and κ and α are material constants. We only deal with the quasi-static case in which the acceleration term is neglected.

$$F = \kappa\left[y_{n+1} - y_n + \alpha(y_{n+1} - y_n)^2\right] - \kappa\left[y_n - y_{n-1} + \alpha(y_n - y_{n-1})^2\right] \tag{2.195}$$

Assuming that $x = nh$ (x: coordinate, n: number of particles, h: the distance between two nearby particles), the continuum model can be derived through the following Taylor expansion to Equation 2.195:

$$y_{n\pm1} = y_n \pm hy_x + \frac{h^2}{2!}y_{xx} \pm \frac{h^3}{3!}y_{xxx} + \frac{h^4}{4!}y_{xxxx} + \cdots \tag{2.196}$$

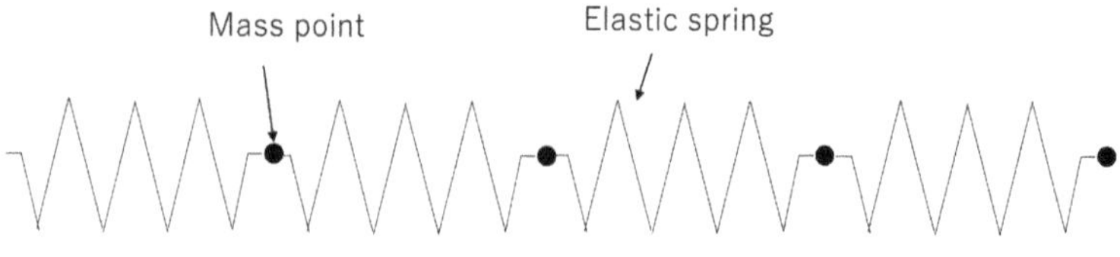

Figure 2.33 Discrete model with non-linear spring.

where subscript x denotes the differentiation with respect to the coordinate x.

Substituting Equation 2.196 into Equation 2.195 and truncating the higher-order terms, we have

$$\frac{\partial \sigma}{\partial x} = F \tag{2.197}$$

$$\sigma = \kappa h^2 \varepsilon + \frac{1}{12} \kappa h^4 \varepsilon_{xx} + \alpha \kappa h^3 \varepsilon^2 \tag{2.198}$$

where $\varepsilon = y_x$ is the strain and σ is the stress.

From Equation 2.198, we obtain the following gradient-dependent elastic model:

$$\sigma = a\varepsilon + b\varepsilon_{xx} + c\varepsilon^2 \tag{2.199}$$

where $a = \kappa h^2$, $b = \kappa h^4/12$, and $c = \kappa \alpha h^3$ are positive constants.

It is interesting to note that in the derivation of Equation 2.199, the gradient term comes from the discrete nature of the physical model. In addition, it becomes evident that the material parameters depend on the distance between particles h, in other words, the material parameters include a characteristic length scale. We call this model the first-gradient model. It is worth noting that the gradient term in Equation 2.199 works as a destabilizer because the sign of this term is negative at the localized zone of the strain. On the other hand, the last term in the right hand side of Equation 2.199 plays a hardening role.

2.6.2 Second gradient-dependent elastic model

We derive the other type of gradient-dependent model. We consider the chain with two neighbor interactions (see Figure 2.34) that are similar to the interactions of the elastic model with a micro-structure by Kunin (1982).

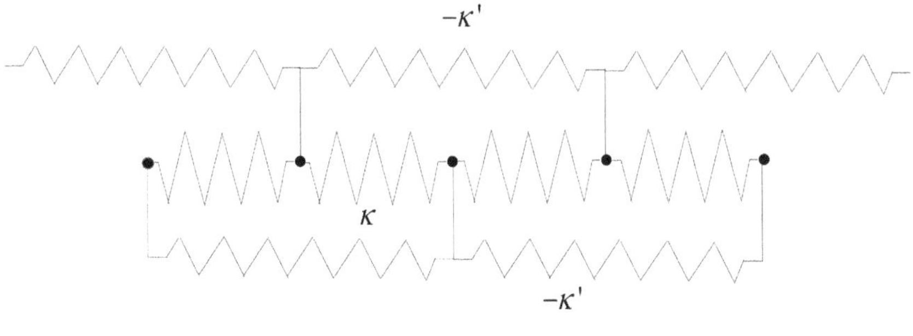

Figure 2.34 Nearest and next-nearest neighbor interactions.

In this model, we assume that the spring for the next-nearest neighbor interactions has an opposite sign to the one for the nearest neighbor interactions.

$$F_1 = \kappa r_i + \kappa \alpha r_i^2 \tag{2.200}$$

$$F_2 = -\kappa' r_i' \tag{2.201}$$

where F_i (i = 1, 2) is the interaction force and κ' is a constant of the interaction between the mass point and the next-nearest neighbor mass point as $r_i' = y_i - y_{i-2}$.

Considering two interactions, the interaction force becomes

$$F = \kappa(y_{n+1} - 2y_n + y_{n-1})\left[1 + \alpha(y_{n+1} - y_{n-1})\right] - \kappa'(y_{n+2} - 2y_n + y_{n-2}) \tag{2.202}$$

From the Taylor series expansion:

$$y_{n\pm1} = y_n \pm h y_x + \frac{h^2}{2!} y_{xx} \pm \frac{h^3}{3!} y_{xxx} + \frac{h^4}{4!} y_{xxxx} + \cdots \tag{2.203}$$

$$y_{n\pm m} = y_n \pm H y_x + \frac{H^2}{2!} y_{xx} \pm \frac{H^3}{3!} y_{xxx} + \frac{H^4}{4!} y_{xxxx} + \cdots \tag{2.204}$$

where $H = mh$, $m > 1$.

The case in Figure 2.34 corresponds to m = 2. Substituting Equations 2.203 and 2.204 into Equation 2.202 and truncating the higher-order terms, we have

$$\frac{\partial \sigma}{\partial x} = F \tag{2.205}$$

$$\sigma = (\kappa - \kappa' m^2) h^2 \varepsilon + (\kappa - \kappa' m^4) h^2 \varepsilon_{xx} / 12 + \alpha \kappa h^3 \varepsilon^2 \tag{2.206}$$

Requiring $(\kappa - \kappa' m^2) > 0$, $(\kappa - \kappa' m^4)/12 < 0$, and $\alpha < 0$, Equation 2.206 can be rewritten as

$$\sigma = a\varepsilon - b\varepsilon^2 - c\varepsilon_{xx} \tag{2.207}$$

where a, b, c are positive constants.

Let us discuss the long-range interaction, i.e., the interaction between the point and the next-nearest point. For granular materials, it is known that the stress force chain is more important than the physical configuration of particles (Oda 1972; Oda and Konishi 1974; Daniels 2017; Daniels et al. 2017). The strongest force chain is parallel to the major principal stress direction but this feature deviates near the strain localization area (Taboada et al. 2005). The photo elastic and numerical experiments show that there exists a secondary stronger force chain which is not necessarily parallel to the maximum principal stress direction. Hence, we can assume that the secondary stronger force chain may carry the long-range force interaction and this long-range influence can be modeled by the next-nearest interaction. The experiments show that this main stress force direction deviates near shear bands. This indicates that the long-range interaction may play an important role near the strain localization area.

For cohesive materials such as clayey soil, let us discuss the two-neighbor interactions based on the card house structure of clay (Push 1970; Mitchell 1976; Shirozu 1988). In the card house structure, the attractive force is dominant around intersections with an electrical double layer. On the other hand, between two points of the clay plate far from the intersection point, repulsive force is significant. This is a typical example of the two-neighbor interactions introduced in Equations 2.200 and 2.201. We call this type of model the second-gradient model in the following. In this type of model, the gradient term works as a stabilizer because of its positive sign and the last term of Equation 2.208 plays a softening role. As has been discussed here, it is evident that the introduction of the gradient term into the constitutive model takes into account the physical nature of the interactions among particles, i.e., discreteness and complex structures.

2.6.3 Solutions of gradient-dependent elastic models

2.6.3.1 Solution of the second-gradient elastic model

Coleman and Hodgdon (1985) and Zbib and Aifantis (1989) obtained the analytical solution of shearing motions for the gradient-dependent material mode. If the unloading part is neglected, their model is similar to the second model mentioned above. Their model is for plastic behavior. The models derived in this section are, however, for elastic behavior because of the loss of loading–unloading conditions. Herein, we treat only the elastic model for simplicity.

Although a different method to find the analytical solutions is used in this section, the result for the second model includes those of Coleman and Hodgdon (1985) and Zbis and Aifantis (1988).

Considering the scaling of the coordinate x, Equation 2.207 can be written as Equation 2.208 without loss of generality. Hence, we focus on a special case of the second model in one-dimensional form expressed as

$$\sigma = a\varepsilon - b\varepsilon_{xx} - 3b\varepsilon^2 \tag{2.208}$$

where a and b are positive material constants and subscript x denotes the partial differentiation with respect to the x coordinate.

Now, we derive the solution under the constant stress $\sigma = c_0$.

Since $\sigma = c_0$ is constant, the equilibrium equation is automatically satisfied. We set

$$\varepsilon' = \varepsilon - e_0 \tag{2.209}$$

in which e_0 is a constant to be determined by the values a, b.

Using $\varepsilon' = \varepsilon = e_0$, we have

$$A\varepsilon' - b(\varepsilon'_{xx} + 3\varepsilon'^2) = 0 \tag{2.210}$$

where

$$c_0 = ae_0 - 3be_0{}^2 \tag{2.211}$$

$$A = a - 6be_0 \tag{2.212}$$

In the following, we set $\varepsilon' = \varepsilon$ neglecting the prime. To solve the non-linear theory on partial differential equations, Hirota (1971, 1976) developed a powerful method to obtain the solution in non-linear wave propagation theory. In this section, we use Hirota's derivative to solve the differential equation of the gradient-dependent model (Equation 2.210). Equation 2.210 can be written as

$$(AD_x^2 - bD_x^4)(f \cdot f) = 0 \tag{2.213}$$

where the logarithmic transformation:

$$\varepsilon = 2\frac{\partial^2}{\partial x^2}\ln f \tag{2.214}$$

is used.

In Equation 2.213, Hirota's bi-linear differential operator is defined by

$$D_x^n(f \cdot g) = \left(\frac{\partial}{\partial x} - \frac{\partial}{\partial y}\right)^n f(x)g(y)\Big|_{y=x} \tag{2.215}$$

in which f and g are functions of x.

The key feature of Hirota's bi-linear differential operator is shown in Appendix A2.2. Next, in order to solve Equation 2.213, we formally expand the function f as

$$f = 1 + \ni f^{(1)} + \ni^2 f^{(2)} + \ni^3 f^{(3)} + \cdots \tag{2.216}$$

where $\ni$ is an arbitrary small constant.

Substituting Equation 2.216 into Equation 2.213, we obtain

$$(AD_x^2 - bD_x^4)(1 + \ni f^{(1)} + \ni^2 f^{(2)} + \ni^3 f^{(3)} + \cdots)(1 + \ni f^{(1)} + \ni^2 f^{(2)} + \cdots) = 0 \quad (2.217)$$

From the 0th order of $\ni$:

$$o(1): \ (AD_x^2 - bD_x^4)1 = 0 \quad (2.218)$$

Then, from the first-order term of $\ni$, we have

$$o(\ni): \ (AD_x^2 - bD_x^4)(f^{(1)} \cdot 1) = 2(A\partial_x^2 - b\partial_x^4)f^{(1)} = 0 \quad (2.219)$$

The solution of Equation 2.219 is

$$f^{(1)} = e^{\eta_1}, \ \ \eta_1 = k_1 x + \eta_0 \quad (2.220)$$

where k_1 and η_0 are constants.

Substituting Equation 2.220 into Equation 2.219, the following relation is obtained:

$$k_1 = \pm\sqrt{A/b} \quad (2.221)$$

From the second-order term of $\ni$, we have

$$2(A\partial_x\partial_x - b\partial_x^4)(f^{(2)} \cdot 1) = -(AD_x^2 - bD_x^4)(f^{(1)} \cdot f^{(1)}) = 0 \quad (2.222)$$

From Equation 2.220 and the definition of Hirota's operator, the relation $-(AD_x^2 - bD_x^4)(f^{(1)} \cdot f^{(1)}) = 0$ is always satisfied.

From Equation 2.222, we have

$$f^{(2)} = e^{\eta_1} \text{ or } 0 \quad (2.223)$$

In a similar manner, we obtain the relations for the higher-order term of $\ni$ without loss of generality:

$$f^{(n)} = 0, \ (n \geq 2) \quad (2.224)$$

From the above derivations, f is obtained by taking the value of $\ni = 1.0$ because of the arbitrariness of $\ni$.

$$f = 1 + Ce^{\eta_1}, \eta_1 = k_1 x, \ C = e^{\eta_0} \quad (2.225)$$

Substituting Equation 2.225 into Equation 2.214, the strain ε is obtained:

$$\varepsilon = \frac{2Ck_1^2 e^{k_1 x}}{(1 + Ce^{k_1 x})^2} + e_0 = \frac{2Ck_1^2}{C^2 e^{k_1 x} + 2C + e^{-k_1 x}} + e_0 \quad (2.226)$$

Let us discuss the solution of Equation 2.226. At $x = 0$,

$$\varepsilon = \frac{2Ck_1^2}{(1+C)^2} + e_0 \tag{2.227}$$

On the other hand, at infinity, the strain is constant:

$$\varepsilon \to e_0 \text{ as } x \to \pm\infty$$

Figure 2.35 illustrates numerical examples of the solution of Equation 2.226 and shows that the width of the localized zone and the magnitude of strain depend on the value of k_1.

2.6.3.2 Solution of the first gradient-dependent elastic model

The first gradient-dependent elastic model derived in Section 2.6.3.1 is expressed as

$$\sigma = a\varepsilon + b\varepsilon_{xx} + 3b\varepsilon^2 \tag{2.228}$$

where the material parameters a and b are positive.

From the method used in the previous section, we derive the solution under the constant stress $\sigma = c_0$.

Since $\sigma = c_0$ is constant, the equilibrium equation is automatically satisfied. The equilibrium equation becomes

$$A\varepsilon' + b(\varepsilon'_{xx} + 3\varepsilon'^2) = 0 \tag{2.229}$$

where

$$\varepsilon' = \varepsilon - e_0, \quad c_0 = ae_0 + 3be_0{}^2, \; A = a + 6be_0 \tag{2.230}$$

Then, we set $\varepsilon' = \varepsilon$ neglecting the prime for simplicity.

Using the same method as in the previous section, the transformed governing equation

by Hirota's derivative becomes

$$(AD_x^2 + bD_x^4)(f \cdot f) = 0 \tag{2.231}$$

This equation is solved as

$$f = 1 + Ce^{\eta_1}, \; \eta_1 = k_1 x, \; C = e^{\eta_0} \tag{2.232}$$

$$k_1 = \pm i\sqrt{A/b} \tag{2.233}$$

where k_1 and η_0 are constants and $i = \sqrt{-1}$.

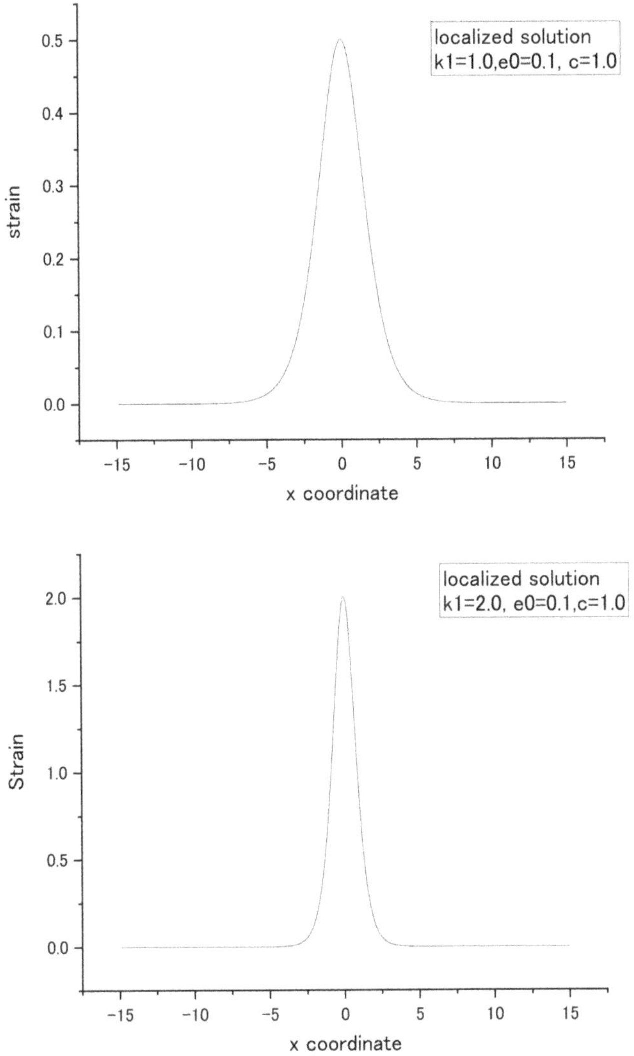

Figure 2.35 Localization solutions.

In order to show the characteristics of the solution for the first model, we can set

$\sqrt{A/b}=1$, $k_1 = \pm i$ without loss of generality.

In this case, the strain ε is obtained as

$$\varepsilon = \frac{-2C}{(C^2 e^{-ix} + 2C + e^{ix})} + e_0 \tag{2.234}$$

Equation 2.234 indicates that the solution is periodic.

2.7 STRAIN-SOFTENING CONSTITUTIVE MODEL CONSIDERING THE MEMORY AND INTERNAL VARIABLES

2.7.1 Introduction

Dense sands, overconsolidated clays, and soft rocks exhibit strain-hardening and strain-softening behavior in a certain range of confining pressure. It is well known that during the straining of soft rock, stress increases at an early stage of loading and reaches a peak stress, after which the stress decreases and approaches a residual strength. As the strain-softening behavior is closely related to the progressive failure phenomenon, it is important to establish a constitutive model that can describe the strain-softening behavior.

Two approaches have been used to describe the strain-softening behavior. One approach calls for the strain-softening behavior to be incorporated into a constitutive model (e.g., Höeg 1972; Nayak and Zienkiewicz 1972; Benerjee and Stipho 1979; Adachi and Oka 1995; Frantzskonis and Desai 1987). However, it has been pointed out that the use of classical rate-independent constitutive models with strain softening leads to the problem of ill-posedness in the numerical calculation due to the loss of the positive-definiteness of the structural tangential stiffness matrix. In addition, a complicated procedure for the numerical solution, such as the arc length method, is needed instead of the usual numerical technique (Riks 1972) and the solution depends highly on the mesh size in the finite element methods. Desai (1974) proposed a consistent finite element technique for work-softening behavior to avoid the negative tangent modulus. In order to resolve the problem of ill-posedness, Bazan et al. (1988) and Aifantis (1984) introduced the non-local approach into constitutive modeling.

The other approach to the strain-softening problem is based on the idea that strain softening is due to the development of strain localization in the specimen. In this context, strain softening is explained by the onset of the bifurcation of the solution for the boundary value problem using a strain-hardening type of constitutive model (e.g., Read and Hegemier 1984; Sandler 1986). Although this type of approach does give some information on the onset conditions of the shear bands (e.g., Yatomi et al. 1989a,b), it is complicated to pursue the numerical analysis in the post-localization and/or the post-failure regime (Zbib and Aifantis 1988). Therefore, this type of approach has been applied in the study of strain localization but has not been fully established for post-localization problems.

Pietruszczak and Mroz (1981) introduced a smear element into the finite element analysis with strain softening to explain the overall strain-hardening and strain-softening behavior of geological materials by assuming a shear zone a priori. In addition, from a physical point of view, it is difficult to exclude the changes in the material properties during the strain-softening process.

On the other hand, it can be assumed that the strain-softening behavior is due to both the changes in material properties, i.e., destructuration of the material, and the bifurcation such as strain localization. Desai et al. (1991) and Desai (1992) proposed a constitutive model for geological materials based on the disturbed concept (DSC) by taking into consideration the degradation of materials. As a result, the derived model can simulate the strain-softening behavior of materials (Shao and Desai 2000). From a conceptual point of view, it seems that there is a similarity between the theory in this section and Desai's theory. Kimoto and Oka (2005) proposed an elasto-viscoplastic model based on the overstress type of viscoplasticity theory, in which the degradation of the material is described by the shrinkage of the boundary and yield surfaces with respect to the evolution of inelastic strain. The theory can reproduce the strain-softening behavior. Bazan et al. (1988) proposed an alternative model that describes the strain-softening behavior based on the micro-plane model. From the above considerations, a model should be derived such that a deformation analysis is possible in the pre-failure and post-failure regimes. The model explained in this section describes the strain-hardening and strain-softening behavior for soft sedimentary rocks.

It is worth noting that several methods have been proposed to regularize the ill-posedness of the boundary value problem: non-local formulation, fluid–solid multiphase formulation, viscoplastic regularization, and dynamics formulation (Oka and Kimoto 2012).

In this section, we will explain the strain-softening model following Adachi and Oka (1995). Oka and Adachi (1985) and Adachi and Oka (1995) derived a constitutive model by introducing a stress history tensor which is defined as a functional of stress history with respect to the strain history. The derived model is evaluated by a comparison of the simulated and test results of sedimentary soft rocks, i.e., porous tuff. In addition, the applicability of the model to numerical analysis is discussed in relation to the uniqueness of the solution in both pre-failure and post-failure regimes for the initial and boundary value problems.

2.7.2 Interpretation of the strain-softening process

Figure 2.36 shows a schematic diagram of the stress–strain relations with both strain hardening and strain softening. It is assumed that the material strength is composed of frictional strength and others due to cementation and/or the cohesion of materials. In addition, it is also presumed that the frictional strength is relatively small at the early stage of the straining process, and the strength due to cementation and/or cohesion of the material decreases with the advances in deformation and eventually frictional strength develops.

Kovari (1982) used a similar concept for his block model in order to explain the failure of rock material. In other words, the strength due to

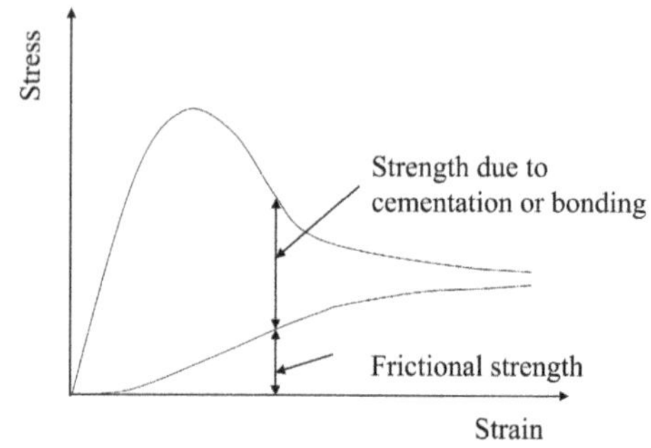

Figure 2.36 Schematic diagram of the stress–strain relations with both strain hardening and strain softening. (After Adachi, T. and Oka, F. 1995. An elasto-plastic constitutive model for soft rock with strain softening. *Int. J. Numer. Anal. Geomech.* 19: 233–247.)

cementation and/or cohesion between particles is released in the strain-softening regime and the frictional strength increases rapidly after the peak strength. Oka (1985) proposed a viscoplastic constitutive theory by introducing the stress history tensor with respect to real time. In the following, a stress history tensor is defined with respect to the strain measure and is introduced into the elasto-plastic model. This stress history tensor will be introduced into the constitutive model that controls the frictional component of the strength in order to describe this physical concept.

2.7.3 Inherent strain measure, stress history tensor, and kernel function

It is assumed that the total strain increment tensor can be decomposed into two parts, i.e., an elastic component, $d\varepsilon_{ij}^e$, and a plastic component, $d\varepsilon_{ij}^p$:

$$d\varepsilon_{ij} = d\varepsilon_{ij}^e + d\varepsilon_{ij}^p \tag{2.235}$$

A new strain measure z is defined as

$$dz = (Q_{ijkl} d\varepsilon_{ij} d\varepsilon_{kl})^{1/2} \tag{2.236}$$

where Q_{ijkl} is generally a fourth-order tensor.

Herein, we assume that Q_{ijkl} is an isotropic tensor. Since Q_{ijkl} is inherent to a specific material, a new measure, z, can be called an inherent strain measure.

The total stress history tensor, σ_{ij}^z, is a union of the two sets of the present stress tensor, σ_{ij}, and the reduced stress history tensor, σ_{rij}^z. The reduced stress history is defined by the history of the stress without the present stress. Following the description method in continuum mechanics (e.g.,

Eringen 1967), the total stress history and the reduced stress history are described as

$$\sigma_{ij}^{z} = \left[\sigma_{ij}(z), \sigma_{rij}^{z}\right] \tag{2.237}$$

$$\sigma_{rij}^{z} = \left[\sigma_{ij}(z - z'), 0 < z' \leq z\right] \tag{2.238}$$

where z is a total strain measure obtained by the integration of a strain measure increment defined by Equation 2.236.

The stress history tensor σ_{ij}^{*} is assumed to be a function of the reduced stress history as

$$\sigma_{ij}^{*} = \sigma_{ij}^{*}(\sigma_{rij}(z - z')) \tag{2.239}$$

In cases when the stress history tensor is given by a function of the reduced stress history with respect to the total strain measure, z, the stress history tensor is expressed by

$$\sigma_{ij}^{*} = \int_0^z K(z - z')\sigma_{ij}(z')dz' \tag{2.240}$$

where K is a continuous bounded kernel function and $\partial K / \partial z < 0$ is assumed.

Although K is generally the tensor function of z, K is assumed to be a scalar here for simplicity. From this assumption, the stress history tensor satisfies the principle of fading memory by which the influence of the past history becomes less significant. In addition, due to the theorem of integration in a wider sense, the interval for the integration can be taken as in $0 \leq z' \leq z$. For soft rock, we adopted the kernel function, $K(z)$, given by a simple exponential function and the strain measure z is defined by the invariant of the deviatoric strain tensor as

$$k(z) = \exp(-z / \tau) / \tau, \quad dz = (de_{ij} de_{ij})^{1/2} \tag{2.241}$$

where τ is a strain-softening material parameter and de_{ij} is the deviatoric strain tensor.

From Equations 2.240 and 2.241, the stress history tensor is derived as

$$\sigma_{ij}^{*} = \frac{1}{\tau}\int_0^z \exp(-(z - z') / \tau)\sigma_{ij}(z')dz' \tag{2.242}$$

When the value of τ is small, the retardation of the development of the stress history tensor decreases, and the stress history tensor coincides with the real stress only when $\tau = 0$. This condition corresponds to the model for a fully destructured and/or failed frictional material.

The not-well known terminology "destructured material" is used to describe materials such as soils and rocks from which the effect of the internal structure as cementation or bonding has been removed (e.g., Leroueil and Vaughan 1990). The destructured state is similar to the disturbed state advocated by Desai (2001).

2.7.4 Flow rule and yield function

The non-associated flow rule is adopted as

$$d\varepsilon_{ij}^{p} = H\frac{\partial f_p}{\partial \sigma_{ij}} df_y \tag{2.243}$$

where f_p is the plastic potential function; f_y is the yield function; and H is the hardening-softening function.

It is assumed that the yield function depends on the stress history tensor and the hardening/softening parameter κ as

$$f_y(\sigma_{ij}^{*}, \kappa) = 0 \tag{2.244}$$

The loading criterion is given by

$$f_y = 0, \ df_f = \frac{\partial f_y}{\partial \sigma_{ij}^{*}} d\sigma_{ij}^{*} > 0; \ d\varepsilon_{ij}^{p} \neq 0 \tag{2.245}$$

$$f_y = 0, \ df_f \leq 0; \ d\varepsilon_{ij}^{p} = 0 \tag{2.246}$$

Applying the assumptions that a material is elastic when the stress ratio is constant and a plastic strain develops only when the stress ratio changes, Adachi and Oka (1984a,b) and Oka et al. (1989) derived an elasto-plastic constitutive model for overconsolidated clay. Following the same concept, the yield function for soft rock is given by

$$f_y = \eta^{*} - \kappa = 0 \tag{2.247}$$

where η^{*} is the stress history ratio defined by

$$\eta^{*} = (S_{ij}^{*} S_{ij}^{*} / \sigma_{m}^{*2})^{1/2} \tag{2.248}$$

where S_{ij}^{*} and σ_{m}^{*} are the deviatoric part and the isotropic part of the stress history tensor, respectively, and δ_{ij} is Kronecker's delta.

2.7.5 Elastic boundary surface

The distinction of the loading and unloading condition can be evaluated by Equations 2.244–2.246. In the case of cyclic loading, a kinematic hardening rule is necessary to rationally describe the cyclic behavior of the material. Since the present formulation is in the context of the isotropic hardening-softening theory, an elastic domain can be introduced to describe the elastic behavior in a simple manner:

$$f_e(\sigma_{ij}, \kappa_e) = 0 \tag{2.249}$$

In the elastic domain, where $f_e \leq 0$, the material is elastic and the value η^* is frozen. The functional form of f_e is similar to the form of the yield function. When the material state enters the elastic domain, the value of κ_e never changes within the elastic domain. However, κ_e may shrink during the softening process.

2.7.6 Plastic potential function and overconsolidation boundary surface

The plastic potential function is assumed to be given by the present real stress, a plastic potential parameter L:

$$f_p(\sigma_{ij}, L) = 0 \tag{2.250}$$

The plastic potential function for soft rock is given by

$$f_p = \bar{\eta} + \bar{M} \ln\{(\sigma_m + b) / \sigma_{mb}\} = 0 \tag{2.251}$$

where σ_{mb} and b are the material parameters that describe the material structure and $\bar{\eta}$ is the stress ratio defined by

$$\bar{\eta} = \{S_{ij}S_{ij} / (\sigma_m + b)^2\}^{1/2} \tag{2.252}$$

in which S_{ij} is the deviatoric part of the stress tensor; σ_m is the mean stress; b is a material parameter of the shape of the plastic potential surface; and $\bar{M}$ is discussed in the following.

The plastic potential function in the stress invariant space is schematically shown in Figure 2.37. Next, an overconsolidation boundary surface is introduced that controls the changes in the shape of the plastic potential function. From a physical point of view, the material is in the overconsolidated region when $f_b < 0$, and in the normally consolidated region when $f_b \geq 0$. For soft rocks, the overconsolidated boundary surface is given by

$$f_b = \bar{\eta} + \bar{M}_m \ln\{(\sigma_m + b) / \sigma_{mb}\} = 0 \tag{2.253}$$

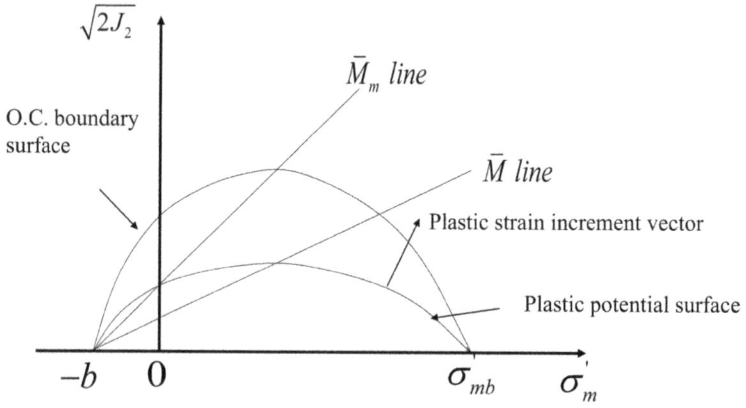

Figure 2.37 Plastic potential function in the stress invariant space. (After Adachi, T. and Oka, F. 1995. An elasto-plastic constitutive model for soft rock with strain softening. *Int. J. Numer. Anal. Geomech.* 19: 233–247.)

where $\bar{M}_m$ is the value of $\bar{\eta}$ when maximum compression takes place.

By introducing the overconsolidated boundary surface, the value of $\bar{M}$ in Equation 2.251 is given in what follows.

In the overconsolidated region $f_b < 0$:

$$\bar{M} = -\bar{\eta} / \ln\{(\sigma_m + b) / \sigma_{mb}\} \tag{2.254}$$

In the normally consolidated region $f_b \geq 0$:

$$\bar{M} = \bar{M}_m \tag{2.255}$$

From Equation 2.254, the value of $\bar{M}$ can be determined by the value of the present stress and by the material parameters b and σ_{mb}.

2.7.7 Strain-hardening and strain-softening parameter

It is assumed that the evolutional equation of the strain hardening/softening parameter κ is given by

$$\dot{\kappa} = \frac{G'(M_f^* - \kappa)^2 \dot{\gamma}^p}{M_f^{*2}} \tag{2.256}$$

in which $\dot{\gamma}^p$ is the second invariant of the deviatoric plastic strain rate tensor, given by

$$\dot{\gamma}^p = (\dot{e}_{ij}^p \dot{e}_{ij}^p)^{1/2} \tag{2.257}$$

where $\dot{e}_{ij}^p$ is the deviatoric plastic strain rate tensor.

Then, κ is given by

$$\kappa = \int_0^t \dot{\kappa} dt \tag{2.258}$$

When κ is zero at the initial state under proportional loading conditions, the hyperbolic function is given by integration as

$$\kappa = \frac{M_f^* G' \gamma^p}{M_f^* + G' \gamma^p} \tag{2.259}$$

in which M_f^* is the value of η^* at the residual state and G' is an initial tangent of Equation 2.259.

2.7.8 Elasto-plastic constitutive model with strain softening

For the elastic strain rates, the isotropic Hooke's law is used as

$$d\varepsilon_{ij}^p = dS_{ij} / (2G) + d\sigma_m \delta_{ij} / 3K \tag{2.260}$$

where G is the elastic shear modulus and K is the elastic bulk modulus.

The plastic strain increment tensor, $d\varepsilon_{ij}^p$, is derived by using Prager's compatibility condition in Equation 2.261 with Equations 2.243, 2.247, and 2.251.

$$df_y = d(\eta^* - \kappa) = 0 \tag{2.261}$$

$$d\varepsilon_{ij}^p = \Lambda \left[\frac{\eta_{ij}}{\bar{\eta}} + (\bar{M} - \bar{\eta}) \frac{\delta_{ij}}{3} \right] \left[\frac{\eta_{kl}^*}{\eta^*} - \eta^* \frac{\delta_{kl}}{3} \right] \frac{d\sigma_{kl}^*}{\sigma_m^*}, \quad \Lambda = \frac{M_f^{*2}}{G'(M_f^* - \eta^*)^2} \tag{2.262}$$

where $\eta_{ij} = S_{ij} / (\sigma_m + b)$ and $\eta_{kl}^* = S_{kl}^* / \sigma_m^*$.

Eight parameters are included in the constitutive model, i.e., G (elastic shear modulus), K (elastic bulk modulus), τ (stress history parameter), b and σ_{mb} (plastic potential parameters), M_f^* (stress ratio at a residual state), $\bar{M}_m$ (parameter for overconsolidation boundary surface), and G' (strain-hardening/softening parameter).

2.7.9 Uniqueness of the solution for the initial value problem

The uniqueness of the solution is an important characteristic for the initial and/or boundary value problems. In this section, the uniqueness of the solution for the initial value problem is discussed for the strain-softening constitutive equations mentioned above following the work by Valanis (1985) and Willam et al. (1986).

The increment of the stress history tensor is given by

$$d\sigma_{ij}^{*} = (\sigma_{ij} - \sigma_{ij}^{*})dz / \tau \tag{2.263}$$

Employing Equation 2.263, therefore, the total strain increment tensor can be rewritten as

$$d\varepsilon_{ij} = D_{ijkl}d\sigma_{kl} + A_{ij} \tag{2.264}$$

where D_{ijkl} is the elastic compliance tensor, i.e.,

$$D_{ijkl}d\sigma_{kl} = \frac{d\sigma_{ij}}{2G} - \frac{\lambda d\sigma_{kk}}{2G(3\lambda + 2G)}\delta_{ij}, \ \lambda = K - 2G/3 \tag{2.265}$$

A_{ij} is given by

$$A_{ij} = \Lambda\left[\frac{\eta_{ij}}{\bar{\eta}} + (\bar{M} - \bar{\eta})\frac{\delta_{ij}}{3}\right]\left[\frac{\eta_{kl}^{*}}{\eta^{*}} - \eta^{*}\frac{\delta_{kl}}{3}\right]\frac{(\sigma_{kl} - \sigma_{kl}^{*})dz}{\tau\sigma_{m}^{*}} \tag{2.266}$$

Following the work of Valanis (1985), we examine the uniqueness of the solution at time $t + \Delta t$ when the solution at time t is unique. Now, let two different solutions exist at time $t + \Delta t$. If the difference between the two solutions satisfies the following inequality, the material can be called positive material and the solution to the initial value problem is unique.

$$\Delta\dot{\varepsilon}_{ij}\Delta\dot{\sigma}_{ij} > 0 \tag{2.267}$$

Two different solutions are given by

$$\Delta\dot{\sigma}_{ij} = \dot{\sigma}_{ij(1)} - \dot{\sigma}_{ij(2)} = \left[\frac{d\sigma_{ij}}{dz}_{(1)} - \frac{d\sigma_{ij}}{dz}_{(2)}\right]\frac{dz}{dt} \tag{2.268}$$

$$\Delta\dot{\varepsilon}_{ij} = \dot{\varepsilon}_{ij(1)} - \dot{\varepsilon}_{ij(2)} = \left[\frac{d\varepsilon_{ij}}{dz}_{(1)} - \frac{d\varepsilon_{ij}}{dz}_{(2)}\right]\frac{dz}{dt} \tag{2.269}$$

From Equation 2.264, we obtain

$$\Delta\dot{\varepsilon}_{ij}\Delta\dot{\sigma}_{ij} = (\dot{\sigma}_{ij(1)} - \dot{\sigma}_{ij(1)})D_{ijkl}(\dot{\sigma}_{ij(1)} - \dot{\sigma}_{ij(1)})(dz/dt)^2 \tag{2.270}$$

Since $A_{ij}dz$ is equal for the two solutions and $dz/dt > 0$, Equation 2.267 holds when D_{ijkl} is positive definite. From Equation 2.265, D_{ijkl} is positive definite and $(\dot{\sigma}_{ij(1)} - \dot{\sigma}_{ij(2)})$ is symmetric and not zero, and Equation 2.267

holds for the proposed model. As a result, the uniqueness of the solution for the initial problem holds for the above softening model. It is worth noting that if A_{ij} depends on the rate of strain or stress rate, i.e., the model is rate independent in a classical sense, Equation 2.267 does not always hold.

2.7.10 Simulation of triaxial compression test results of sedimentary soft rock

First, we explain the properties of the model before showing the simulated results of triaxial compression tests using the softening model. Using Figure 2.38, the stress–strain relations are discussed. With the advance of deformation, the strain measure (axial strain) increases monotonically, and q (deviator stress = $\sigma_{11} - \sigma_{33}$) increases at the early stages of loading when τ is not zero and reaches a peak value. Then, q decreases and eventually reaches a residual strength. During the straining process, q^*, which corresponds to the stress history tensor, increases at a high rate and finally approaches the residual value of stress. Therefore, at the early stages of loading, $q - q^*$ increases and then decreases to zero after its peak stress. Since q^* is equal to q when $\tau = 0$, the stress (q)–the strain (ε_{11}) relation can be modeled by the hyperbolic curve without a peak value. As mentioned above, q^* corresponds to the frictional strength and $q - q^*$ corresponds to the strength due to the cementation and/or cohesion of the material. On the one hand, therefore, cases in which $\tau = 0$ describe the stress–strain relations of normally consolidated clay and loose sand, and the behavior of

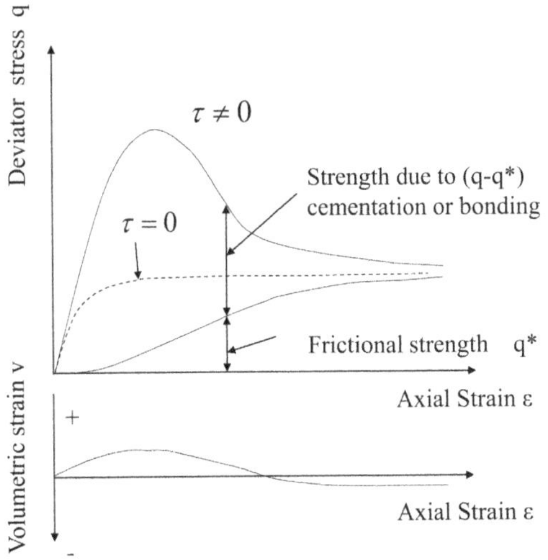

Figure 2.38 Strain hardening-softening behavior and the growth of frictional strength. (After Adachi, T. and Oka, F. 1995. An elasto-plastic constitutive model for soft rock with strain softening. *Int. J. Numer. Anal. Geomech.* 19: 233–247.)

soft rock at high confining pressures. On the other hand, the models with $\tau \neq 0$ describe the behavior of overconsolidated clay, dense sand, and soft rock at low confining pressures.

Figures 2.39 and 2.40 show simulated results for the behavior of sedimentary soft rock (porous tuff) under drained triaxial compression conditions (Adachi and Ogawa 1981) by the model. Table 2.9 shows the material parameters used in the calculation. The material parameters were determined as follows: the elastic shear modulus, G, and Poisson's ratio, ν, are determined either from the unloading slope of the stress–strain curve or by the initial tangent moduli, $\Delta q / \Delta e_{11}$ and $\Delta v / \Delta e_{11}$, respectively, through the following formula:

$$G = \Delta q / (3\Delta e_{11}), \quad \nu = (2\Delta v / \Delta e_{11} - 3) / (\Delta v / \Delta e_{11} + 3) \tag{2.271}$$

In a classical elasto-plastic theory, the initial tangential modulus includes both elastic and inelastic moduli when the initial elastic domain is taken into account in the model. To the contrary, in the present model, the initial modulus is purely elastic because the inelastic strain increment depends on the increment of the invariant of the stress history tensor, and the initial increment of the stress history is always zero.

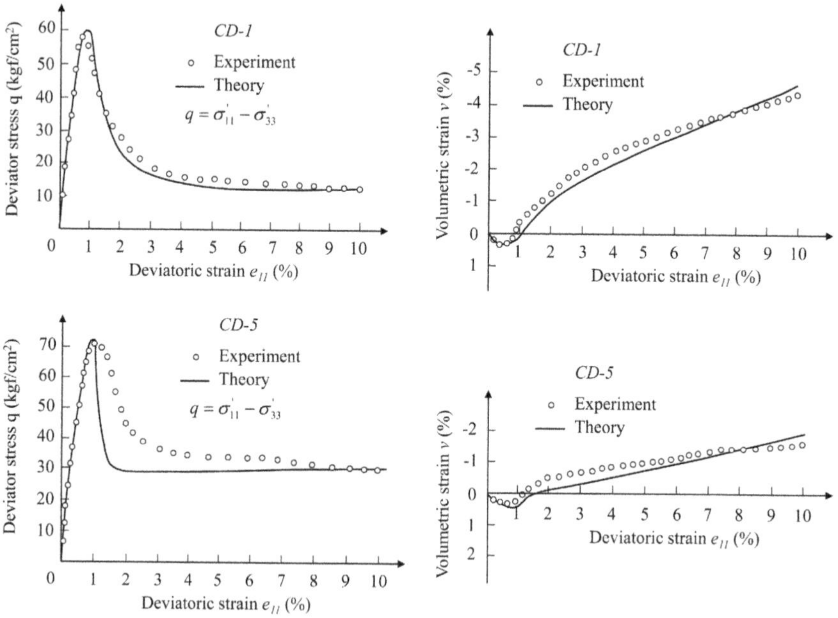

Figure 2.39 Stress–strain relations of soft rock. (After Adachi, T. and Oka, F. 1995. An elasto-plastic constitutive model for soft rock with strain softening. *Int. J. Numer. Anal. Geomech.* 19: 233–247.)

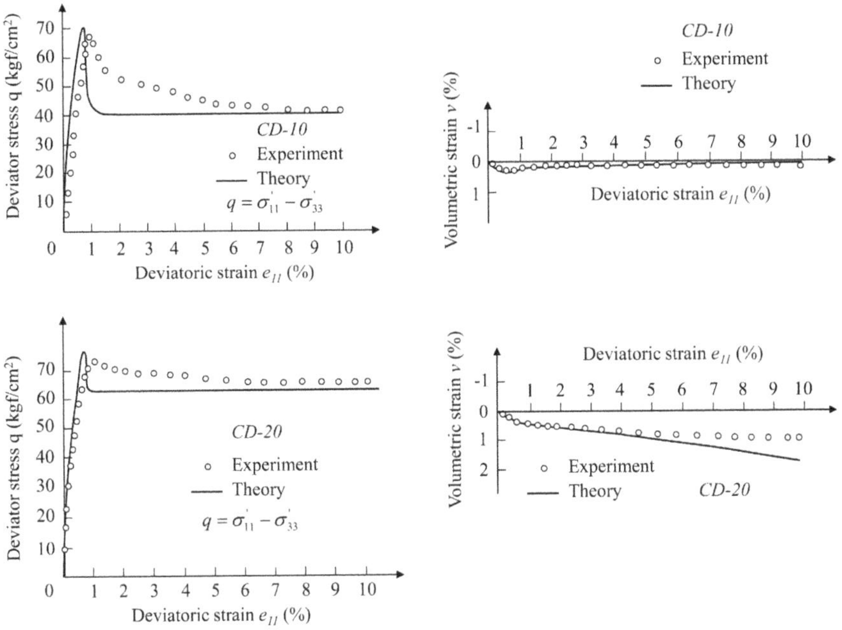

Figure 2.40 Stress–strain relations of soft rock. (After Adachi, T. and Oka, F. 1995. An elasto-plastic constitutive model for soft rock with strain softening. *Int. J. Numer. Anal. Geomech.* 19: 233–247.)

Table 2.9 Material parameters and initial conditions for soft rock

Test no.	*CD-1*	*CD-5*	*CD-10*	*CD-20*
σ'_{m0}: Initial mean effective stress (×98 kPa)	1	5	10	20
e_0: Initial void ratio	0.72	0.72	0.72	0.72
ν: Poisson ratio	−0.108	0.119	0.144	0.0323
E: Young's modulus (×98 kPa)	13500	10410	13500	16500
K: Elastic bulk modulus (×98 kPa)	3700	4550	5260	5880
M_f^*: Stress ratio at failure, $M_f^* = \overline{M}_m$	1.97	1.61	1.42	1.26
G': Hardening parameter	1000	3000	5000	10000
b: Parameter for plastic potential (×98 kPa)	12	30	40	40
σ'_{mb}: (×98 kPa)	150	150	150	150
T	0.09	0.019	0.0007	0.0025

Source: Data from Adachi, T. and Oka, F. 1995. An elasto-plastic constitutive model for soft rock with strain softening. *Int. J. Numer. Anal. Geomech.* 19: 233–247.

The increment of the stress history tensor is given by Equation 2.263. From Equation 2.263, it is easily seen that the increment of the stress history tensor, $d\sigma^*_{ij}$, is zero when the real stress is zero since the stress history tensor is zero at the initial state. Therefore, Young's modulus and the elastic bulk modulus are obtained with the following well-known relations:

$$E = 2(1+\nu)G, \quad K = E/3(1-2\nu) \tag{2.272}$$

where M^*_f is the stress ratio for the residual stress and G' is obtained from the initial tangent modulus of the stress–strain relation of the destructured material in which τ is zero.

Alternatively, G' could be determined by applying the curve fitting method to an overall stress–strain curve. The stress ratio at maximum compression is $\bar{M}_m$. The plastic potential parameters, b and σ_{mb}, are determined such that the model can reasonably describe the dilatancy at large deformations. Finally, τ is determined by the curve adjusting method so that the stress–strain curves of experiments and calculated results reasonably correspond. All the material parameters can be determined by the drained triaxial compression test results (see Table 2.9). Figures 2.39 and 2.40 indicate that the proposed model can well describe the stress–strain relations of sedimentary soft rock. Let us discuss the confining pressure dependency of τ. Figure 2.41 is the relationship between τ and the confining pressure. It is easily seen that τ decreases with an increase in the initial confining pressure, σ_{m0}. This means that the strain-softening characteristic becomes less significant at large values for the confining pressure. In other words, the value of τ is closely related to the mechanism of the development of the strength.

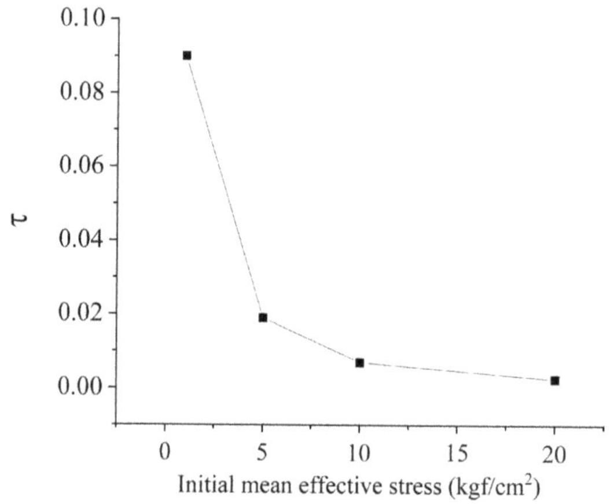

Figure 2.41 Parameter τ versus initial mean effective stress. (After Adachi, T. and Oka, F. 1995. An elasto-plastic constitutive model for soft rock with strain softening. *Int. J. Numer. Anal. Geomech.* 19: 233–247.)

We will explain the reason why Poisson's ratio for the CD-1 case is negative. The value −0.108 is in the interval of $-1 \le \nu \le 0.5$ for the ideal elastic material. It is very unusual for the value of ν to be negative for geological materials. This point can be interpreted in the following. In the present formulation, the total strain rate is decomposed into two parts for simplicity: the elastic and the plastic strain rates. For a more elaborate formulation, the total elastic strain rate can be divided into two parts. One is a purely elastic strain rate and the other is an elastic strain due to changes in the material structure. Although Poisson's ratio of the latter part might be negative, Poisson's ratio of the purely elastic strain component is considered to be positive. In general, it is not so easy to separate the elastic strain. One possible method may be to use an ultra-measurement of the wave velocity for determining pure elasticity. This consideration is consistent with the concept of damage mechanics (Chaboche 1988). Oka et al. (1989) divided the elastic strain component into two parts in the formulation of the constitutive model for natural clay.

2.8 STRAIN-SOFTENING CONSTITUTIVE MODEL FOR FROZEN SOIL

2.8.1 Introduction

Experimental works have revealed that frozen soil exhibits rate-sensitive, strain hardening and at large strain levels, strain-softening behavior (e.g., Ladanyi 1980). Many models have been proposed to describe the behavior of frozen soil (Vialov 1963; Andersland an Al-Nouri 1970; Fish 1980; Assure 1979; Ting 1983; Wang et al. 2014). Most of these studies were conducted under uniaxial loading conditions. Oka (1985) proposed a new type of viscoplastic model with memory and internal variables. Based on this model, Oka and Adachi (1985, 1995) constructed an elasto-plastic constitutive model with strain softening introducing a stress history tensor, which is presented in Section 2.7. In that model, the time measure is used instead of real time, which is similar to the endochronic time by Valanis (1971). Adachi et al. (1990) proposed a constitutive model for frozen sand using a generalized measure of time. In this section, we present the strain-softening viscoplastic model for frozen sand based on Adachi et al. (1990) and Oka and Adachi (1995).

2.8.2 Elasto-viscoplastic softening model for frozen sand

We consider the infinitesimal strain fields for simplicity. A new time measure z is given by

$$dz = F(\text{strain rate})dt \tag{2.273}$$

where dz is an increment of a new time measure; t is the real time; and the function F is to be determined experimentally for the particular medium.

The stress history tensor σ_{ij}^{*} is given as a functional of the reduced stress history with respect to the new time measure z.

$$\sigma_{ij}^{*} = \sigma_{ij}^{*}\left[\sigma_{rij}^{z}(z)\right] \tag{2.274}$$

$$\sigma_{rij}^{z} = \left[\sigma_{ij}(z-z'); 0 < z' \le z\right] \tag{2.275}$$

where σ_{rij}^{z} is the reduced stress history.

Herein, the stress history tensor is expressed by a single exponential type of kernel function as

$$\sigma_{ij}^{*} = \int_0^z \frac{1}{\tau}\exp[-(z-z')/\tau](\sigma_{ij}(z') - \sigma_{ij}(0))dz' \tag{2.276}$$

where τ is a material parameter expressing the retardation of the stress with respect to the time measure, and $\sigma_{ij}(0)$ is the value of the stress tensor at $z = 0$; z is a measure defined by

$$dz = g(\dot{\varepsilon}_{ij})dt \tag{2.277}$$

where $\dot{\varepsilon}_{ij}$ denotes the strain rate tensor.

A full description of the constitutive model is obtained with the flow rule, the yield function, the elastic strain rate tensor, etc.

The following new time measure is adopted for frozen sand:

$$dz = F_0\left(\frac{\dot{\varepsilon}_{11}}{\dot{\varepsilon}_0}\right)^a dt \tag{2.278}$$

A full description of the constitutive model is obtained with the flow rule, yield function, hardening rule, the elastic strain rate tensor, etc., described in Section 2.7.

2.8.3 Instability analysis of the model

We will discuss the stability of the proposed model in one-dimensional form. In one-dimensional form, the yield condition can be written as

$$f = \sigma^{*} - \kappa = 0 \tag{2.279}$$

where σ^{*} is the stress history and κ is the hardening/softening parameter.

The evolutional equation of the hardening/softening parameter κ is written as

$$\dot{\kappa} = \frac{(a-\kappa)^2}{ab}\dot{\varepsilon}^{vp} \tag{2.280}$$

where a and b are positive material parameters.

Since the stress history is written as

$$\sigma^* = \frac{1}{\tau}\int_0^z \exp\left(-\frac{z-z'}{\tau}\right)\sigma(z')dz' \tag{2.281}$$

and we assume $dz = dt$ and $\sigma(0) = 0$ for simplicity, the total strain rate is given by

$$\frac{d\varepsilon}{dt} = \frac{1}{E}\frac{d\sigma}{dt} + \frac{ab}{\tau(a-\sigma^*)^2}(\sigma - \sigma^*) \tag{2.282}$$

where $d\varepsilon$ and $d\sigma$ are the strain and stress increments, respectively.

Under the condition in which the initial viscoplastic increment is zero, the viscoplastic strain is given by the integration as

$$\varepsilon^{vp} = \frac{b\sigma^*}{a-\sigma^*} \tag{2.283}$$

From Equations 2.282 and 2.283, the viscoplastic strain rate is given by

$$\frac{d\varepsilon^{vp}}{dt} = f(\varepsilon^{vp}, \sigma) \tag{2.284}$$

where

$$f(\varepsilon^{vp}, \sigma) = \frac{1}{ab\tau}\left[b^2\sigma + (2b\sigma - ab)\varepsilon^{vp} + (\sigma - a)\varepsilon^{vp2}\right] \tag{2.285}$$

Using linear transformation, Equation 2.282 can be transformed into the well-known logistic model ($dx/dt = Ax(1 - Bx)$, x: variable, A, $B > 0$: constants) used in mathematical biology (May 1974). In addition, it is known that the discretized equation of the logistic model may produce bifurcation, chaotic behavior, and pattern formation (e.g., Drazin 1992).

The function f has two equilibrium points in general. From $f = 0$, the two equilibrium points are

$$\varepsilon^{vp} = -b \tag{2.286}$$

and

$$\varepsilon^{vp} = -\frac{b\sigma}{\sigma - a} \tag{2.287}$$

Differentiating f with respect to the viscoplastic strain gives

$$\frac{\partial f}{\partial \varepsilon^{vp}} = \frac{1}{ab\tau}\left[2\varepsilon^{vp}(\sigma - a) + b(2\sigma - a)\right] \tag{2.288}$$

Figure 2.42 shows a graph of the function $f(\varepsilon^{vp},\sigma)$. The behavior of the function f depends on the value of the stress σ. In the case where σ is less than a, two-equilibrium points exist: points B and A in Figure 2.42. Point B is unstable and point A is stable. Since point A is asymptotically stable and the sign of $df/d\varepsilon^{vp}$ is negative, the behavior of the model is stable when $\sigma < a$. On the other hand, Figure 2.42 shows that in the case of $\sigma = a$, f has one equilibrium point at point B and it is unstable. In this case, the behavior of the model is unstable. When $\sigma > a$, f has two equilibrium points (points B and C). Point C is stable and point B is unstable.

$$\frac{\partial f}{\partial \varepsilon^{vp}} = \frac{1}{\tau} > 0 \text{ at point } B$$

This means that the material is unstable in the region where $\sigma \geq a$. From the consideration mentioned above, the behavior of the model loses stability at the point where $\sigma = a$. The point of $\sigma = a$ is a bifurcation point.

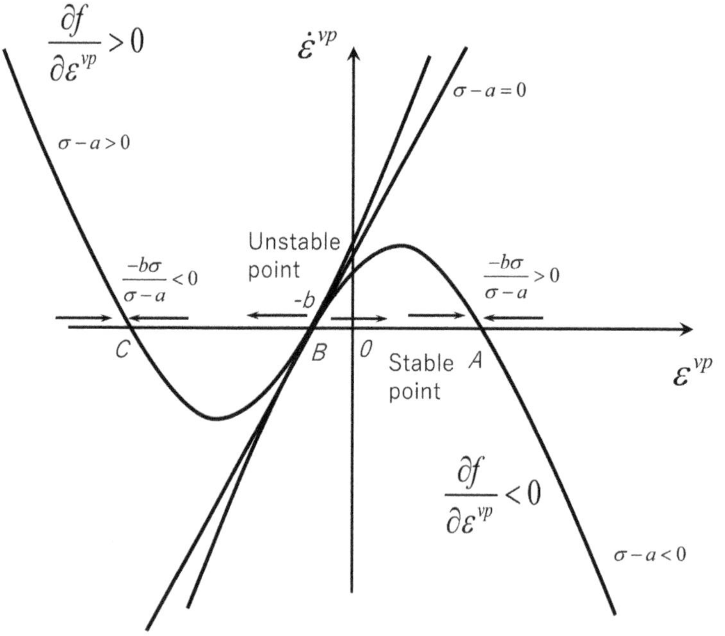

Figure 2.42 Graph of function f in Equation 2.285. (After Oka, F., Adachi, T., Yashima, A. andChu,L.L.1994.Strainlocalizationanalysisbyelasto-viscoplasticsofteningmodelfor frozen sand. *Int. J. Numer. Anal. Geomech.* 18: 813–832.)

Next, we will discuss the uniaxial compression test condition with constant strain rate. In the case of compression with constant strain rate C, the stress rate is given by

$$\frac{d\sigma}{dt} = E\left(C - \frac{ab}{(a-\sigma^*)^2}(\sigma - \sigma^*)\frac{1}{\tau} \right) \tag{2.289}$$

In the early stage of straining, $d\sigma/dt$ is positive due to the small value of the viscoplastic strain rate. With an increase in strain, σ^* asymptotically reaches a and $a - \sigma^*$ becomes very small. Subsequently, the viscoplastic strain rate rapidly grows and $d\sigma/dt$ becomes negative. In this situation:

$$d\varepsilon d\sigma < 0 \tag{2.290}$$

Hence, the model can reproduce the strain-softening behavior, actually, in the numerical simulation shown in Figure 2.43.

Let us discuss the stability of the solution in the sense of Liapunov (e.g., Perko 1991).

Consider the two solutions, ε_1^{vp} and ε_2^{vp}, for slightly different constant stresses under creep conditions in which the stresses are $\sigma_1 = \sigma$ and $\sigma_2 = \sigma + \alpha$ where α is a positive small constant. Since the stress is constant during the creep process, we obtain

$$\varepsilon^{vp} = \frac{b\sigma^*}{a-\sigma^*} = \frac{b\sigma(1-\exp(-t/\tau))}{a-\sigma+\sigma\exp(-t/\tau)} \tag{2.291}$$

At time t, the difference between the viscoplastic strains can be evaluated by the following relation:

$$\left|\varepsilon_2^{vp} - \varepsilon_1^{vp}\right| = \frac{ab\alpha(1-\exp(-t/\tau))}{[a-\sigma-\alpha+(\sigma+\alpha)\exp(-t/\tau)][a-\sigma+\sigma\exp(-t/\tau)]} \tag{2.292}$$

When $a \geq \sigma$, the denominator is always positive because α is an arbitrary positive small constant, and the difference is limited. On the other hand, when $a < \sigma$, the denominator becomes zero at a finite time, t_c, and the difference becomes infinite.

$$t_c = \tau \ln\left(\frac{\sigma+\alpha}{\sigma+\alpha-a} \right) \tag{2.293}$$

Equations 2.292 and 2.293 indicate that fluctuation grows at a finite time even if the fluctuation is very small, and it means that the solution is not stable in the sense of Liapunov. From the above discussion, it is shown that the model has the potential for describing material instability.

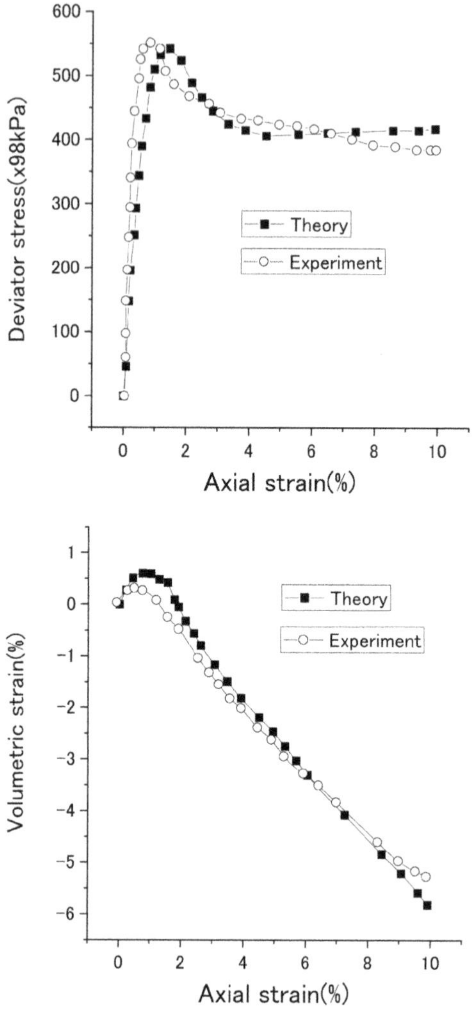

Figure 2.43a Simulated and experimental stress–strain relations of frozen sand. (After Oka, F., Adachi, T., Yashima, A. and Chu, L.L. 1994. Strain localization analysis by elasto-viscoplastic softening model for frozen sand. *Int. J. Numer. Anal. Geomech.* 18: 813–832.). (a) $T = -49°C$, $\sigma_3 = 100 \times 98$ kPa, $\dot{\varepsilon}_{11} = 2.7\% / \text{min}$.

2.8.4 Simulation of the triaxial experimental results

The experimental results of the frozen Toyoura silica sand (Shibata et al. 1985) are numerically simulated by the model described in Section 2.7 with the new time measure of Equation 2.278. The material parameters and the initial conditions are listed in Table 2.10. Figure 2.43a shows the typical deviator stress–deviatoric strain and the volumetric strain–deviatoric strain relations of dense sand at a very low temperature with a strain rate

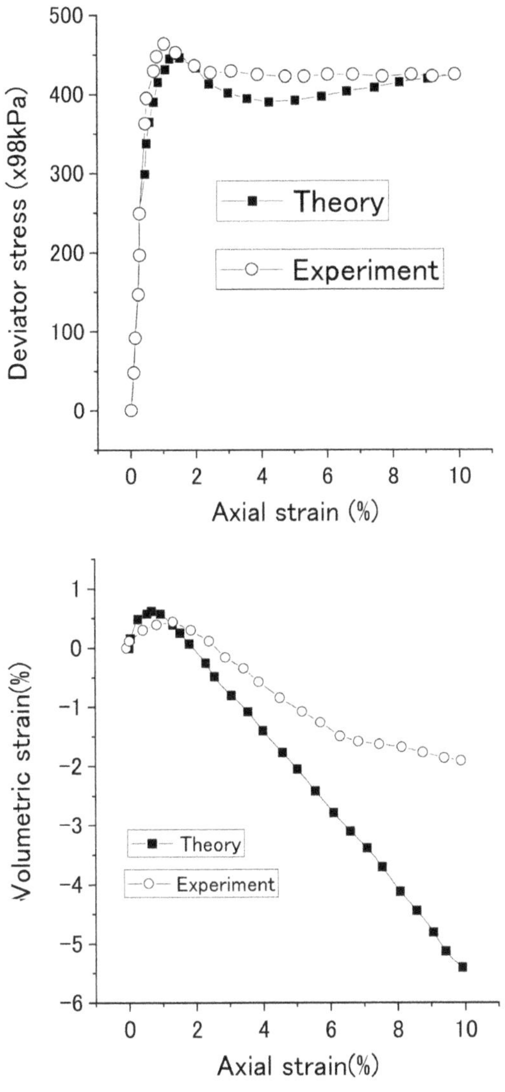

Figure 2.43b Simulated and experimental stress–strain relations of frozen sand. (After Oka, F., Adachi, T., Yashima, A. and Chu, L.L. 1994. Strain localization analysis by elasto-viscoplastic softening model for frozen sand. *Int. J. Numer. Anal. Geomech.* 18: 813–832.). (b) $T = -50°C$, $\sigma_3 = 100 \times 98$ kPa, $\dot{\varepsilon}_{11} = 0.29\%$ / min.

of 2.7%/min. The deviator stress increases until it reaches the peak stress, after which it exhibits strain softening as it approaches the residual value of strength. Under the low strain rate of 0.027%/min (Figure 2.43c), the strain softening is less significant. The simulated results are in good agreement with those of the experiments.

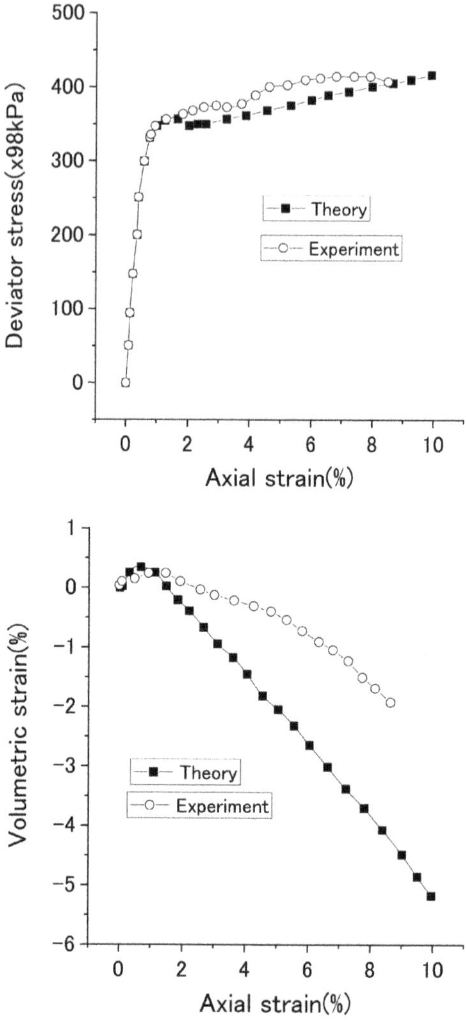

Figure 2.43c Simulated and experimental stress–strain relations of frozen sand. (After Oka, F., Adachi, T., Yashima, A. and Chu, L.L. 1994. Strain localization analysis by elasto-viscoplastic softening model for frozen sand. *Int. J. Numer. Anal. Geomech.* 18: 813–832.). (c) $T = -52°C$, $\sigma_3 = 100 \times 98$ kPa, $\dot{\varepsilon}_{11} = 0.027\%$ / min

APPENDIX A2.1 CONVEXITY OF FAILURE SURFACE

When Ysufuku's failure criterion is expressed by

$$f(\theta) = \frac{D}{\left[B\cos^2 A + \sin^2 A\right]^{1/2}}$$

Table 2.10 Material parameters and initial conditions for frozen sand

Test no.	*C-1*	*C-2*	*C-3*
σ'_{m0}: Initial mean effective stress (×98 kPa)	100	100	100
Strain rate $\dot{\varepsilon}_{11}$	2.7	0.29	0.00272.7
Temperature (°C)	−49	−50	−52
E: Young's modulus (×98 kPa)	80000	80000	80000
K: Elastic bulk modulus (×98 kPa)	21060	21060	21060
M_f^*: Stress ratio at failure	1.5	1.5	1.5
M_m^*: Stress ratio at failure	1.4	1.4	1.4
G'	200	200	200
b: (×98 kPa)	400	400	400
σ_{mb}: (×98 kPa)	10000	10000	10000
τ	75	75	75
a	0.92	0.92	0.92
F_0	1.0	1.0	1.0

Source: Data from Oka, F., Adachi, T., Yashima, A. and Chu, L.L. 1994. Strain localization analysis by elasto-viscoplastic softening model for frozen sand. *Int. J. Numer. Anal. Geomech.* 18: 813–832.

$$D = M_{fc}^* \omega_f^*,\ B = \omega_f^{*2},\ A = \frac{3}{2}\left(\theta + \frac{\pi}{6}\right) \tag{A2.1}$$

where $f(\theta)$ with respect to θ on the π plane is not always convex. Hence, the convexity of the failure surface on plane π has to be confirmed using the necessary and sufficient conditions for the convexity of $f(\theta)$ given by Hashiguchi (2009) as

$$F = \frac{1}{f} + \frac{\partial^2}{\partial \theta^2}\left(\frac{1}{f}\right) > 0 \tag{A2.2}$$

APPENDIX A2.2 HIROTA'S BI-LINEAR DIFFERENTIAL OPERATOR (HIROTA 1976)

The key feature of Hirota's bi-linear differential operator is

$$D_x^n(f \cdot g) = \left(\frac{\partial}{\partial x} - \frac{\partial}{\partial y}\right)^n f(x)g(y)\Big|_{x=y}$$

For example, in the case of $n = 2$:

$$D_x^2(f \cdot g) = \left(\frac{\partial}{\partial x} - \frac{\partial}{\partial y}\right)^2 f(x)g(y)\Big|_{x=y} = \frac{\partial^2 f}{\partial x^2} g - 2\frac{\partial f}{\partial x}\frac{\partial g}{\partial x} + f\frac{\partial^2 g}{\partial x^2}$$

REFERENCES

Adachi, T. and Ogawa, T. 1981. Mechanical behaviour of sedimentary soft rocks and failure criteria. *Proc. JSCE* 295: 51–63 (in Japanese).

Adachi, T. and Oka, F. 1984a. Constitutive equations for sands and overconsolidated clays and assigned works for sand. In: *Results of the International Workshop on Constitutive Relations for Soils*, ed. G. Gudehus, F. Darve and I. Vardoulakis, 141–157. Balkema, Rotterdam.

Adachi, T. and Oka, F. 1984b. Constitutive equations for sands and overconsolidated clays. In: *Results of International Workshop on Constitutive Equations for Soils*. ed. G. Gudehus and F. Darve, 111–122. Balkema, Rotterdam.

Adachi, T., Oka, F. and Poorooshasb, H.B. 1990. A constitutive model for frozen sand. *J. Energy Resour. Technol. ASME* 112: 208–212.

Adachi, T., Oka, F., Hirata, T., Hashimoto, T., Pratan, T.B.S., Nagaya, J. and Mimura, M. 1991. Triaxial and torsional hollow cylinder tests of sensitive natural clay and an elasto-viscoplastic constitutive model. In: *Proceedings of 10th ECSME*, 1: 3–6. Balkema, Rotterdam.

Adachi, T. and Oka, F. 1995. An elasto-plastic constitutive model for soft rock with strain softening. *Int. J. Numer. Anal. Methods Geomech.* 19: 233–247.

Adachi, T., Oka, F., Hirata, T., Hashimoto, T., Pratan, T.B.S., Nagaya, J. and Mimura, M. 1995. Stress-strain behavior and yielding characteristics of eastern Osaka clay. *Soils Found.* 35(3): 1–13.

Aifantis, E.C. 1984. On the microstructural origin of certain in elastic models. *J. Engr. Mater. Tech. ASME* 106(4): 326–330.

Aifantis, E.C. 1987. The physics of plastic deformation. *Int. J. Plast.* 3(3): 211–247.

Akaki, T., Kimoto, S. and Oka, F. 2016. Chemo-thermo-mechanically coupled seismic analysis of methane hydrate-bearing sediments during a predicted Nankai Trough Earthquake. *Int. J. Numer. Anal. Meth. Geomech.* 40(16): 2207–2237. doi: 10.1002/nag.2527.

Anand, L. 1983. Plane deformations of ideal granular materials. *J. Mech. Phys. Solids* 31(2): 105–122.

Andersland, O.B. and AlNouri, I. 1970. Time dependent strength behaviour of frozen soils. *J. Soil Mech. Found. ASCE* 96, SM4: 1249–1265.

Armstrong, P.J. and Frederick, C.O. 1966. A mathematical representation of the multiaxial Bauschinger effect. *Berkeley Nuclear Laboratories 1966, Technical Report RD/B/N 731.*

Asaoka, A., Noda, T., Yamada, E., Kaneda, K. and Nakano, M. 2002. An elasto-plastic description of two distinct volume change mechanism of soils. *Soils Found.* 42(5): 47–57.

Asaro, R.J. 1979. Geometrical effects in the inhomogeneous deformation of ductile single cristals, *Acta Metallurgia* 27(3): 445–453.

Assure, A. 1979. Some promising trends in ice mechanics. In: *Proc. of the IUTAM Symp. Phys. and Mechanics of Ice*, ed. P. Tryde, 1–15. Springer-Verlag, Berlin.

Aurthr, J.R.F., Chua, K.S., Dunstan, T. and Rodriguez, C.J.I. 1980. Principal stress rotation: A missing parameter. *J. Geotech. Eng. ASCE* 106: 419–433.

Banerjee, P.K. and Stipho, A.S. 1979. An elasto-plastic model for undrained behaviour of heavily over-consolidated clays, short communication. *Int. J. Numer. Anal. Meth. Geomech.* 3(1): 97–103.

Bazant, Z. and Prat, P.C. 1988. Micro-plane model for brittle-plastic material, I: Theory and II: verification. *J. Eng. Mech. ASCE* 114(10): 1672–1702.

Bazant, Z.P. and Pijaudier-Cabot, G. 1988. Non-local continuum damage, localization, instability and convergence. *J. Appl. Mech. ASME* 55(2): 287–293.

Bigoni, D. 2012. *Nonlinear Solid Mechanics.* Cambridge University Press, Cambridge.

Boehler, J.P. and Sawczuk, A. 1977. On yielding of oriented solids. *Acta Mech.* 27(1–4): 185–206.

Butterfield, R. and Harkness, R.M. 1972. The kinetics of Mohr–Coulomb material. In: *Stress-Strain Behavior of Soils*, ed. R.H.G. Parry, 220–233. Foulis Henley-on Thames, Oxfordshire.

Chaboche, J.L. and Rousselier, G. 1983. On the plastic and viscoplastic constitutive equations; part I and part II. *J. Press. Vessel Technol. Trans. ASME* 105(2): 153–164.

Chaboche, J.L. 1988. Continuum damage mechanics part I-general concept. *J. Appl. Mech. ASME* 55(1): 59–64.

Chiaro, C., Koseki, J. and Nalin De Silva, L.I. 2013. An elasto-plastic model for liquefiable sands subjected to torsional shear loadings. In: *Constitutive Modeling of Geomaterials*, ed. Q. Yang, J.-M. Zhang, H. Zheng and Y. Yao, 519–526. Springer.

Coleman, B.D. and Hodgdon, M.L. 1985. On shar bands in ductile materials. *Arch. Rat Mech. Anal.* 90(3): 219–249.

Dafalias, Y.F., Herrman, L.R. and Anandarajah, A. 1981. Cyclic loading response of cohesive soils using a bounding surface plasticity model. In: *Proc. of International Conferences on Recent Advances in Geotechnical Earthquake Engineering and Soil Dynamics.* Paper No. 16: 139–144.

Dafalias, Y. 1983. Corotational rates for kinematical hardening at large strains. *J. Appl. Mech. ASME* 52: 561–565.

Dafalias, Y. 1985. The plastic spin. *J. Appl. Mech. ASME* 52(4): 865–871.

Daniels, K.E. 2017. The role of force networks in granular materials. In *Powders and Grains 2017. EPJ Web Conf.* 140: 01006. doi: 10.1051/epjconf/2017140 01006.

Daniels, K.E., Kollmer, J.E. and Pucke, J.G. 2017. Photoelastic force measurements. In *Granular materials. Rev. Sci. Instrum.* 88(5): 051808. doi: 10.1063/1.4983049.

De Jong, G.D.J. 1959. *Statics and Kinematics in the Failable Zone of a Granular Material*, thesis. Delft University.

De Jong, G.D.J. 1971. The double sliding free rotating model for granular assemblies. *Geotechnique* 21(2): 155–163.

Desai, C.S. 1974. A consistent finite element technique for work-softening behaviour. In: *Proc. Int. Conf. on Computational Methods in Non-Linear Mechanics*, ed. J.T. Oden et al., 969–978. University of Texas, Austin, TX.

Desai, C.S., Armaleh, S., Katti, D. and Ma, Y. 1991. Disturbed state concept for modelling of soils and joints. In: *Proc. 7th IACMAG Conf.*, ed. G. Beer, J.R. Booker and J.P.C. Carter, 1: 321–326. Balkema, Rotterdam.

Desai, C.S. 1992. The disturbed state concept for modelling of mechanical response of materials and interfaces. *Report of Dept. of Civil Eng. and Eng. Mech.* University of Arizona, Tucson, AZ.

Desai, C.S. 2001. *Mechanics of Materials and Interfaces: The Disturbed State Concept*. CRC Press/Taylor & Francis, Boca Raton, FL.

Drazin, P.G. 1992. *Nonlinear Systems*, 81–94. Cambridge University Press, Cambridge.

Drescher, A. and de Josselin de Jong, G. 1972. Photo elastic verification of a mechanical model for the flow of a granular material. *J. Mech. Phys. Solids* 20(5): 337–351.

Eringen, E.C. 1967. *Mechanics of Continua*. Chapter 9. John Wiley & Sons, New York.

Fahey, M. 1992. Shear modulus of cohesionless soil: Variation of stress-strain level. *Can. Geotech. J.* 29(1): 157–161.

Fermi, E. 1965. *Collected Papers of Enrico Fermi*. Vol. II, 978. Chicago University Press, Chicago, IL.

Fish, A.M. 1980. Kinetic nature of the long term strength of frozen soils. In: *Proc. Int. Symp. on Ground Freezing*, Tronheim, Norway, 95–108.

Frantzskonis, G. and Desai, C.S. 1987. Analysis of a strain softening constitutive model. *Int. J. Solids Struct.* 21(6): 751–767.

Gallipoli, D., Wheeler, S. and Karstunen, M. 2003. Modelling the variation of degree of saturation in a deformable unsaturated soil. *Geotechnique* 53(1): 105–112.

Gallipoli, D., Gens, A., Chen, G. and D'Onza, F. 2008. Modelling unsaturated soil behavior during normal consolidation. *Comput. Geotech.* 35(6): 825–834.

Garcia, E.F., Oka, F. and Kimoto, S. 2010. Instability analysis and simulation of water infiltration into an unsaturated elasto-viscoplastic material. *Int. J. Solids Struct.* 47(25–26): 3519–3536.

Ghaboussi, J. and Momen, H. 1979. Plasticity model for cyclic behaviour of sands. In: *Proc. of 3rd Int. Conf. on Numer. Methods in Geomech.*, Aachen, ed. W. Wittke, 423–434. Balkema, Rotterdam.

Graham, J. and Houlsby, G.T. 1983. Anisotropy elasticity of a natural clay. *Geotechnique* 33(2): 165–180.

Gutierrez, M., Ishihara, K. and Towhata, I. 1991. Flow theory for sand during rotation of principal stress direction. *Soils Found.* 31(4): 121–132.

Hardin, B.O. and Drnevich, V.P. 1972. Shear modulus and damping in soils: Design equation and curves. *J. Soil Mech. Found. ASCE* 98 (SM7): 667–692.

Hashiguchi, K. 1998. The tangential plasticity. *Met. Mater.* 4(4): 652–656.

Hashiguchi, K. and Chen, Z.-P. 1998. Elastoplastic constitutive equations of soils with subloading surface and the rotational hardening. *Int. J. Numer. Anal. Meth. Geomech. Num. Ana. Meth. Geomech.* 22: 197–227.

Hashiguchi, K. 2009. *Elastoplasticity Theory*, 390–391. Springer.

Hill, R. 1962. Acceleration waves in solids. *J. Mech. Phys. Solids* 10(1): 1–16.

Hirota, R. 1971. Exact solution of the Korteweg-de Vries equation for multiple collisions of solitons. *Phys. Rev. Lett.* 27(18): 1192–1194.

Hirota, R. 1976. Direct methods of finding exact solutions of non-linear evolution equations. In: *Backlund Transformations*, ed. R.M. Miura. *Lecture. Notes in Mathematics*, 515. Springer-Verlag, New York.

Höeg, K. 1972. Finite element analysis of strain softening. *J. Soil Mech. Found. Engng. ASCE* 98(1): 43–58.

Ishihara, K. and Kabilamany, K. 1990. Stress dilatancy and hardening laws for rigid granular model of sand. *Soil Dyn. Earthquake Eng.* 9(2): 66–77.

Kato, R., Oka, F., Kimoto, S., Kodaka, T. and Sunami, S. 2009. A method of seepage-deformation coupled analysis of unsaturated ground and its application to river embankment. *Trans. C. JSCE* 65(1): 226–240 (in Japanese).

Kato, R., Oka, F. and Kimoto, S. 2014. A numerical simulation of seismic behavior of highway embankments considering seepage flow. In: *Computer Methods and Recent Advances in Geomechanics*, ed. F. Oka, A. Murakami, R. Uzuoka and S. Kimoto, 755–760 (*Proc. of the 14th Int. Conf. of Int. Assoc. Comp. Meth. Recent Adv. Geomech.*, Kyoto, Japan, 22–25 September).

Kimoto, S. and Oka, F. 2005. An elasto-viscoplastic model for clay considering destructuralization and consolidation analysis of unstable behavior. *Soils Found.* 45(2): 29–42.

Kimoto, S., Oka, F., Fukutani, J., Yabuki, T. and Nakashima, K. 2011. Monotonic and cyclic behavior of unsaturated sandy soil under drained and fully undrained conditions. *Soils Found.* 51(4): 663–681.

Kimoto, S., Shahbodagh, B., Mirjalili, M. and Oka, F. 2015. A cyclic elastoviscoplastic constitutive model for clay considering nonlinear kinematic hardening rules and structural degradation. *Int. J. Geomech. ASCE* 15(5): A4014005-1–A4014005-14.

Kimoto, S., Ishikawa, R. and Akaki, T. 2017. Behavior of unsaturated sandy soil during triaxial compression tests under fully undrained conditions and its modelling. In: *Proc. 19th Int. Conf. on Soil Mechanics and Geotechnical Engineering 2017*, Seoul, Korea, September, 17–22, 1187–1190.

Kolymbas, D. 1991. A outline of hypoplasticity. *Arch. Appl. Mech.* 61: 143–151.

Kovari, K. 1982. Rock mass behaviour and its mathematical modelling. In: *Numerical Methods in Geomechanics, Nato Advances Study Institute Series*, ed. J.B. Martin, 145–164. D. Reidel, Dordrecht.

Kunin, I.A. 1982. *Elastic Media with Microstructure I, One-Dimensional Models.* Solid State Sciences 26. Springer-Verlag, Berlin.

Ladanyi, B. 1980. Mechanical behaviour of frozen soils. In: *Mechanics of Structured Media.* Part B, ed. A.S.P. Selvadurai: 203–245.

Lade, P.V. and Duncan, J.M. 1975. Elastoplastic stress-strain theory for cohesionless soil. *J. Geotech. Engng. ASCE* 101(10): 1037–1053.

Lade, P.V. and Musante, H.M. 1978. Three-dimensional behavior of remolded clay. In: *Proc. ASCE* 104(GT2): 193–209.

Lade, P.V. 1992. Static instability and liquefaction on loose fine sandy slopes. *J. Geotech., Eng. ASCE* 118(1): 51–71.

Lee, C.W. 2012. *A Study on Dynamic Stability of Unsaturated Road Embankments Using Dynamic Centrifugal Model Tests*, PhD thesis. Kyoto University.

Leroueil, S. and Vaughan, P.R. 1990. The general and congruent effects of structure in natural soils and weak rocks. *Geotechnique* 40(3): 467–488.

Loret, B. 1983. On the effect of plastic rotation in the finite deformation of anisotropic elastoplastic materials. *Mech. Meter* 2: 287–304.

Maleki, M. and Cambou, B. 2009. A cyclic elastoplastic-viscoplastic constitutive model for soils. *Geomech. Geoeng. An Int. J.* 4(3): 209–220.

Malvern, L.E. 1969. *Introduction to the Mechnics of Continuous Media.* New York: Prentice Hall.

Matsuoka, H. and Nakai, T. 1977. Stress-strain relationship of soil based on SMP. In: *Proceedings of Speciality Session 9 of 9th ICSMFE*, Tokyo, 153–162.

May, R.M. 1974. Populations with non-overlapping generation stable points, stable cycles and chaos. *Science* 186(4164): 645–647.
Mehrabadi, M.M. and Cowin, S.C. 1978. Initial planer deformation of dilatant granular materials. *J. Mech. Phys. Solids* 26(4): 269–284.
Mehrabadi, M.M. 1979. *Planer Deformation of Granular Materials*, PhD thesis. Tulane University.
Mirjalili, M. 2010. *Numerical Analysis of a Large-Scale Levee on Soft Soil Deposits Using Two-Phase Finite Deformation Theory*, PhD thesis. Kyoto University.
Mitchell, J.K. 1976. *Fundamentals of Soil Behavior.* John Wiley & Sons.
Modaressi, H. and Laloui, L. 1997. A thermo-viscoplastic constitutive model for clays. *Int. J. Numer. Anal. Meth. Geomech.* 21(5): 313–335.
Morio, B., Kusakabe, S., Yasufuku, N. and Hyodo, M. 1994. Evaluation of undrained behavior of anisotropic sand based on elastoplastic constitutive model. In: *Proc. JSCE*, 505, III-29, 287–296.
Mühlhaus, H.-B.. and Aifantis, E.C. 1991. A variational principle for gradient plasticity. *Int. J. Solids Struct.* 28(7): 845–857.
Mühlhaus, H.-B.. and Aifantis, E.C. 1991. The influence of microstructure-induced gradients on the Localization of deformation in viscoplastic materials. *Acta Mech.* 89(1–4): 217–231.
Mühlhaus, H.-B. and Oka, F. 1996. Dispersion and wave propagation in discrete and continuum models for granular materials. *Int. J. Solids Struct.* 33(19): 2841–2858.
Murayama, S. 1964. A theoretical consideration on a behavior of sand. In: *Proc. IUTAM Symp. on Rheology and Soil Mechanics*, 146–159. Springer, Berlin.
Nakai, T. and Mihara, Y. 1984. A new mechanical quantity for soils and its application to elastoplastic constitutive models. *Soils Found.* 24(2): 82–94.
Nakai, T. and Matsuoka, H. 1986. A generalized elastoplastic constitutive model for clay in three-dimensional stresses. *Soils Found.* 26(3): 81–98.
Nakai, T. 2012. *Constitutive Modeling of Geomaterials.* CRC Press/Taylor & Francis, Boca Raton, FL.
Nayak, G.C. and Zienkiewicz, O.C. 1972. Elasto-plastic stress analysis. *Int. J. Numer. Anal. Meth. Geomech.* 5: 113–135.
Nemat-Nasser, S., Mehrabadi, M.M. and Iwakuma, T. 1981. Three-dimensional constitutive relations and ductile fracture. In Proc. of IUTAM Symposium, ed. S. Nemat-Nasser: 157–172. North-Holland, Amsterdam.
Nemat-Nasser, S. 1983. On finite plastic flow of crystalline solids and geomaterials. *J. Appl. Mech. ASME* 50(4b): 1114–1126.
Niemunis, A. and Herle, I. 1997. Hypoplastic model for cohesionless soils with elastic range. *Mech. Cohesive Mater.* 2: 279–299.
Oda, M. 1972. The mechanism of fabric changes during compressional deformation of sand. *Soils Found.* 12(2): 1–18.
Oda, M. and Konishi, J. 1974. Microscopic deformation mechanism of granular material in simple shear. *Soils Found.* 14(4): 25–38.
Oda, M. and Nakayama, H. 1988. Yield function for soil with anisotropic fabric. *J. Eng. Mech. ASCE* 115(1): 89–104.
Oka, F. 1985. Elasto/viscoplastic constitutive equations with memory and internal variables. *Comput. Geotech.* 1(1): 59–69.

Oka, F. and Adachi, T. 1985. An elasto-plastic constitutive equation of geologic materials with memory. In: *Proc. 5th Int. Conf. on Numerical Methods in Geomechanics*, ed. T. Kawamoto and Y. Ichikawa, 1: 293–300. Balkema, Rotterdam.

Oka, F., Leroueil, S. and Tavenas, F. 1989. A constitutive model for natural clays with strain softening. *Soils Found.* 29(3): 54–66.

Oka, F., Yashima, A., Adachi, T. and Aifantis, E.C. 1991. A gradient dependent viscoplastic model for clay and its application to FEM consolidation analysis. In: *Proc. 3rd Int. Conf. on Constitutive Laws for Engineering Materials-Theory and Application*, ed. C.S. Desai et al., 313–316. ASME Press, New York.

Oka, F. 1992. A cyclic elasto-viscoplastic constitutive model for clay based on the non-linear-hardening rule. In: *Proc. 4th Intl. Symp. on Numerical Models in Geomechanics*, ed. Swansea Pande, G.N. and Pietruszczak, S., 1, 105–114. Balkema, Rotterdam.

Oka, F., Yashima, A., Adachi, T. and Aifantis, E.C. 1992. Instability of gradient dependent viscoplastic model for clay saturated with water and FEM analysis. *Appl. Mech. Rev. ASME* 45(3S)(3-2): 103–111.

Oka, F. 1993a. Anisotropic and pseudo-anisotropic elasto-viscoplastic constitutive models for clay. In: *Modern Approaches to Plasticity (Proc. Workshop in Horton 1992)*, ed. D. Kolymbas, 505–526. Elsevier Sci. Pub. B.V., Amsterdam.

Oka, F. 1993b. An elasto-viscoplastic constitutive model for clay using a transformed stress tensor. *Mech. Mater.* 16(1–2): 47–53.

Oka, F., Adachi, T., Yashima, A. and Chu, L.L. 1994. Strain localization analysis by elasto-viscoplastic softening model for frozen sand. *Int. J. Numer. Anal. Meth. Geomech.* 18(12): 813–832.

Oka, F. 1995. A gradient-dependent elastic model for granular materials and strain localization. In: *Continuum Models for Materials with Microstructure*, ed. H.B. Mühlhaus, 145–158. John Wiley & Sons.

Oka, F., Yashima, A., Tateishi, A., Taguchi, Y. and Yamashita, S. 1999. A cyclic elasto-plastic constitutive model for sand considering a plastic-strain dependence of the shear modulus. *Geotechnique* 49(5): 661–680.

Oka, F., Kodaka, T. and Kim, Y.S. 2004. A cyclic viscoelastic–viscoplastic constitutive model for clay and liquefaction analysis of multi-layered ground. *Int. J. Numer. Anal. Meth. Geomech.* 28(2): 131–179.

Oka, F., Kodaka, T., Kimoto, S., Kim, Y.S. and Yamasaki, N. 2006. An elasto-viscoplastic model and multiphase coupled FE analysis for unsaturated soil. In: *Unsaturated Soils 2006*, 2039–2050. ASCE.

Oka, F., Kimoto, S., Kato, R., Sunami, R. and Kodaka, T. 2008. Unsaturated river embankment due to seepage flow and overflow. In *Proc. 12th IACMAG*, ed. D.N. Singh, Goa, India: 2029–2041.

Oka, F., Kodaka, T., Suzuki, H., Kim, Y.S., Nishimatsu, N. and Kimoto, S. 2010. Experimental study on the behavior of unsaturated compacted silt under triaxial compression. *Soils Found.* 50(1): 27–44.

Oka, F. and Kimoto, S. 2012. *Constitutive Modeling of Multiphase Geomechanics.* CRC Press/Taylor and Francis, Boca Raton, FL.

Oka, F. and Kimoto, S. 2016. Effect of the non-associativity of plasticity model on the cyclic behavior of geomaterials. *Key Eng.* 725: 322–326 (*Proc. 13th Asia-Pacific Symposium on Engineering Plasticity and Its Applications*, December 4–8).

Oka, F. and Kimoto, S. 2018. A cyclic elastoplastic constitutive model and effect of non-associativity on the response of liquefiable sandy soils. *Acta Geotech.* 13(6): 1283–1297.
Oka, F., Shahbodagh, B. and Kimoto, S. 2019. A computational model for dynamic strain localization in unsaturated elasto-viscoplastic soils. *Int. J. Numer. Anal. Meth. Geomech.* 43(1): 138–165. doi: 10.1002/nag.2857.
Pastro, M., Zienkiewicz, O.C. and Chan, A.H.C. 1990. Generalized plasticity and the modeling of soil behavior. *Int. J. Numer. Ana. Meth. Geomech.* 14(3): 151–190.
Perko, L. 1991. *Differential Equations and Dynamical Systems*, 128–134. Springer, Berlin.
Perzyna, P. 1963. The constitutive equations for work-hardening and rate sensitive plastic materials. In: *Proc. of Vibration Problems*, Warsaw, 3(4): 281–290.
Pestana, J.M. and Whittle, A.J. 1999. Formulation for unified constitutive model for clays and sands. *Int. J. Numer. Anal. Meth. Geomech.* 23: 1215–1243.
Pietruszczak, S. and Mroz, Z. 1981. Finite element analysis of deformation of strain softening materials. *Int. J. Numer. Meth. Engng.* 7: 327–334.
Prevost, J.H. 1978. Plasticity theory for soils stress-strain behavior. *J. Eng., Mech. ASCE* 104(5): 1177–1194.
Push, R. 1970. Microstructural changes in soft quick clay at failure. *Can. Geotech. J.* 7(1): 1–7.
Read, H.E. and Hegemier, G.P. 1984. Strain-softening of rocks, solid and concrete: A review article. *Mech. Mater.* 3(4): 271–294.
Riks, E. 1972. The application of Newton's methods to the problem of elastic stability. *J. Appl. Mech. ASME* 39(4): 1060–1066.
Roscoe, K.H., Basset, R.H. and Cole, E.R. 1967. Principal axes observed during simple shear of a sand. In: *Proc. Geotech. Conf.*, Oslo, 1: 231–237.
Roscoe, K.H. 1970. The influence of strains in soil mechanics, The 10th Rankine Lecture. *Geotechnique* 20(2): 129–170.
Rudnicki, J.W. and Rice, J.R. 1975. Conditions for the localization of deformation in pressure sensitive dilatant materials. *J. Mech. Phys. Solids* 23(6): 371–394.
Sadeghi, H., Kimoto, S., Oka, F. and Shahbodagh, B. 2014. Dynamic analysis of river embankments during earthquakes using a finite deformation FE analysis method. In: *Proc. 14th Int. Conf. of IACMAG*, ed. F. Oka, A. Murakami, R. Uzuoka and S. Kimoto, 637–642. CRC Press/Taylor & Fancis.
Sandler, I.S. 1986. Strain softening for static and dynamic problems. In: *Proc. Symp. on Constitutive Equations; Macro and Computational Aspects, ASME Winter Ann. Meeting*, 217–231.
Shahbodagh, K.B. 2011. *Large Deformation Dynamic Analysis Method for Partially Saturated Elasto-Viscoplastic Soils*, PhD thesis. Kyoto University.
Shahbodagh, B., Mirjalili, M., Kimoto, S. and Oka, F. 2014. Dynamic analysis of strain localization in water-saturated clay using a cyclic elasto-viscoplastic model. *Int. J. Numer. Anal. Meth. Geomech.* 38(8): 771–793.
Shao, C. and Desai, C.S. 2000. Implementation of DSC model and application of analysis of field pile load tests under cyclic loading. *Int. J. Numer. Anal. Meth. Geomech.* 44(6): 601–624.
Shibata, T., Adachi, T., Yashima, A., Takahashi, T. and Yoshioka, I. 1985. Time dependence and volume characteristic change of frozen sand under triaxial condition. In: *Proc. 4th Int. Symp. on Ground Freezing*, ed. S. Kinoshita and Fukuda, 173–179. Balkema, Rotterdam.

Shirozu, H. 1988. *Introduction to Clay Mineralogy*. Asakura Pub. Co., Tokyo (in Japanese).
Spencer, A.J.M. 1964. A theory of the kinematics of ideal soils under plane strain conditions. *Geotechnique* 21: 190–192.
Stören, S. and Rice, J.R. 1975. Localized necking in thin sheets. *J. Mech. Phys. Solids* 23(6): 421–441.
Taboada, A., Chang, K.J., Radjaï, F. and Bouchette, F. 2005. Rheology, force transmission, and shear instabilities in frictional granular media from biaxial numerical tests using the contact dynamics method. *J Geophys Res. Solid Earth*, 110(B9): B09202.
Teunissen, J.A.M. and Vermeer, P.A. 1988. Analysis of double shearing in frictional materials. *Int. J. Numer. Anal. Meth. Geomech.* 12(3): 323–340.
Thomas, T.Y. 1961. *Plastic Flow and Fracture of Solids*. Academic Press, New York.
Ting, J.M. 1983. Tertiary creep model for frozen sands, *J. Getech. Engng. ASCE*. 109(7): 932–945.
Tobita, Y. 1988. Yield condition of anisotropic granular materials. *Soils Found.* 28(2): 113–126.
Tsutsumi, S. and Hashiguchi, K. 2005. General non-proportional loading behaviour of soils. *Int. J. Plast.* 21(10): 1941–1969.
Valanis, K.C. 1971. A theory of viscoplasticity without a yield surface, *Arch. Mech. Stos*. 23(4): 517–533.
Valanis, K.C. and Read, H.E. 1982. A new endochronic plasticity model for soils. In: *Soil Mechanics-Transient and Cyclic Loads*, ed. G.N. Pande and O.C. Zienkiewicz, 375–417. John Wiley and Sons.
Valanis, K.C. 1985. On the uniqueness of solution of the initial value problem in softening materials. *J. Appl. Mech. ASME* 52(3): 649–653.
van Genuchten, M.Th. 1980. A closed-form equation for predicting the hydraulic conductivity of unsaturated soils. *Soil Sci. Soc. Am. J.* 44(5): 892–898.
Vardoulakis, I. and Aifantis, E.C. 1991. A gradient flow theory of plasticity for granular materials. *Acta Mech.* 87(3–4): 197–217.
Vermeer, P.A. 1981. A formulation and analysis of granular flow. In: *Proc. Int. Conf. Symp. Mech. Behaviour of Struc. Media, B*, 325–339. Elsevier, Amsterdam.
Vialov, S.S. 1963. Rheology of frozen soils. In: *Proc. 1st Permafrost Conf.*, Lafayette, 332–342. Nat. Acad. Sci.: Nat. Res. Council, Washington, DC.
Wang, C.C. 1970. A new representation theorem for isotropic functions, part I and II. *Arch. Rat Mech. Anal.* 36: 166–197, 198–223.
Wang, C.C. 1971. Corrigendum to my recent papers on "Representations for isotropic functions". *Arch. Rat Mech. Anal.* 43(5): 392–395.
Wang, S., Qi, J., Yin, Z., Zhang, J. and Ma, W. 2014. A simple rheological element based creep model for frozen soils. *Cold Reg. Sci. Technol.* 106–107: 47–54.
Wichtmann, T. and Triantafyllidis, Th. 2015. An experimental data base for the development, calibration and verification of constitutive models for sand with focus to cyclic loading. Part II: tests with strain cycles and combined loading. *Acta Geotech.* 11(4): 763–774. doi: 10.1007/s11440-015-0412-x.
Willam, K.J., Pramono, E. and Sture, S. 1986. Stability and uniqueness of strain-softening computations. In: *Finite Element Methods for Non-Linear Problems*, Europe–US Symp., Tronheim, 1985, ed. P. Bergan, K.-J. Bathe and W. Wunderlich, 119–142. Springer-Verlag, Berlin.

Yamada, Y. and Ishihara, K. 1979. Anisotropic deformation characteristics of sand under three dimensional stress conditions. *Soils Found.* 19(2): 79–94.
Yang, Y. and Yu, H.S. 2006a. Numerical simulation of simple shear with non-coaxial soil models. *Int. J. Numer. Anal. Meth. Geomech.* 30(1): 1–19.
Yang, Y. and Yu, H.S. 2006b. A non-coaxial critical state model and its application to simple shear simulations. *Int. J. Numer. Anal. Meth. Geomech.* 30(13): 1369–1390.
Yasufuku, N. 1990. *Yielding Characteristics of Anisostropically Consolidated Sand in the Wide Range of Stress Level and Elastoplastic Constitutive Equation*, Dr thesis. Kyushu University (in Japanese).
Yatomi, C., Yashima, A., Iizuka, A. and Sano, I. 1989a. Shear bands formation numerically simulated by a non-coaxial Cam-clay model. *Soils Found.* 29(4): 1–13.
Yatomi, C., Yashima, A., Iizuka, A. and Sano, I. 1989b. General theory of shear bands formation by a non-coaxial Cam-clay models. *Soils Found.* 29(3): 41–53.
Yong, R.N. and Mckyes, E. 1971. Yield and failure of clay under triaxial stresses. In: *Proc. ASCE* 97(SM1), 159–176.
Yu, H.-S. 1998. CASM: A unified state parameter model for clay and sand. *Int. J. Numer. Anal. Meth. Geomech.* 22(8): 621–753.
Yu, H.-S. 2006. Noncoaxial plasticity. In: *Plasticity and Geotechnics*, Chapter 8. Springer, New York.
Zbib, H.M. and Aiantis, H.C. 1988. On the localization and post-localization behaviour of plastic deformation, I on the initiation of shear bands. *Res Mec.* 23: 261–277.
Zbib, H.M. and Aifantis, E.C. 1988. On the localization and post localization behavior of plastic deformation II, on the evolution and thickness of shear bands. *Res. Mech.* 23: 279–292.
Zbib, H.M. and Aifantis, E.C. 1989. A gradient dependent flow theory of plasticity, Application to metal and soil instability. *Appl. Mech. Rev.* 42: 295–304.

Chapter 3

Governing equations and finite element formulation for large deformation of three-phase materials

3.1 GOVERNING EQUATIONS FOR THREE-PHASE GEOMATERIALS

The governing equations for the dynamic analysis of multiphase geomaterials, i.e. air-water-soil three-phase materials, are described within the framework of the theory of porous media which was presented in Section 1.1. In the theory, the governing equations are derived for multiphase mixtures of solids and immiscible fluids by many researchers (Coussy 1995; de Boer 1996; Schrefler 2002; Laloui et al. 2003; Ehlers et al. 2004; Borja 2004; Oka and Kimoto 2012). The u-p formulation is derived in the finite deformation regime using the updated Lagrangian method for the weak form of the equation of motion. An elasto-viscoplastic constitutive model, considering nonlinear kinematic hardening, is used to describe the viscoplastic behavior of the soil skeleton during dynamic loading. In addition, a soil-water characteristic curve is used as the constitutive equation describing the dependency of the volume fraction occupied by each phase to suction.

The governing equations are obtained from the extension of the formulations of the finite deformation analysis method of Shahbodagh (2011) and Oka et al. (2019) and the three-phase porous theory proposed by Kato et al. (2009, 2014) and Oka et al. (2019).

3.1.1 Partial stress tensors for the mixture

Partial stresses and the total stress concept for three-phase materials are adopted as presented in Chapter 1 and in the computational modeling of multiphase geomaterials by Oka and Kimoto (2012).

In the present study, the skeleton stress tensor is used as the basic stress variable in the constitutive model of the soil skeleton, and the suction, p^C (see Equation 3.27), is used in the constitutive model describing the bonding effect.

DOI: 10.1201/9781003200031-3

3.1.2 Conservation of mass

The conservation of mass for each phase is given by

$$\frac{d^{\alpha}\bar{\rho}^{\alpha}}{dt}+\bar{\rho}^{\alpha}v_{i,i}^{\alpha}=0\ \ (\alpha=S,W,G) \tag{3.1}$$

where v_i^{α} is the velocity vector of phase α; d^{α}/dt is the material time derivative with respect to phase α; and $\bar{\rho}^{\alpha}$ is the average density for phase α:

$$\bar{\rho}^{S}=n^{S}\rho^{S} \tag{3.2}$$

$$\bar{\rho}^{W}=n^{W}\rho^{W} \tag{3.3}$$

$$\bar{\rho}^{G}=n^{G}\rho^{G} \tag{3.4}$$

where ρ^S, ρ^W, and ρ^G are the mass density of the soil particles, water, and air, respectively, and n^{α} $(\alpha = S,W,G)$ is the volume fraction. Considering Equations 3.1–3.4 and using porosity $n(=n^W+n^G)$, the mass conservation laws for the solid, water, and gas phases are expressed as

$$-\dot{n}\rho^{S}+(1-n)\rho^{S}v_{i,i}^{S}=0 \tag{3.5}$$

$$\dot{n}S_r\rho^{W}+n\dot{S}_r\rho^{W}+nS_r\rho^{W}v_{i,i}^{W}=0 \tag{3.6}$$

$$\dot{n}(1-S_r)\rho^{G}-n\dot{S}_r\rho^{G}+n(1-S_r)\dot{\rho}^{G}+n(1-S_r)\rho^{G}v_{i,i}^{G}=0 \tag{3.7}$$

in which the superimposed dot denotes the material time derivative with respect to the whole mixture, and the spatial derivatives of the saturation and the porosity are negligibly small. Note that the soil particles and the water are assumed to be incompressible in the above equations. Dividing both sides of Equation 3.5 by ρ^S/S_r, and Equation 3.6 by ρ^W, and adding these equations together yield the continuity equation for water as

$$n\dot{S}_r+S_rD_{ii}=-\dot{w}_{i,i}^{W} \tag{3.8}$$

in which D_{ij} is the stretching tensor given as $D_{ij}=(L_{ij}+L_{ji})/2$ and $L_{ij}=v_{i,j}^{S}$ is the velocity gradient of the solid phase and $\dot{w}_i^{\beta}$ is the average velocity vector of the fluid to the solid skeleton defined by

$$\dot{w}_i^{\beta}=n^{\beta}\left(v_i^{\beta}-v_i^{S}\right)\ \ \beta=W,G \tag{3.9}$$

Similarly, the continuity equation for gas is obtained from Equations 3.5 and 3.7. Dividing both sides of Equation 3.5 by $\rho^S/(1-S_r)$, and Equation 3.7 by ρ^G, and adding these equations together yield

$$-n\dot{S}_r + n(1-S_r)\frac{\dot{\rho}^G}{\rho^G} + (1-S_r)v_{i,i}^S + n(1-S_r)\left(v_i^G - v_i^S\right)_{,i} = 0 \tag{3.10}$$

Substituting Equation 3.9 into Equation 3.10 and noting $D_{ii} = v_{i,i}^S$, the mass conservation equation for the gas phase can be expressed as

$$(1-S_r)D_{ii} - n\dot{S}_r + n(1-S_r)\frac{\dot{\rho}^G}{\rho^G} = -\dot{w}_{i,i}^G \tag{3.11}$$

3.1.3 Conservation of linear momentum

The conservation of linear momentum for each phase is given by

$$\bar{\rho}^\alpha a_i^\alpha = \sigma_{ji,j}^\alpha + \bar{\rho}^\alpha b_i + \sum_\gamma h_i^{\alpha\gamma} \quad (\alpha, \gamma = S, W, G) \tag{3.12}$$

where a_i^α is the acceleration vector for phase α; b_i is the gravity force vector; and $h_i^{\alpha\gamma}$ is the interaction force vector between phases α and γ. The interaction force vector satisfies $h_i^{\alpha\gamma} = -h_i^{\gamma\alpha}$ and $h_i^{\alpha\gamma} = 0$ when $\alpha = \gamma$. Assuming that the interaction forces through the relative velocity between the water and the gas are sufficiently small, the interaction force vector between the solid and water phases and that between the solid and gas phases are given by

$$h_i^{SW} = \frac{n^W \rho^W g}{k^W}\dot{w}_i^W, \quad h_i^{SG} = \frac{n^G \rho^G g}{k^G}\dot{w}_i^G \tag{3.13}$$

in which k^W and k^G are the permeability coefficients of the water and gas phases, respectively, and g is the gravity acceleration. Using Equations 3.3, 3.4, and 3.9, and considering that the spatial gradient of the porosity is negligible, the conservation law of linear momentum for the water and gas phases becomes

$$\rho^\beta n^\beta\left(a_i^S + \frac{\ddot{w}_i^\beta}{n^\beta}\right) + h_i^{S\beta} = \sigma_{ji,j}^\beta + \rho^\beta n^\beta b_i \quad (\beta = W, G) \tag{3.14}$$

Assuming that the relative accelerations of the fluid phases are much smaller than the acceleration of the solid phase, i.e.,

$$a_i^S >> \ddot{w}_i^\beta \quad (\beta = W, G) \tag{3.15}$$

Equation 3.14 can be approximated as follows:

$$\rho^\beta n^\beta a_i^S + h_i^{S\beta} = \sigma_{ji,j}^\beta + \rho^\beta n^\beta b_i \quad (\beta = W, G) \tag{3.16}$$

This approximation is valid for most practical geotechnical engineering problems such as the seismic analysis of porous media, in particular for unsaturated multiphase porous media due to their lower permeability compared to saturated media. Substituting Equations 1.26, 1.27, and 3.13 into Equation 3.16 and assuming that the spatial gradients of the volume fractions are small, the average relative velocity vectors of the fluid phases to the solid skeleton are expressed as

$$\dot{w}_i^\beta = \frac{k^\beta}{\rho^\beta g}\left(-p_{,i}^\beta - \rho^\beta a_i^S + \rho^\beta b_i\right) \quad (\beta = W, G) \tag{3.17}$$

3.1.4 Equation of motion for the whole mixture

The sum of the conservation of linear momentum (Equation 3.12) for the three phases gives

$$\rho^S n^S a_i^S + \rho^W n^W a_i^W + \rho^G n^G a_i^G = \sigma_{ji,j} + \rho b_i \tag{3.18}$$

where $\rho = \sum_\alpha \rho^\alpha n^\alpha$ $(\alpha = S, W, G)$. This equation can be written in the following alternative form:

$$\rho a_i^S + \rho^W n^W \left(a_i^W - a_i^S\right) + \rho^G n^G \left(a_i^G - a_i^S\right) = \sigma_{ji,j} + \rho b_i \tag{3.19}$$

When the relative accelerations are disregarded, Equation 3.19 becomes

$$\rho a_i^S = T_{ji,j} + \rho b_i \tag{3.20}$$

in which $T_{ij} \equiv \sigma_{ij}$ is the total Cauchy stress tensor. This equation represents the equation of motion for the whole multiphase mixture.

Using the first Piola–Kirchhoff stress tensor, Π_{ij}, Equation 3.20 can be rewritten as

$$\rho_0\left(a_i - b_i\right) - \Pi_{ji,j} = 0 \tag{3.21}$$

in which $a_i \equiv a_i^S$ is the acceleration vector of the soil skeleton; ρ_0 is the mass density with respect to the reference configuration; and Π_{ij} is the first Piola–Kirchhoff stress tensor, which is also known as the nominal stress tensor. In the updated Lagrangian formulation, the current configuration (at time t) is chosen as the reference state.

3.1.5 Continuity equations for fluid phases

The continuity equations for the fluid phases, i.e., liquid and gas, are obtained using their mass conservation and linear momentum equations.

For the liquid phase, substituting Equation 3.17 into Equation 3.8 yields

$$\rho^{W}\dot{D}_{ii} + p^{W}_{,ii} - \frac{\gamma^{W}}{k^{W}}\left(S_r D_{ii} + \dot{S}_r n\right) = 0 \tag{3.22}$$

in which γ^W is the unit weight of the liquid phase; $b_{i,j} = 0$ when the body force is spatially constant; D_{ii} is the volumetric stretching; and $\dot{D}_{ii} = a^S_{i,i}$.

Similarly, substituting Equation 3.17 into Equation 3.11 results in the continuity equation for the gas phase:

$$\rho^{G}\dot{D}_{ii} + P^{G}_{,ii} - \frac{\gamma^{G}}{k^{G}}\left((1-S_r)D_{ii} - \dot{S}_r n + (1-S_r)n\frac{\dot{P}^{G}}{P^{G}}\right) = 0 \tag{3.23}$$

in which γ^G is the unit weight of the gas phase and the following relation is used if the gas is the ideal gas ($\rho^G = M^G P^G / R\theta$; M^G: molecular weight of the gas, R: gas constant, θ: temperature) under the isothermal condition.

$$\frac{\dot{\rho}_G}{\rho_G} = \frac{\dot{P}^G}{P^G} \tag{3.24}$$

3.1.6 Suction–saturation characteristic curve of unsaturated soil

The interaction between water and air induced by the surface tension can be described by the soil-water characteristic curve. In the present study, the van Genuchten model (1980) is adopted as

$$S_{re} = \left(1 + (\alpha P^C)^{n'}\right)^{-m} \tag{3.25}$$

where S_{re} is the effective saturation, i.e.,

$$S_{re} = \frac{S_r - S_{r\min}}{S_{r\max} - S_{r\min}} \tag{3.26}$$

The maximum and minimum saturation values are $S_{r\max}$ and $S_{r\min}$, respectively; α, m, and n' are the material parameters where the relation $m = 1 - 1/n'$ holds; and P^C is the suction:

$$P^C = P^G - P^W \tag{3.27}$$

For the permeability of unsaturated soil, the following equation is used to avoid numerical instability (Garcia et al. 2010):

$$k^W = k_s^W S_{re}{}^a\left(1-(1-S_{re}^{1/m})^{n'}\right),\ k^G = k_s^G(1-S_{re})^b\left(1-(S_{re}^{1/m})^{n'}\right) \tag{3.28}$$

where k_s^W is the water permeability under saturated condition; k_s^G is the gas permeability under fully dry condition; and a, b are the material parameters. Both k_s^W and k_s^G depend on the void ratio e in the following form:

$$k_s^W = k_{s0}^W \exp[(e-e_0)/C_k],\ k_s^G = k_{s0}^G \exp[(e-e_0)/C_k] \tag{3.29}$$

where k_{s0}^W and k_{s0}^G are the initial values at $e = e_0$, and C_k is the material parameter controlling the volume change dependency of the soil permeability coefficients.

3.2 FINITE ELEMENT DISCRETIZATION OF GOVERNING EQUATIONS

In this section, the weak forms of the governing equations, i.e., the continuity equations for the water and gas phases and the equation of motion for the whole multiphase mixture, are derived. The spatial discretization of the governing equations is achieved using the finite element method in the finite deformation regime with the updated Lagrangian description, whereas the time integration is conducted using the Newmark method.

3.2.1 Discretization of equation of motion for the whole mixture

In non-linear dynamic finite element analysis involving large deformations with incremental law, it is necessary to linearize the equation of motion. Solving the total form of the equation of motion is more fundamental than solving its incremental form. It is possible to use an incremental formulation if the calculation error is completely zero, otherwise the errors will accumulate leading to numerical instability. In the present formulation, the no rate-form equation of motion is adopted and discretized in the updated Lagrangian scheme.

In the Lagrangian formulation, as the reference state refers to the configuration at time t, the equation of motion at time $t+\Delta t$ becomes

$$[\rho(a_i - b_i) - \Pi_{ji,j}]_{t+\Delta t} = 0 \tag{3.30}$$

For simplicity, we disregard the notation of time $t+\Delta t$ in the following. The boundary conditions for the mixture are considered as

$$\Pi_{ji}N_j = \bar{t}_i \text{ on } \Gamma_t, \quad v_i = \bar{v}_i \text{ on } \Gamma_u \tag{3.31}$$

where Γ_t is the traction boundary surface; Γ_u is the displacement boundary surface; N_j is an outward unit normal vector of surface Γ_t; $\bar{t}_i$ is the traction vector at time $t+\Delta t$; v_i is the velocity vector of the solid phase; and the superimposed bars indicate the prescribed values. Moreover, the total boundary is obtained as $\Gamma = \Gamma_t + \Gamma_u$. A weak form of Equation 3.30 can be given by

$$\int_V \rho a_i \delta v_i dV + \int_V \Pi_{ji} \delta L_{ij} dV = \int_{\Gamma_t} \bar{t}_i \delta v_i d\Gamma + \int_V \rho b_i \delta v_i dV \tag{3.32}$$

The nominal stress Π_{ij} at time $t+\Delta t$ can be approximated by $\Pi_{ji|t+\Delta t} \cong \Pi_{ji|t} + (\Delta t)\dot{\Pi}_{ji}$, where the superimposed dot indicates time differentiation.

From the time derivation of Nanson's formula (Malvern 1969), we have

$$\dot{\Pi}_{ji} = JF_{jk}^{-1}\dot{\hat{S}}_{ki} \tag{3.33}$$

in which $F_{kj}(= (\partial x_k / \partial X_j))$ is the deformation gradient tensor; J is its Jacobian; and $\dot{\hat{S}}_{ki}$ is the nominal stress rate with respect to the configuration at time t:

$$\dot{\hat{S}}_{ki} \equiv \dot{T}_{ki} - L_{kp}T_{pi} + L_{pp}T_{ki} \tag{3.34}$$

In the updated Lagrangian method, the configuration at time t, i.e., the latest known configuration, is considered as a reference configuration, while the configuration at time $t+\Delta t$ is unknown. Therefore, at time t, $JF_{ij}^{-1} = \delta_{ij}$, $\Pi_{ji} = T_{ji}$, and $\dot{\Pi}_{ji} = \dot{\hat{S}}_{ji}$.

Using Equations 3.33 and 3.34, Equation 3.32 becomes

$$\int_V \rho a_i \delta v_i dV + \int_V T_{ij|t} \delta L_{ij} dV + (\Delta t) \int_V \dot{\hat{S}}_{ji} \delta L_{ij} dV = \int_{\Gamma_t} \bar{t}_i \delta v_i d\Gamma + \int_V \rho b_i \delta v_i dV \tag{3.35}$$

The relations among the nominal stress rate tensor $\dot{\hat{S}}_{ij}$, the nominal skeleton stress rate tensor $\dot{\hat{S}}'_{ij}$, and the Cauchy skeleton stress rate tensor $\dot{T}'_{ij}$ are

$$\dot{\hat{S}}_{ij} = \dot{\hat{S}}'_{ij} - \dot{P}^F\delta_{ij} - L_{kk}P^F\delta_{ij} + P^F L_{ji} \tag{3.36}$$

$$\dot{\hat{S}}'_{ij} = \dot{T}'_{ij} + L_{kk}T'_{ij} - L_{ik}T'_{kj} \tag{3.37}$$

Using Equations 3.36 and 3.37, Equation 3.35 can be written in matrix form as

$$\int_V \rho\{\delta v\}^T\{a\}dV + (\Delta t)\int_V \{\delta D\}^T\{\dot{T}'\}dV + (\Delta t)\int_V \{\delta L\}^T[D_S]\{L\}dV$$

$$+(\Delta t)\int_V \{\delta L\}^T\{T'\}tr(D)dV - (\Delta t)\int_V tr(\delta D)^T\dot{P}^F dV - (\Delta t)\int_V \{\delta L\}^T[U]\{L\}dV$$

$$= \int_{\Gamma_t}\{\delta v\}^T\{\bar{t}\}d\Gamma + \int_V \rho\{\delta v\}^T\{b\}dV - \int_V \{\delta L\}^T\{T_{|t}\}dV \tag{3.38}$$

where tr denotes the trace; $[D_S]\{L\} = -T'_{ik}L_{jk}$; and $[U]\{L\} = L_{kk}P^F\delta_{ij} - P^F L_{ji}$.

For elasto-viscoplastic materials, the constitutive equation is described using the Jaumann rate of the Cauchy stress tensor and the stretching tensor as

$$\hat{T}'_{ij} = C^e_{ijkl}(D_{kl} - D^{vp}_{kl}) \tag{3.39}$$

where C^e_{ijkl} is the elastic tangential stiffness matrix; D^{vp}_{ij} is the viscoplastic stretching tensor; and $\hat{T}'_{ij}$ is the objective Jaumann rate of the Cauchy skeleton stress tensor given by

$$\hat{T}'_{ij} = \dot{T}'_{ij} + T'_{ik}W_{kj} - W_{ik}T'_{kj} \tag{3.40}$$

in which $W_{ij}(=(v_{i,j} - v_{j,i})/2)$ is the spin tensor.

It is worth noting that D^e_{ij} (elastic stretching) and D^{vp}_{ij} (viscoplastic or plastic stretching) are used instead of $\dot{\varepsilon}^e_{ij}$ (elastic strain rate) and $\dot{\varepsilon}^{vp}_{ij}$ (viscoplastic or plastic strain rate), respectively, for a finite or large strain analysis such as an updated Lagrangian analysis.

The tangent modulus method (Pierce et al. 1984) is implemented here to determine the viscoplastic stretching tensor. In the following, we use the elasto-viscoplastic constitutive equation derived in Section 2.4 for unsaturated soils. Finite element method is used for the discretization of the governing equations. The relation between the Jaumann rate of the Cauchy stress and the stretching tensor of elasto-vicoplastic material is obtained by the tangent modulus method.

The tangent modulus method, developed by Pierce et al. (1984) for rate-dependent materials, results in more stability and accuracy with time steps much larger than that required by the Euler method. Using this method, the tangential stiffness matrix $[C^{\tan}]$ and the relaxation stress $\{Q\}$ in Equation 3.44 are obtained as

$$C^{\tan}_{ijkl} = C^e_{ijkl} - C^e_{ijrs}C_{rsmn}\frac{\partial f_p}{\partial \sigma'_{mn}}\frac{1}{1+\xi'}(\theta\Delta t)\frac{\partial \Phi}{\partial \sigma'_{pq}}C^e_{pqkl} \tag{3.41}$$

$$Q_{ij} = C^e_{ijkl} C_{klmn} \frac{\partial f_p}{\partial \sigma'_{mn}} \frac{1}{1+\xi'} \Phi_t \tag{3.42}$$

where

$$\begin{aligned}\xi' = (\theta \Delta t) \Bigg\{ & \frac{\partial \Phi}{\partial \sigma'_{ij}} C^e_{ijkl} C_{klmn} \frac{\partial f_p}{\partial \sigma'_{mn}} - \frac{\partial \Phi}{\partial y^*_m} B^*_2 \left(A^*_2 C_{kkij} \frac{\partial f_p}{\partial \sigma'_{ij}} - y^*_m \left| C_{kkij} \frac{\partial f_p}{\partial \sigma'_{ij}} \right| \right) \\ & - \frac{\partial \Phi}{\partial \chi^*_{ij}} B^* \left(A^* C_{ijkl} \frac{\partial f_p}{\partial S_{kl}} - \chi^*_{ij} \left(C_{mnpq} \frac{\partial f_p}{\partial S_{pq}} C_{mnrs} \frac{\partial f_p}{\partial S_{rs}} \right)^{\frac{1}{2}} \right) \Bigg\}\end{aligned} \tag{3.43}$$

where Φ_t is the material function $\Phi(f_y)$ at time t, and θ is the tangent stiffness parameter ranging from 0 to 1, with $\theta = 0$ corresponding to the Euler time integration scheme. Through several numerical examples, Pierce et al. (1984) showed that the method is stable and accurate for θ in the range of 0.5 and 1.0.

Substituting Equation 3.40 into Equation 3.39 yields

$$\{\dot{T}'\} = [C^{\tan}]\{D\} - \{Q\} + \{W'\} \tag{3.44}$$

in which $\left[C^{\tan}\right]$ is the tangential stiffness matrix; $\{Q\}$ is the relaxation stress vector; and $\{W'\} = W_{ik} T'_{kj} - T'_{ik} W_{kj}$.

Taking the acceleration vector $\{a_N\}$, the velocity vector $\{v_N\}$, and the pore pressures of the fluids $\{p^\beta_N\}$ as unknown variables at the nodal points, the following definitions can be set:

$$\begin{aligned}&\{D\} = [B]\{v_N\},\ \{v\} = [N]\{v_N\},\ \{L\} = [N_L]\{v_N\},\ \{a\} = [N]\{a_N\}, \\ &tr(D) = \{B_v\}^T \{v_N\},\ \dot{P}^F = \{N_h\}^T \{\dot{P}^F_N\}\end{aligned} \tag{3.45}$$

in which $[N]$ and $\{N_h\}$ represent the displacement shape functions for the solid phase and the shape functions for the pore pressures of the fluid phases, respectively. Substituting Equation 3.45 into Equation 3.38 and eliminating the arbitrary virtual velocity vector $\{\delta v_N\}$ from the equation yield

$$\begin{aligned}&\int_V \rho [N]^T [N] dV \{a_N\} + (\Delta t) \int_V [B]^T [C^{\tan}][B] dV \{v_N\} + (\Delta t) \int_V [B]^T \{W'\} dV \\ &- (\Delta t) \int_V [B]^T \{Q\} dV + (\Delta t) \int_V [N_L]^T [D_s][N_L] dV \{v_N\} \\ &+ (\Delta t) \int_V [N_L]^T [T'] \{B_v\}^T dV \{v_N\} - (\Delta t) \int_V \{B_v\}\{N_h\}^T dV \{\dot{P}^F_N\} \\ &- (\Delta t) \int_V [N_L]^T [U][N_L] dV \{v_N\} = \int_{\Gamma_t} [N]^T \{\bar{t}\} d\Gamma + \rho \int_V [N]^T \{b\} dV - \int_V [N_L]^T \{T_{|t}\} dV\end{aligned} \tag{3.46}$$

From Equation 1.67, the time derivative of p^F is obtained as

$$\dot{P}^F = \frac{\partial}{\partial t}\left\{S_r P^W + (1-S_r)P^G\right\} = \left\{A_s + (1-S_r)\right\}\dot{P}^G + \{-A_s + S_r\}\dot{P}^W \tag{3.47}$$

where

$$A_s = \frac{\partial S_r}{\partial p^C}(P^W - P^G) = -\frac{\partial S_r}{\partial p^C}P^C \tag{3.48}$$

Finally, we have

$$\begin{aligned}&[M]\{a_N\} + \left(\Delta t\left([K]+[K_L]\right)+[R]\right)\{v_N\} - (-A_s + S_r)(\Delta t)[K_v]\{\dot{P}_N^W\}\\&-(A_s + (1-S_r))(\Delta t)[K_v]\{\dot{P}_N^G\} + (\Delta t)\{T_W\} = \{F\} - \{T^*\}_t\end{aligned} \tag{3.49}$$

where the following definitions are used:

$$[M] = \int_V \rho[N]^T[N]dV,\quad [K] = \int_V [B]^T\left[C^{\tan}\right][B]dV$$

$$[K_L] = \int_V [N_L]^T[D_S][N_L]dV - \int_V [N_L]^T[U][N_L]dV + \int_V [N_L]^T\{T'\}\{B_v\}^T dV$$

$$[K_v] = \int_V \{B_v\}\{N_h\}^T dV,\quad \{T_W\} = \int_V [B]^T\{W'\}dV - \int_V [B]^T\{Q\}dV$$

$$\{F\} = \int_{\Gamma_t} [N]^T\{\bar{t}\}d\Gamma + \rho\int_V [N]^T\{b\}dV,\quad \{T^*\}_t = \int_V [N_L]^T\{T_{lt}\}dV \tag{3.50}$$

The Rayleigh damping $[R]$ considered in the formulation is described by the linear combination of the mass matrix $[M]$ and the stiffness matrix $[K]$ as

$$[R] = \alpha_0[M] + \alpha_1[K] \tag{3.51}$$

in which α_0 and α_1 are the constant values determined by specifying the damping ratios for the lowest and the highest modes of vibration which are expected to significantly contribute to the response (Rayleigh and Lindsay 1945, Clough and Penzien 1975).

3.2.2 Discretization of continuity equations

As described in Section 3.1.5, the continuity equations for the fluid phases are derived from their conservation of mass and their equations of motion. The boundary conditions for the fluid phase are assumed as follows:

$$P^G = \bar{P}^G \text{ on } \Gamma_p^G, \quad P_{,i}^G = \bar{q}_i^G \text{ on } \Gamma_q^G \tag{3.52}$$

$$P^W = \bar{P}^W \text{ on } \Gamma_p^W, \quad P_{,i}^W = \bar{q}_i^W \text{ on } \Gamma_q^W \tag{3.53}$$

where $\bar{q}_i^{\beta}$ $(\beta = G, W)$ is the gradient vector of the pore pressure on boundary Γ_q^{β}, and the superimposed bars indicate the prescribed values. In addition, Γ_p^{β} and Γ_q^{β} satisfy $\Gamma^{\beta} = \Gamma_p^{\beta} + \Gamma_q^{\beta}$.

Considering the test functions for the gas and liquid phases as δP^G and δP^W, respectively, the weak forms of the continuity equations are given as follows:

For the gas phase:

$$\begin{aligned} &\int_V \left(\rho_G \dot{D}_{ii} + P_{,ii}^G - \frac{\gamma^G}{k^G}\left((1-S_r)D_{ii} - \dot{S}_r n + (1-S_r) n \frac{\dot{P}_G}{P_G} \right) \right) \delta P^G dV \\ &- \int_{\Gamma_q^G} (P_{,i}^G - \bar{q}_i^G) \delta P^G n_i d\Gamma = 0 \end{aligned} \tag{3.54}$$

For the liquid phase:

$$\begin{aligned} &\int_V \left(\rho_W \dot{D}_{ii} + P_{,ii}^W - \frac{\gamma^W}{k^W}\left(S_r D_{ii} + \dot{S}_r n \right) \right) \delta P^W dV \\ &- \int_{\Gamma_q^W} (P_{,i}^W - \bar{q}_i^W) \delta P^W n_i d\Gamma = 0 \end{aligned} \tag{3.55}$$

Taking the acceleration vector $\{a_N\}$, the velocity vector $\{v_N\}$, the pore water pressure $\{P_N^W\}$, and the pore gas pressure $\{P_N^G\}$ as unknown variables at nodal points, we have

$$\begin{aligned} &\rho_G \int_V \{N_h\}\{B_v\}^T dV \{a_N\} - \int_V [N_{h,i}]^T [N_{h,i}] dV \{P_N^G\} \\ &- \frac{\gamma^G}{k^G} \int_V (1-S_r)\{N_h\}\{B_v\}^T dV \{v_N\} + \frac{\gamma^G}{k^G} \int_V \{N_h\} n \frac{\partial S_r}{\partial p^C} \{N_h\}^T dV \left(\{\dot{P}_N^G\} - \{\dot{P}_N^W\} \right) \\ &- \frac{\gamma^G}{k^G} \int_V (1-S_r) n \frac{\{N_h\}\{N_h\}^T}{p_t^G} dV \{\dot{P}_N^G\} = - \int_{\Gamma_q^G} \{N_h\}\{n\}^T \{\bar{q}^G\} d\Gamma \end{aligned} \tag{3.56}$$

$$\rho_W \int_V \{N_h\}\{B_v\}^T dV\{a_N\} - \int_V [N_{h,i}]^T [N_{h,i}] dV \{P_N^W\}$$
$$- \frac{\gamma^W}{k^W} \int_V S_r \{N_h\}\{B_v\}^T dV \{v_N\} - \frac{\gamma^W}{k^W} \int_V \{N_h\} n \frac{\partial S_r}{\partial P^C} \{N_h\}^T dV \left(\{\dot{P}_N^G\} - \{\dot{P}_N^W\}\right)$$
$$= -\int_{\Gamma_q^W} \{N_h\}\{n\}^T \{\bar{q}^W\} d\Gamma \tag{3.57}$$

where P_t^G is the absolute pore gas pressure at time t. The discretized continuity equations can be written as follows:

For the gas phase:

$$\rho^G [K_v]^T \{a_N\} - \left(\frac{\gamma^G}{k^G}\right)(1-S_r)[K_v]^T \{v_N\} - [K_h]\{P_N^G\}$$
$$- \left(\frac{\gamma^G}{k^G}\right) n \frac{\partial S_r}{\partial p^C} [K_n]\{\dot{P}_N^W\} + \left(\frac{\gamma^G}{k^G}\right) n \frac{\partial S_r}{\partial p^C} [K_n]\{\dot{P}_N^G\} \tag{3.58}$$
$$- \left(\frac{\gamma^G}{k^G}\right)(1-S_r) n [K_c^P]\{\dot{P}_N^G\} = -\{q^G\}$$

For the liquid phase:

$$\rho^W [K_v]^T \{a_N\} - \left(\frac{\gamma^W}{k^W}\right) S_r [K_v]^T \{v_N\} - [K_h]\{P_N^W\}$$
$$+ \left(\frac{\gamma^W}{k^W}\right) n \frac{\partial S_r}{\partial p^C} [K_n]\{\dot{P}_N^W\} - \left(\frac{\gamma^W}{k^W}\right) n \frac{\partial S_r}{\partial p^C} [K_n]\{\dot{P}_N^G\} = -\{q^W\} \tag{3.59}$$

In Equations 3.58 and 3.59, the following definitions are used:

$$P^\beta = \{N_h\}^T \{P_N^\beta\}, \quad \{P_{,j}^\beta\} = [N_{h,j}]\{P_N^\beta\}$$

$$D_{ii} = \{B_v\}^T \{v_N\}, \quad \dot{D}_{ii} = \{B_v\}^T \{a_N\}, \quad [K_v]^T = \int_V \{N_h\}\{B_v\}^T dV$$

$$[K_h] = \int_V [N_{h,i}]^T [N_{h,i}] dV, \quad \{q^\beta\} = \int_{\Gamma_q^\beta} \{N_h\}\{n\}^T \{\bar{q}^\beta\} d\Gamma \quad (\beta = G, W)$$

$$[K_n] = \int_V \{N_h\}\{N_h\}^T dV, \quad \left[K_c^P\right] = \int_V \frac{\{N_h\}\{N_h\}^T}{P_t^G} dV \tag{3.60}$$

3.2.3 Discretization of governing equations in time domain

For the time discretization, Newmark's β method is adopted as

$$\{u_N\}_{t+\Delta t} = \{u_N\}_t + \Delta t\{v_N\}_t + \frac{(\Delta t)^2}{2}\{a_N\}_t + \beta(\Delta t)^2(\{a_N\}_{t+\Delta t} - \{a_N\}_t) \tag{3.61}$$

$$\{\mathrm{v}_N\}_{t+\Delta t} = \{v_N\}_t + \Delta t\{a_N\}_t + \gamma\Delta t(\{a_N\}_{t+\Delta t} - \{a_N\}_t) \tag{3.62}$$

where Δt is the time increment; β and γ are Newmark's constants; and $\{u_N\}$ is the nodal displacement.

3.2.4 Final form of discretized governing equations

Finally, the discretized equations in matrix form are given as follows:

$$[A]\{X\} = \{B\} \tag{3.63}$$

where

$$[A] = \begin{bmatrix} \begin{matrix}[M]_{t+\Delta t} \\ +\gamma(\Delta t)\left\{\begin{matrix}([K]_{t+\Delta t} + [K_L]_{t+\Delta t})\Delta t \\ +[R]_{t+\Delta t}\end{matrix}\right\}\end{matrix} & -(-A_s + S_r)[K_v]_{t+\Delta t} & -(A_s + (1 - S_r))[K_v]_{t+\Delta t} \\ \Delta t\left(\rho_W - \frac{\gamma^W(\Delta t)\gamma}{k^W}S_r\right)[K_v]^T_{t+\Delta t} & \begin{matrix}-(\Delta t)[K_h]_{t+\Delta t} \\ +\left(\frac{\gamma^W}{k^W}\right)n\frac{\partial S_r}{\partial P^C}[K_n]_{t+\Delta t}\end{matrix} & -\left(\frac{\gamma^W}{k^W}\right)n\frac{\partial S_r}{\partial P^C}[K_n]_{t+\Delta t} \\ \Delta t\left(\rho_G - \frac{\gamma^G(\Delta t)\gamma}{k^G}(1 - S_r)\right)[K_v]^T_{t+\Delta t} & -\frac{\gamma^G}{k^G}n\frac{\partial S_r}{\partial P^C}[K_n]_{t+\Delta t} & \begin{matrix}-\Delta t[K_h]_{t+\Delta t} \\ +\frac{\gamma^G}{k^G}n\left(\begin{matrix}\frac{\partial S_r}{\partial P^C}[K_n]_{t+\Delta t} \\ -(1 - S_r)[K_c^P]_{t+\Delta t}\end{matrix}\right)\end{matrix} \end{bmatrix}$$

$$\{X\} = \left\{\{a_N\}^T_{t+\Delta t}, \{P_N^W\}^T_{t+\Delta t}, \{P_N^G\}^T_{t+\Delta t}\right\}^T$$

$$\{B\} = \left\{ \begin{array}{c} \{F\}_{t+\Delta t} - \{T^*\}_t - (\Delta t)\{T_W\}_t - \{([K]_{t+\Delta t} + [K_L]_{t+\Delta t})\Delta t + [R]_{t+\Delta t}\} \\ [(\Delta t)(1-\gamma)\{a_N\}_t + \{v_N\}_t] - (-A_s + S_r)[K_v]_{t+\Delta t}\{p_N^W\}_t \\ -(A_s + (1-S_r))[K_v]_{t+\Delta t}\{p_N^G\}_t \\ \\ -(\Delta t)\{q^W\}_{t+\Delta t} + (\Delta t)\left(\dfrac{\gamma^W}{k^W}\right) S_r [K_v]^T_{t+\Delta t}\left(\Delta t(1-\gamma)\{a_N\}_t + \{v_N\}_t\right) \\ +\left(\dfrac{\gamma^W}{k^W}\right) n \dfrac{\partial S_r}{\partial P^C}[K_n]_{t+\Delta t}\left(\{P_N^W\}_t - \{P_N^G\}_t\right) \\ \\ -(\Delta t)\{q^G\}_{t+\Delta t} + \Delta t \dfrac{\gamma^G}{k^G}(1-S_r)[K_v]^T_{t+\Delta t}\left(\Delta t(1-\gamma)\{a_N\}_t + \{v_N\}_t\right) \\ +\dfrac{\gamma^G}{k^G} n \dfrac{\partial S_r}{\partial P^C}[K_n]_{t+\Delta t}\left(\{P_N^G\}_t - \{P_N^W\}_t\right) - \dfrac{\gamma^G}{k^G} n(1-S_r)[K_c^P]_{t+\Delta t}\{P_N^G\}_t \end{array} \right\}$$

REFERENCES

Borja, R.I. 2004. Cam-clay plasticity part V: A mathematical framework for three-phase deformation and strain localization analysis of partially saturated porous media. *Comput. Meth. Appl. Mech. Eng.* 193(48–51): 5301–5338.

Clough, R.W. and Penzien, J. 1975. *Dynamics of Structures*. McGraw-Hill, New York.

Coussy, O. 1995. *Mechanics of Porous Continua*. John Wiley & Sons, Chichester/New York (originally published in French as *Mécanique de Milieux Poreux*, 1991 Editions Technip).

de Boer, R. 1996. Highlights in the development of the theory of porous media-toward a consistent macroscopic theory. *Appl. Mech. Rev. ASME* 49(4): 201–262.

Ehlers, W., Graf, T. and Ammann, M. 2004. Deformation and localization analysis of partially saturated soil. *Com. Meth. Appl. Mech. Engng.* 193(27–29): 2885–2910.

Garcia, E.F., Oka, F. and Kimoto, S. 2010. Instability analysis and simulation of water infiltration into an unsaturated elasto-viscoplastic material. *Int. J. Solids Struct.* 47(25–26): 3519–3536.

Kato, R., Oka, F. and Kimoto, S. 2014. A numerical simulation of seismic behavior of highway embankments considering seepage flow. In: *Computer Methods and Recent Advances in Geomechanics*, ed. F. Oka, A. Murakami, R. Uzuoka and S. Kimoto, Proc. of 14th Int. Conf. of Int. Assoc. Comp. Meth. Recent Adv. Geomech., Kyoto, Japan, 22–25 September, CRC Press/Balkema, Leiden: 755–760.

Kato, R., Oka, F., Kimoto, S., Kodaka, T. and Sunami, S. 2009. A method of seepage-deformation coupled analysis of unsaturated ground and its application to river embankment. *Trans. JSCE, C* 65(1): 226–240 (in Japanese).

Laloui, L., Klubertanz, G. and Vulliet, L. 2003. Solid-liquid-air coupling in multiphase porous media. *Int. J. Numer. Anal. Meth. Geomech.* 27(3): 183–206.
Malvern, L.E. 1969. *Introduction to the Mechanics of a Continuous Medium.* Prentice-Hall, New York.
Oka, F. and Kimoto, S. 2012. *Computational Modeling of Multiphase Geomaterials.* CRC Press/Taylor & Francis, Boca Raton, FL.
Oka, F., Shahbodagh, B. and Kimoto, S 2019. A computational model for dynamic strain localization in unsaturated elasto-viscoplastic soils. *Int. J. Numer. Anal. Meth. Geomech.* 43(1): 138–165.
Pierce, D., Shih, C.F. and Needleman, A. 1984. A tangent modulus method for rate dependent solids. *Comput. Struct.* 18(5): 875–887.
Rayleigh, J.W.S. and Lindsay, R.B. 1945. *The Theory of Sound.* Dover Publications, New York.
Schrefler, B.A. 2002. Mechanics and thermodynamics of saturated/unsaturated porous materials and quantitative solutions. *Appl. Mech. Rev. ASME* 55(4): 351–388.
Shahbodagh, B. 2011. *Large Deformation Dynamic Analysis Method for Partially Saturated Elasto-Viscoplastic Soils*, PhD thesis. Kyoto University.
van Genuchten, M.Th. 1980. A closed-form equation for predicting the hydraulic conductivity of unsaturated soils. *Soil Sci. Soc. Am. J.* 44(5): 892–898.

Chapter 4

Strain localization in geomaterials

4.1 STRAIN LOCALIZATION MODES IN POROUS MATERIAL: SHEAR AND COMPACTION BANDS

Many researchers know that the strain localization phenomenon, such as shear band formation, is a precursor to failure in geomaterials. Numerous contributions have been made to the analysis of strain localization over the past few decades (e.g., Rudnicki and Rice 1975). Another type of strain localization is the formation of "compaction bands", which are planar deformation bands (almost) perpendicular to the maximum compressive principal stress with no shear offset, as shown in Figure 4.1. Compaction bands in geomaterials were first observed in the field by Mollema and Antonellini (1996) in natural outcrops of sandstone, and were as such defined as planar bands without shear offset.

Using Mohr's stress circle to re-examine the strain localization theory by Rudnicki and Rice (1975), Perrin and Leblond (1993) made some corrections to Rudnicki and Rice's formulation (1975). Based on the work by Perrin and Leblond (1993), Olsson (1999) pointed out that compaction bands can be described within the framework of the bifurcation theory proposed by Rudnicki and Rice (1975). In this section, the conditions under which compaction bands occur will be reviewed based on the approach (theory) proposed by Rudnicki and Rice (1975) and Perrin and Leblond (1993).

4.1.1 Strain localization analysis using Mohr's stress circle

The principal values of deviatoric stresses are σ_i' ($i = I, II, III$ or $= 1,2,3$), $\sigma_I' > \sigma_{II}' > \sigma_{III}'$, $\tau = \left(\frac{1}{2}\sigma_{ij}'\sigma_{ij}'\right)^{1/2}$, $N_i = \sigma_i' / \tau$, and $\sigma_{ij}' = \sigma_{ij} - \sigma_{kk} / 3$.

Hence, $\sigma_I' + \sigma_{II}' + \sigma_{III}' = 0$, $2\tau^2 = \sigma_I'^2 + \sigma_{II}'^2 + \sigma_{III}'^2$, and $\sigma = -\frac{1}{3}\sigma_{kk}$ is the compressive isotropic component.

DOI: 10.1201/9781003200031-4

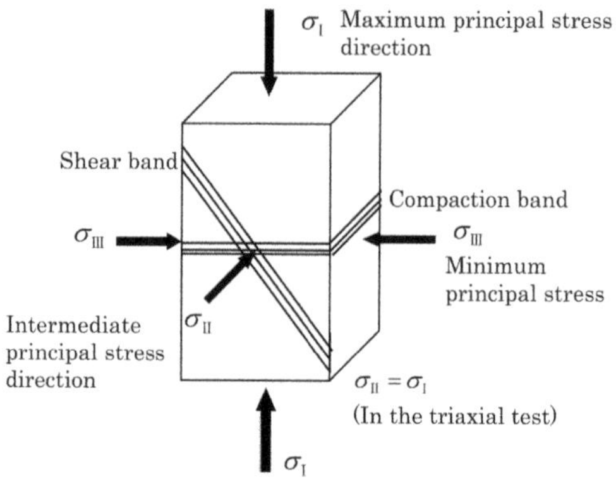

Figure 4.1 Shear and compaction bands.

The localization conditions of the pressure-sensitive dilatant material model by Rudnicki and Rice (1975) are given as follows:

The plastic loading condition is expressed by

$$\dot{\tau} - \mu\dot{\sigma} = \frac{\sigma'_{kl}}{2\tau}\hat{\sigma}'_{kl} + \frac{\beta'}{3}\hat{\sigma}_{kk} > 0 \tag{4.1}$$

where $\hat{\sigma}_{ij}$ is the Jaumann rate of Cauchy's stress.

The deviatoric and isotropic components of stretching, D'_{ij} and D_{kk}, are given by

$$\begin{aligned} D'_{ij} &= \frac{\hat{\sigma}'_{ij}}{2G} + \frac{1}{h}\frac{\sigma'_{ij}}{2\tau}\left[\frac{\sigma'_{kl}}{2\tau}\hat{\sigma}_{kl} + \beta'\frac{\hat{\sigma}_{kk}}{3}\right] \\ D_{kk} &= \frac{\hat{\sigma}_{kk}}{3K} + \frac{\beta}{h}\left[\frac{\sigma'_{kl}}{2\tau}\hat{\sigma}_{kl} + \beta'\frac{\hat{\sigma}_{kk}}{3}\right] \end{aligned} \tag{4.2}$$

where β' is the internal friction coefficient and β is the dilatancy factor.

The stress rates are continuous across the localization band in the direction of x_2, and the gradients of the strain rates are discontinuous (see Figures 4.2 and 4.3).

Since $\Delta\dot{\sigma}_{2j} = L_{2jk2}\Delta D_{k2} = 0$ but $\Delta D_{k2} \neq 0$ and considering $L_{2jk2} = L_{2j2k}$, we have

$$Det[L_{2jk2}] = 0 \tag{4.3}$$

Then, the corresponding hardening coefficient h is given by

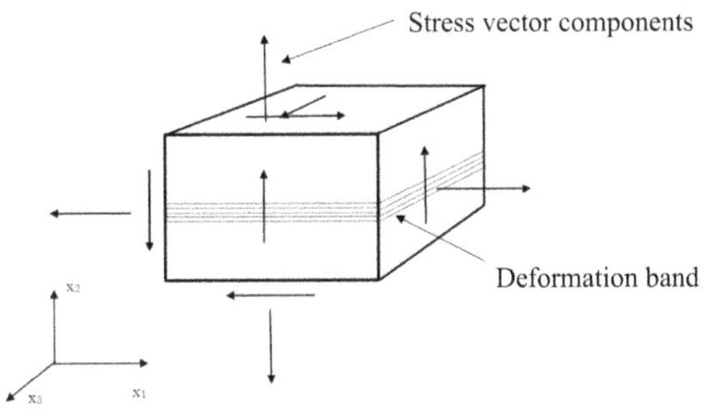

Figure 4.2 Coordinates and stress components.

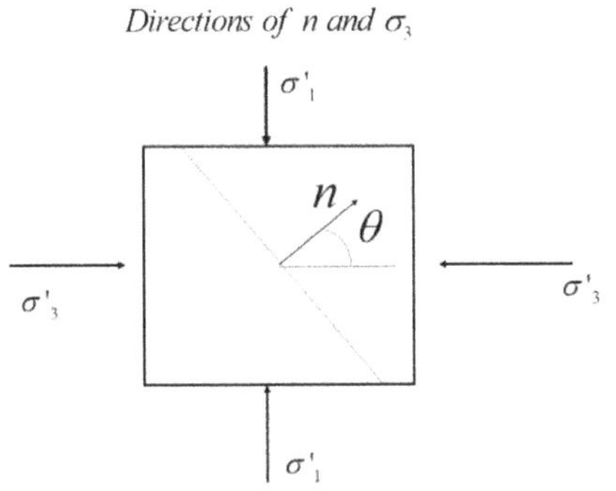

Figure 4.3 Direction of the normal to the localization band.

$$h = \frac{(G\sigma'_{22} + K\beta\tau)(G\sigma'_{22} + K\beta'\tau)}{(K + 4G/3)\tau} + \frac{G(\sigma'^2_{21} + \sigma'^2_{23})}{\tau^2} - G - K\beta\beta' \tag{4.4}$$

where the normal component of the stress on a surface of the unit area perpendicular to the x_2 direction is $\Pi = \sigma'_{22}$, and the square of the tangential component is $T^2 = \sigma'^2_{21} + \sigma'^2_{23}$.

Therefore, Equation 4.2 is rewritten in elliptic form as

$$h = A(\Pi - \Pi_0)^2 + BT^2 + c \tag{4.5}$$

$$A = \frac{G^2}{(K + 4G/3)\tau^2} = \frac{(1 - 2\nu)G}{2(1 - \nu)\tau^2}, \quad \Pi_0 = -\frac{K(\beta + \beta')\tau}{2G} = -\frac{(1 + \nu)(\beta + \beta')\tau}{3(1 - 2\nu)}$$

$$B = \frac{G}{\tau^2}$$

$$\begin{aligned} c &= -\frac{K^2(\beta - \beta')^2}{4(K + 4G/3))} - G - K\beta\beta' \\ &= -G\left[1 + \frac{2(1+\nu)}{3(1-2\nu)}\beta\beta' + \frac{(1+\nu)^2(\beta-\beta')^2}{18(1-\nu)(1-2\nu)}\right] \end{aligned} \tag{4.6}$$

where ν is Poisson's ratio and $A < B$.

Herein, we consider the normal to the planar band n_i, $i = 1, 2, 3$, and the three-dimensional Mohr stress circle shown in Appendix A4.1.

Case (1): General case (shear bands)
The maximum Mohr's circle (Figure 4.4) is

$$(\Pi - \sigma'_I)(\Pi - \sigma'_{III}) + T^2 = 0 \tag{4.7}$$

From Equations 4.5 and 4.7, we obtain

$$(B - A)\Pi^2 + (B\sigma'_{II} + 2A\Pi_0)\Pi + B(\sigma'_{II} - \tau^2) - A\Pi_0^2 + h_c - c = 0 \tag{4.8}$$

where the relations $\sigma_I'^2 + \sigma_{II}'^2 + \sigma_{III}'^2 = 2\tau^2$ and $\sigma'_I + \sigma'_{II} + \sigma'_{III} = 0$ are used.

Since Mohr's circle is tangent to the ellipse of the localization condition, this equation has a double solution in Π.

$$\Pi = \frac{-(B\sigma'_{II} + 2A\Pi_0) \pm \sqrt{(B\sigma'_{II} + 2A\Pi_0)^2 - 4(B-A)(B(\sigma'_{II} - \tau^2) - A\Pi_0^2 + h_c - c)}}{2(B-A)}$$

Then,

$$h_c = c + B(\tau^2 - \sigma'_{II}) + A\Pi_0^2 + \frac{(B\sigma'_{II} + 2A\Pi_0)^2}{4(B-A)} \tag{4.9}$$

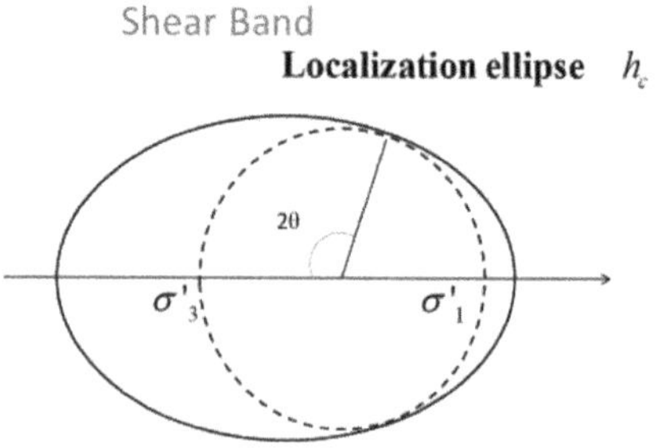

Figure 4.4 Localization ellipse and Mohr's circle for general case.

Using Equation 4.6, Equation 4.9 is rewritten as

$$\frac{h_{cr}}{G} = \frac{1+\nu}{9(1-\nu)}(\beta - \beta')^2 - \frac{1+\nu}{2}\left(N_{II} + \frac{\beta + \beta'}{3}\right)^2 \tag{4.10}$$

where $N_{II} = \sigma'_{II} / \tau$.

Equation 4.10 corresponds to equation (20) of Rudnicki and Rice (1975).

Alternatively, Equation 4.10 can be obtained by differentiating Equation 4.4 with respect to the orientation of the stresses in Equation 4.3.

Using the condition for a double solution, since $\sigma'_{III} \leq \Pi \leq \sigma'_I$, we have

$$\sigma'_{III} < \Pi = \frac{-(B\sigma_{II} + 2A\Pi_0)}{2(B-A)} < \sigma'_I \tag{4.11}$$

Rewriting Equation 4.11, we obtain

$$N_{III} + (1-\nu)N_{II} < \frac{(1+\nu)(\beta+\beta')}{3} < N_I + (1-\nu)N_{II} \tag{4.12}$$

Note that the definitions give

$$\begin{aligned} \Pi &= \sigma'_{22} = n_1^2\sigma'_I + n_2^2\sigma'_{II} + n_3^2\sigma'_{III} \\ T^2 &= \sigma'^2_{21} + \sigma'^2_{23} = n_1^2\sigma'^2_I + n_2^2\sigma'^2_{II} + n_3^2\sigma'^2_{III} - \sigma'^2_{22} \end{aligned} \tag{4.13}$$

Hence, we obtain the preferred direction for the localization of $n_1 = \sin\theta$, $n_2 = 0$, and $n_3 = \cos\theta$, where θ is the angle between the normal and σ_{III} direction.

Differentiating h in Equation 4.4 with respect to θ:

$$\tan^2\theta_0 = \frac{\xi - N_{III}}{N_I - \xi}, \quad \xi = \frac{(1+\nu)(\beta+\beta')}{3} - (1-\nu)N_{II} \tag{4.14}$$

Thus, the localization direction θ_0 is obtained.

Alternatively, from Mohr's circle, we have

$$\cos 2\theta = -\frac{\Pi + \sigma'_{II}/2}{(\sigma'_I - \sigma'_{III})/2} \tag{4.15}$$

Note that the center of Mohr's circle is given by $\sigma'_I - \sigma'_{III}/2 = -\sigma'_{II}/2$.

Case (2): Compaction band $n_3 \neq 0$, $n_1 = n_2 = 0$

This is the case of equation 12 in Perrin and Leblond's paper (1993).

In this case, since $n_1 = n_2 = 0$, from Equation A4.9, we have

$$(\Pi - \sigma'_2)(\Pi - \sigma'_1) + T^2 = 0, \quad (\Pi - \sigma'_1)(\Pi - \sigma'_3) + T^2 = 0$$

Then, $\sigma'_2 = \sigma'_3$; the Mohr circle is tangential at $(\sigma'_3, 0)$ in Figure 4.5.

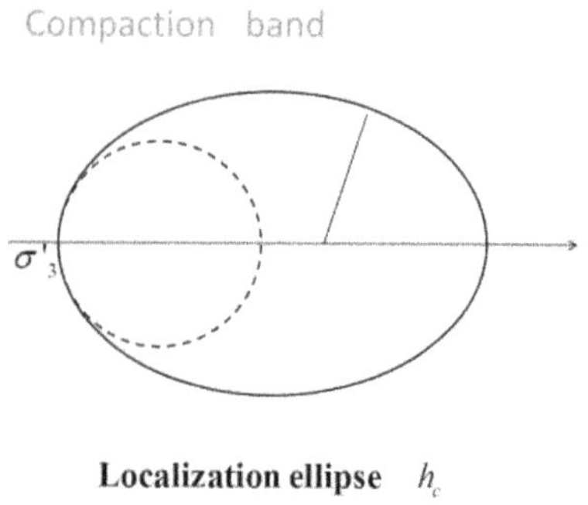

Figure 4.5 Localization ellipse and Mohr's circle for compaction band.

This case does not satisfy the left-hand side of the inequality equation (7) in Perrin and Leblond's paper (1993). Then, the curvature of the Mohr circle at the point $(\sigma_3', 0)$ is larger than that of the localization ellipse Equation 4.4. Hence,

$$\frac{A}{B}(\Pi_0 - \sigma_3') \geq \frac{1}{2}(\sigma_1' - \sigma_3') \tag{4.16}$$

The curvature of the localization ellipse is given as follows:

$$A(\Pi - \Pi_0)^2 + BT^2 = h - c$$

is rewritten as

$$\frac{(\Pi - \Pi_0)^2}{\left(\sqrt{\frac{h-c}{A}}\right)^2} + \frac{T^2}{\left(\sqrt{\frac{h-c}{B}}\right)^2} = 1$$

In general, the curvature and the curvature radius are obtained as

We set the long axis $2a$ and short axis $2b$.

$x = a\ (\cos\theta)$, $y = b\ (\sin\theta)$, θ: angle of the circle and the curvature is $k = ab/((a\sin\theta)^2 + (b\cos\theta)^2)^{3/2}$, then the curvature radius: $R = 1/k$.

The curvature at the point $(a,0)$ is $\rho = b^2/a$ and the curvature is a/b^2.

The curvature of the localization ellipse is less than that of Mohr's circle. Hence,

$$\text{Curvature}\left(=\frac{a}{b^2}\right) = \frac{\sqrt{\frac{h-c}{A}}}{\frac{h-c}{B}} = \frac{B}{\sqrt{A(h-c)}} < \frac{1}{R}\ \text{(curvature of Mohr's circle)}$$

Since at the point $(\sigma_3', 0)$ the shear component is zero, $h = c + A(\sigma_3 - \Pi_0)^2$:

$$\frac{1}{2}(\sigma_1' - \sigma_3') \leq \frac{1-2\nu}{2(1-\nu)}\left(-\frac{(1+\nu)(\beta+\beta')\tau}{3(1-2\nu)} - \sigma_3'\right) \tag{4.17}$$

By rearranging Equation 4.17, we have

$$N_3 + (1-\nu)N_2 \geq \frac{(1+\nu)(\beta+\beta')}{3} \tag{4.18}$$

Rewriting Equation 4.18:

$$-(1-\nu)N_1 + \nu N_3 \geq \frac{(1+\nu)(\beta+\beta')}{3} \tag{4.19}$$

Since $N_1 + N_2 + N_3 = 0$, N_3 is negative. In the case that $\beta + \beta'$ is negative, i.e., in the volumetric compression case, compaction bands occur.

Case (3): Dilation bands ($n_1 \neq 0, n_2 = n_3 = 0$)

In the case in which Mohr's circle (Figure 4.6) is tangential at the point $(\sigma_1', 0)$, from the Mohr circles. Hence, $(\Pi - \sigma_1')(\Pi - \sigma_3') + T^2 = 0$, $\sigma_1' = \sigma_2'$.

Similarly as in Case (2), we have

$$N_{max} + (1-\nu)N_2 \leq \frac{(1+\nu)(\beta+\beta')}{3} \tag{4.20}$$

This case corresponds to the dilational bands which are often observed when necking occurs during the extension of the metal with air bubbles.

4.1.2 Summary

The following localization bands were confirmed as $n_2 = 0$, $n_1, n_3 \neq 0$ (shear bands), $n_3 \neq 0$, $n_1 = n_2 = 0$ (compaction bands), $n_1 \neq 0$, $n_2 = n_3 = 0$ (dilational band), $n_3 = 0$, $n_1, n_2 \neq 0$ (impossible), and $n_1, n_2, n_3 \neq 0$ (impossible).

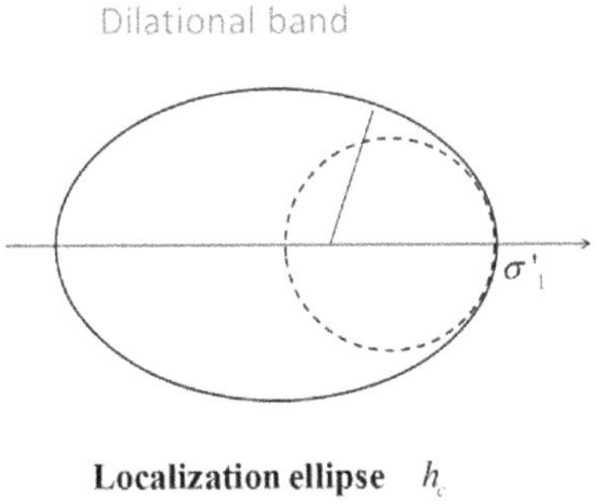

Figure 4.6 Localization ellipse and Mohr's circle for dilational band.

4.2 ELASTO-VISCOPLASTIC NUMERICAL ANALYSIS OF COMPACTION BANDS OF DIATOMACEOUS MUDSTONE

4.2.1 Introduction

Shear strain localization phenomena, such as shear band formation, have been well studied by many researchers as a precursor to failure in geomaterials (e.g., Rudnicki and Rice 1975). On the other hand, other forms of strain localization called "compaction bands" have also been studied, which are planar deformation bands that are (or almost) perpendicular to the maximum compressive principal stress with no shear offset, as shown in Figure 4.1. Compaction bands in the geomaterials of sandstone were first observed in the field by Mollema and Antonellini (1996) in natural outcrops, and as such were defined as planar bands without shear offset. In laboratory tests, Olsson (1999) studied compaction bands in Castlegate sandstone, while Wong et al. (1997) observed similar bands in Darley Dale sandstone. For soils, Arroyo et al. (2005) observed compaction bands in natural calcarenite and some cemented soils. To date, compaction bands have been observed in the laboratory mainly in porous sandstones at high confining pressure levels (e.g., Wong et al. 1997; Olsson 1999), in cemented soils (Arroyo et al. 2005), and in soft rocks (Kodaka et al. 2006). However, Garbrecht (1973) also experimentally observed compaction bands in glass beads as a result of sudden grain crushing. Based on various studies in the current literature, it has been found that one cause of compaction bands is the collapse of the material structure (Issen 2002), while another plausible cause is particle crushing (Papamichos et al. 1993; Vardoulakis and Sulem 1995). Perrin and Leblond (1993) pointed out that Rudnicki and Rice's theory (1975) is applicable to deformation bands such as compaction bands.

A theoretical study on compaction bands by Olsson (1999) pointed out that compaction bands can be described within the framework of the bifurcation theory proposed by Rudnicki and Rice (1975). Furthermore, Issen and Rudnicki (2000) extended Olsson's approach and obtained the onset conditions together with the orientation for both shear and compaction bands by re-examining Rudnick and Rice's theory. In addition, they introduced a cap-type yield function in conjunction with a shear yield function and found the conditions for compaction, shear, and dilation bands (Issen 2002; Issen and Rudnicki 2000). Issen and Rudnicki (2001) obtained the values for the material parameters that control band formation; however, the theoretically obtained parameters were not very consistent with the parameters obtained from the experiments (Olsson 1999).

Arroyo et al. (2005) adopted an elasto-plastic bonded soil model to describe the behavior of bonded and/or cemented soil and explored the onset of compaction bands. They showed compaction band formation under oedometric conditions considering the bifurcation conditions set out by Rudnicki and Rice (1975), and thus emphasized the importance of softening in the material. Borja (2004) studied the computational modeling of deformation bands, including compaction bands, within the framework of the bifurcation theory and numerically analyzed the onset conditions for deformation bands using an advanced constitutive model. To date, very little work has been done on the numerical simulation of compaction bands although many numerical predictions have been made for the formation of shear bands in geomaterials. In general, the proper numerical reproduction of compaction bands is quite challenging.

The purpose of this section is to demonstrate the formation of compaction bands in laboratory tests and subsequently numerically simulate them using an elasto-viscoplastic model in the finite element analysis following Oka and Kimoto (2005) and Oka et al. (2011). All discussions and numerical analyses of compaction bands will pertain to a diatomaceous mudstone called "Noto diatomaceous mudstone" that is very porous due to its rich composition of diatoms.

4.2.2 Behavior of diatomaceous mudstone

Drained triaxial compression tests were carried out on Noto diatomaceous mudstone in order to study deformation bands, including compaction bands, under several confining pressure conditions (Oka et al. 2011). Rectangular prismatic specimens were used for all test cases in order to facilitate the optical measurement of instabilities such as strain localization.

Noto diatomaceous mudstone occurs on the Noto Peninsula in Ishikawa prefecture, Japan. While the mudstone contains rich diatomite and clay minerals, its exact soil composition and index properties have been reported by Maekawa and Miyatake (1983) and Maekawa (1992) as follows: sand fraction = 1%, silt fraction = 66%, clay fraction = 33%, liquid limit = 139.1%, plastic limit = 93.2%, plastic index = 45.9, and organic content = 9.3%. Other physical properties of the mudstone are given as specific gravity = 2.36, natural water content = 115.1%, initial void ratio = 2.72, dry density = 0.63 g/cm^3, saturated density = 1.37 g/cm^3, and consolidation yield stress = 2.55 MPa.

Although diatomaceous mudstone is rather stiff, its porosity is very high compared with other types of soft rocks; the initial porosity is 0.731. A small block sample was trimmed to precise dimensions with a straight edge. The prismatic specimens used in the experiments were 8 cm high with a square section of 4 cm a side. The longitudinal direction of the specimens was made perpendicular to the plane of sedimentation to avoid the effects of the initial anisotropy. The distributions of the local strain on the surface of the specimens were observed by means of two high-resolution charge-coupled device (CCD) cameras, and the results were then analyzed via image analysis based on the particle tracking velocimetry (PTV) technique (see Kodaka et al. 2001).

4.2.2.1 Triaxial testing procedure

A high-pressure, steel, triaxial chamber, equipped with two digital CCD cameras, was used in the tests. A series of conventional consolidated-drained (CD) shear tests were performed. The test cases listed the effective confining pressures as Case CD1 = 0.25 MPa, Case CD2 = 0.50 MPa, Case CD3 = 0.75 MPa, Case CD4 = 1.0 MPa, Case CD5 = 1.5 MPa, and Case CD6 = 2.0 MPa. All the specimens used in the present study were saturated using the double vacuum method, and triaxial testing was carried out under a back pressure of 1 MPa. A typical CD test consisted of isotropically consolidating the specimen at a prescribed level of effective confining pressure for 6 h, followed by shearing at a constant confining pressure and under drained conditions at an axial strain rate of 0.01%/min.

4.2.2.2 Triaxial test results

The triaxial test results were reported by Oka et al. (2011). From the results, the following points were clear. The deviator stress–axial strain relations were obtained for the CD tests at various effective confining pressures ranging from 0.25 MPa to 2 MPa (denoted as CD1–CD6). At relatively low confining pressures (0.25 MPa [CD1] and 0.5 MPa [CD2]), the stress–strain curves display a strain-softening behavior, whereas at higher confining pressure levels (0.75 MPa [CD3], 1 MPa [CD4], 1.5 MPa [CD5], and 2 MPa [CD6]) only a strain-hardening trend is observed. It is noteworthy that the material exhibits mainly contractive behavior, even at low levels of confining pressure. The yield points can be identified from inflection points marking the conditions when volume changes increase rapidly with an increasing mean effective stress. For simplicity, the inflection points can be loosely identified as "yield points", although these would have to be more formally checked by loading-unloading loops. Interestingly, these yield points also correspond to inflection points on the corresponding stress–strain curves. On these curves, the deviator stress at which the inflection points occur decreases, while the degree of strain-hardening increases with confining pressure.

When connecting all the yield points in the deviator stress–mean effective stress space, a cap-shaped yield locus is obtained. The cap yield locus intersects the mean effective stress axis, giving an isotropic yield stress of 2.55 MPa that is consistent with the yield point on the isotropic consolidation curve. As a result, it is apparent that the initial isotropic consolidation stress levels for all CD tests are in the overconsolidated region. Although volume expansion is usually observed for heavily overconsolidated reconstituted soils, contractive behavior was observed for this material, even in the range of a smaller confining pressure than the consolidation yield stress. The highly structured character of the tested material is due to its rich diatom content. Thus, in order to observe volumetric strain localization behavior such as compaction band formation in geomaterials, brittle and contractive properties are a prerequisite.

4.2.2.3 *Image analysis results*

Figure 4.7 shows the "accumulated viscoplastic deviatoric strain" and the "volumetric strain" distributions at 20% global axial strain for the CD tests. These two types of strain quantities are locally distributed on the surface of the specimens and have been obtained through PTV image analysis by assuming a two-dimensional deformation state on the surface. In the CD1 (0.25 MPa) test, steep shear bands can be clearly observed. The local axial strain levels are concentrated in the same region where the shear strain levels localize. It can be seen that one of the shear bands splits the specimen into two parts. For the CD1 test, the shear strain is clearly concentrated from the top-right boundary to the middle-left boundary when the global axial strain is 6%. However, a significant shear band subsequently emerges from the top-left boundary dividing the specimen into two parts in the final stage of axial compression.

Turning to the CD2 (0.5 MPa) test, both the local axial strain and the shear strain are concentrated in the middle of the specimen. By observing the specimen without a membrane at the end of the test, shear bands can be clearly seen inside the middle portion of the specimen. At higher confining

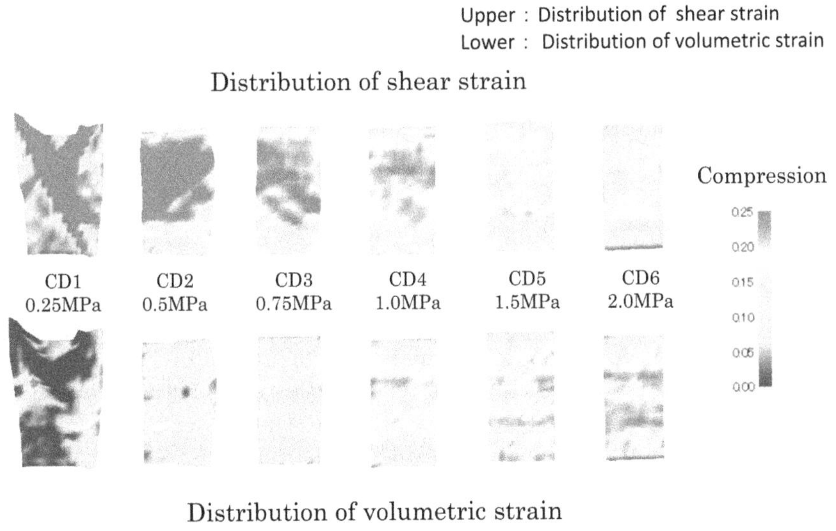

Figure 4.7 Distributions of shear strain and volumetric strain at the surface of the specimens by PTV image analysis. (After Oka, F., Kimoto, S., Higo, Y., Ohta, H., Sanagawa, T. and Kodaka, T. 2011. An elasto-viscoplastic model for diatomaceous mudstone and a numerical simulation of compaction bands. *Int. J. Numer. Anal. Meth. Geomech.* 35(2): 244–263.)

pressures, namely CD3 (0.75 MPa) and CD4 (1.0 MPa), the shear bands become relatively less apparent, with the volumetric strain concentrating perpendicular to the direction of the principal stress. The horizontal axial strain and the shear strain concentrations can also be observed, and as such these cases exhibit mixed deformation patterns. Particularly in the CD4 test, the volumetric strain is concentrated horizontally inside a band. The stress–strain behavior displayed in Figure 4.10b for CD4 denotes that the specimen has already yielded at a global axial strain level of 4%. In other words, compaction bands potentially form in the neighborhood of yield states.

At even higher confining pressures, i.e., CD5 (confining pressure = 1.5 MPa) and CD6 (confining pressure = 2.0 MPa), the shear strain is not significant compared with the axial strain and the volumetric strain. Rather, the volumetric strain levels predominantly concentrate horizontally and periodically throughout the specimen. In addition, shear strain localizes horizontally. Interestingly, the strain localization pattern does not indicate a shear band type, but one of compaction type. The strain concentrations are mainly caused by the volumetric compression of the material to form so-called compaction bands. In particular, periodical compaction bands are clearly observed in the CD6 test. Furthermore, as shown in figure 2 of Oka et al. (2011), the gradient of the stress–strain curve for CD6 clearly changes around a global axial strain of 7%. Thus, localized compressive deformations become significant after a global axial strain of 7% beyond the yield state.

4.2.3 Elasto-viscoplastic constitutive equations

After describing the experimental observations of the various deformation and failure behaviors of Noto diatomaceous mudstone under various effective confining pressures, we next provide a numerical analysis of the experiments using an elasto-viscoplastic constitutive model with structural changes developed by Kimoto and Oka (2005) and Oka and Kimoto (2005). The model is presented as an extension of Adachi–Oka's elasto-viscoplastic model (Oka 1981; Adachi and Oka 1982) and is based on an overstress type of viscoplasticity (Perzyna 1963). The material instability characteristics of the model with microstructural considerations have been previously discussed in Kimoto (2002) and Kimoto and Oka (2005). The structural changes are phenomenologically described by shrinking both the overconsolidation boundary surface and the static yield surface with the evolution of viscoplastic strains. In the analysis, the constitutive model presented in chapter 5.7 of Oka and Kimoto (2012) was used with the dilatancy parameter $\tilde{M}_m^*$ of equation 5.54 of Oka and Kimoto (2012) as

$$\tilde{M}^* = \begin{cases} M_m^* & : \quad f_b \geq 0 \\ -\dfrac{\sqrt{\eta_{ij}^* \eta_{ij}^*}}{\ln(\sigma'_m / \sigma'_{mc})} & : \quad f_b < 0 \end{cases} \tag{4.21}$$

where σ'_{mc} denotes the mean effective stress at the intersection of the overconsolidation boundary surface with the σ'_m axis.

Since the large strain analysis is adopted, i.e., the updated Lagrangian analysis method, for the stress rate, the Jaumann rate of Cauchy's stress and stretching are used instead of the strain rate of Cauchy's stress and the strain rate, respectively.

Finally, deviatoric and volumetric viscoplastic stretching are obtained, respectively, as

$$D'^{vp}_{ij} = C_1 \exp\left[m' \left(\bar{\eta}^* + \tilde{M}^* \ln \frac{\sigma'_m}{\sigma'_{mb}} \right) \right] \frac{\eta^*_{ij} - \eta^*_{ij(0)}}{\bar{\eta}^*} \tag{4.22}$$

$$D^{vp}_{kk} = C_2 \exp\left[m' \left(\bar{\eta}^* + \tilde{M}^* \ln \frac{\sigma'_m}{\sigma'_{mb}} \right) \right] \left[\tilde{M}^* - \frac{\eta^*_{mn} \left(\eta^*_{mn} - \eta^*_{mn(0)} \right)}{\bar{\eta}^*} \right] \tag{4.23}$$

where D'^{vp}_{ij} and D^{vp}_{kk} are deviatoric and volumetric components of the viscoplastic stretching D^{vp}_{ij}.

4.2.4 Elasto-viscoplastic finite element analysis

4.2.4.1 *Finite element analysis method*

We adopt a two-phase mixture finite element formulation allowing for large deformations (Kimoto et al. 2004; Higo et al. 2006; Oka and Kimoto 2012) to numerically simulate the triaxial tests described in Chapter 3 and section 6.5 in Oka and Kimoto (2012). More precisely, Biot's biphasic theory is adopted to develop the governing equations for the soil deformation–water flow coupling problem. The finite element formulation is based on an updated Lagrangian description of deformations together with the objective Jaumann rate of Cauchy's stress since the strain level is not too large and less than 100% (Johnson and Bammann 1984). Moreover, the strain rate tensor $\dot{\varepsilon}_{ij}$ in the previous section is replaced with the stretching (or rate of deformation) tensor D_{ij} in the current formulation.

Using the rate-type formulation for the balance of linear momentum with respect to the current configuration (Kimoto et al. 2004; Higo et al. 2006; Oka and Kimoto 2012), the weak form of the rate type of equilibrium equations for the soil-water whole mixture emerges as

$$\int_V \dot{S}^t_{ji,j} \delta v_i dV = 0 \tag{4.24}$$

where δv_i is the virtual velocity vector and $\dot{S}^t_{ij}$ is the nominal stress rate tensor with respect to the current configuration defined as

$$\dot{S}^t_{ij} = \dot{\sigma}_{ij} + \sigma_{ij} L_{kk} - L_{ik} \sigma_{kj} \tag{4.25}$$

where σ_{ij} denotes the Cauchy stress tensor; L_{ij} is the velocity gradient tensor; and the superimposed dots indicate the time differentiation.

For the fluid phase, the weak form of the continuity equation is employed considering the test function $\hat{\boldsymbol{u}}_w$, namely,

$$\frac{k}{\gamma_w}\int_V \nabla^2 \boldsymbol{u}_w \hat{\boldsymbol{u}}_w \, dV + \int_V D_{kk} \hat{\boldsymbol{u}}_w \, dV = 0 \tag{4.26}$$

where $\boldsymbol{u}_w$ is the pore water pressure; k is the coefficient of permeability; γ_w is the unit weight of the pore water; and D_{kk} is the trace of the stretching tensor D_{ij}.

As for the types of elements used in the subsequent three-dimensional numerical analysis, we adopt a 20-node isoparametric element with a reduced Gaussian (8×8) integration for the soil skeleton and an 8-node isoparametric element with a full (8×8) integration for the pore fluid. Detailed formulations of the coupled equilibrium and continuity equations can be found in Higo et al. (2006) and Oka and Kimoto (2012). In addition, it is interesting to note that this type of discretization shows almost no mesh size dependency (Higo et al. 2006).

4.2.4.2 Numerical results and discussions

Oka et al. (2011) used the boundary conditions associated with the drained triaxial compression tests in displacement-controlled loading mode, as shown in Figure 4.8. Due to the symmetry of the geometrical problem and for efficiency in computations, we analyzed only 1/8 of the specimen in the numerical simulations. As far as hydraulic boundary conditions are concerned, the lateral boundaries of the mesh are set to be impermeable while both the top and bottom boundaries are drained so that the internal redistribution of the pore water is allowed. For deforming the sample, the top boundary of the mesh is displaced at a constant strain rate, herein taken as 0.01%/min. For the present simulation, a time step of 20 s is chosen, which corresponds to an incremental average axial strain of $\Delta\varepsilon = 0.0033\%$. In order to account for the end platen friction, a frictional boundary-type condition with a coefficient of friction equal to 0.1 is imposed on the top boundary of the mesh. While this boundary condition mimics the frictional resistance developed between the top platen and the specimen in the actual experimental set-up, the consideration of interface friction also acts as a trigger for strain localization under external loading.

As a starting point, the material parameters of the constitutive model were first numerically determined and calibrated on the basis of element tests and the experimental data (Oka et al. 2011). The determination of the various model parameters is straightforward, except for the viscoplastic parameters which were obtained following the method proposed

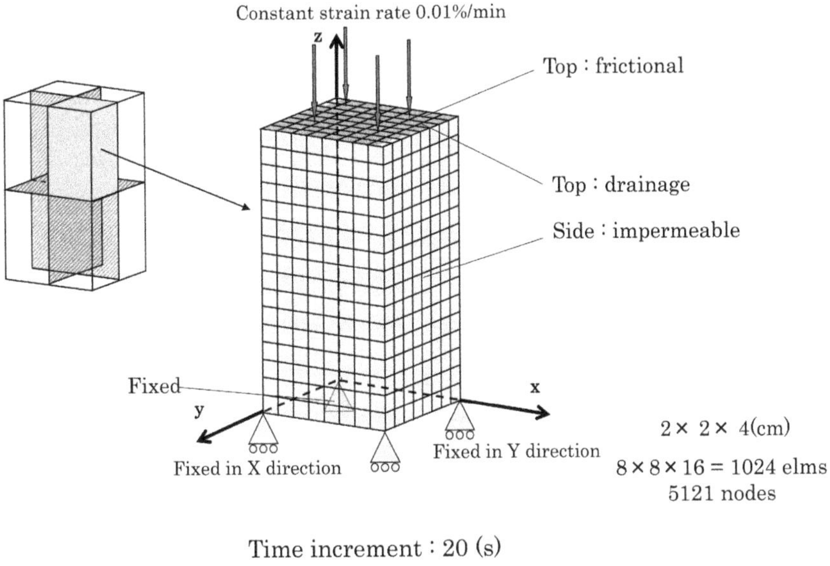

Figure 4.8 Finite element mesh and boundary conditions. (After Oka, F., Kimoto, S., Higo, Y., Ohta, H., Sanagawa, T. and Kodaka, T. 2011. An elasto-viscoplastic model for diatomaceous mudstone and a numerical simulation of compaction bands. *Int. J. Numer. Anal. Meth. Geomech.* 35(2): 244–263.)

by Kimoto and Oka (2005). All material parameters were determined from the triaxial test results. As for the coefficient of permeability for Noto diatomaceous mudstone, we used the value experimentally determined by Maekawa (1992). With the preceding material parameters determined from element tests, the triaxial drained tests are next simulated as a three-dimensional boundary value problem via finite element analysis.

Oka et al. (2011) found that the constant value of the viscoplastic parameter C_2 does not accurately reproduce the experimental results. Hence, Oka et al. (2011) introduced a non-linearity into the viscoplastic parameter C_2 as a function of strains in order to improve the computed stress–strain responses. As such, in the final numerical calculations, C_2 is assumed to depend on the accumulated second invariant of viscoplastic shear strain rate z as follows:

$$C_2 = C_{20}\left[1 + (R-1)\frac{\tanh Wz(1-\tanh Wz_0)}{1-\tanh Wz\tanh Wz_0}\right] \tag{4.27}$$

$$z = \int_0^t \dot{z}dt, \quad \dot{z} = \left(D_{ij}^{vp}D_{ij}^{vp}\right)^{1/2} \tag{4.28}$$

where C_{20} is the value of C_2 at the initial state, and R, W, and z_0 are material parameters. In particular, R is an amplification factor for C_2 whereas z_0 controls the onset of the increase in C_2 with W the rate of increase.

In the analysis, we assumed the values of newly introduced parameters as follows: $R = 10.0$, $z_0 = 0.06$, and $W = 10.0$. The parameters used in the analysis are listed in Table 4.1. Figure 4.9 shows the distributions of the accumulated second invariant of the viscoplastic deviatoric strain defined as $(\gamma^p = \int_0^t (D_{ij}^{\prime vp} D_{ij}^{\prime vp})^{1/2} dt,\ D_{ij}^{\prime vp}$: deviatoric viscoplastic stretching) and the volumetric strain (20%–56%) as computed from the model with the newly introduced parameters. With the appearance of near horizontal strain localization zones with significant volumetric strains, Figure 4.9 clearly shows a patterned formation of compaction bands. Figure 4.10 shows that there is good agreement between simulations and experiments where (a) the average deviator stress versus axial strain, (b) the volumetric strain–mean effective stress, and (c) the volumetric strain–axial strain are presented.

The introduction of the non-linear strain dependency of viscoplastic parameter C_2 invariably improves the numerical simulations, but at the expense of using more material parameters. From Figures 4.9 and 4.10, it is seen that the introduction of the non-linear strain dependency of viscoplastic parameter C_2 well simulates the experimental results.

Table 4.1 Material parameters used in the simulation

Effective confining pressure (MPa) (test no.)	σ'_{m0}	0.25 (CD1), 0.5 (CD2), 0.75 (CD3), 1.00 (CD4), 1.5 (CD5), 2.0 (CD6)
Initial elastic shear modulus (MPa)	G_0	92.0
Compression index	λ	1.292
Swelling index	κ	0.0487
Initial void ratio	e_0	2.72
Compression yield stress (MPa)	σ'_{mai}	2.55
Stress ratio at critical state	M^*_m	1.551
Viscoplastic parameter	m'	28.76
Viscoplastic parameter (1/s)	C_1	1.0×10^{-6}
Viscoplastic parameter (1/s)	C_2	1.2×10^{-5}
Structural parameter (MPa)	σ'_{maf}	2.20
Structural parameter (MPa)	β	2.0
Coefficient of permeability (m/s)	k	1.55×10^{-9}
Material parameter	R	10.0
Material parameter	W	10.0
Material parameter	z_0	0.06

Source: Data from Oka, F., Kimoto, S., Higo, Y., Ohta, H., Sanagawa, T. and Kodaka, T. 2011. An elasto-viscoplastic model for diatomaceous mudstone and a numerical simulation of compaction bands. *Int. J. Numer. Anal. Meth. Geomech.*, 35(2): 244–263.

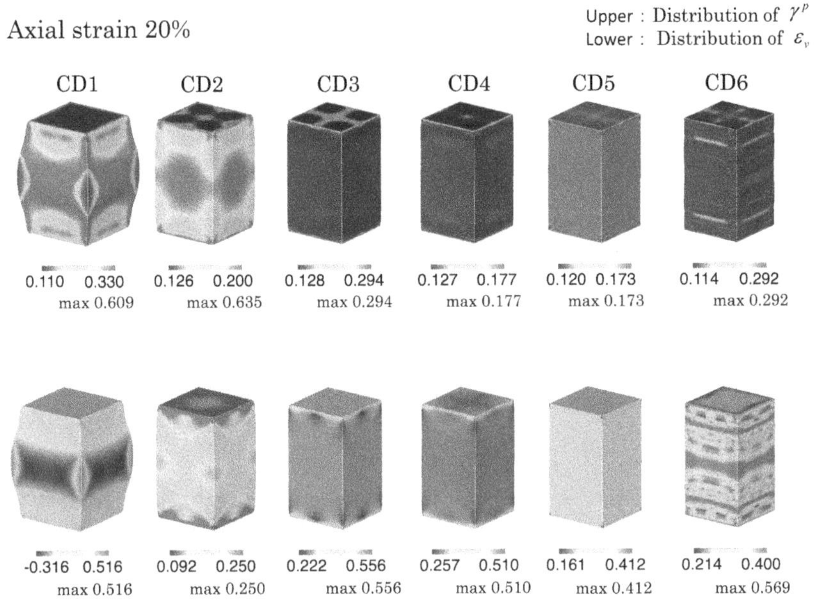

Figure 4.9 Distributions of γ^p and volumetric strain of the numerical simulation with the strain dependency of viscoplastic parameter C_2 for the cases CD1–CD6. (After Oka, F., Kimoto, S., Higo, Y., Ohta, H., Sanagawa, T. and Kodaka, T. 2011. An Elasto-viscoplastic model for diatomaceous mudstone and a numerical simulation of compaction bands. *Int. J. Numer. Anal. Meth. Geomech.* 35(2): 244–263.)

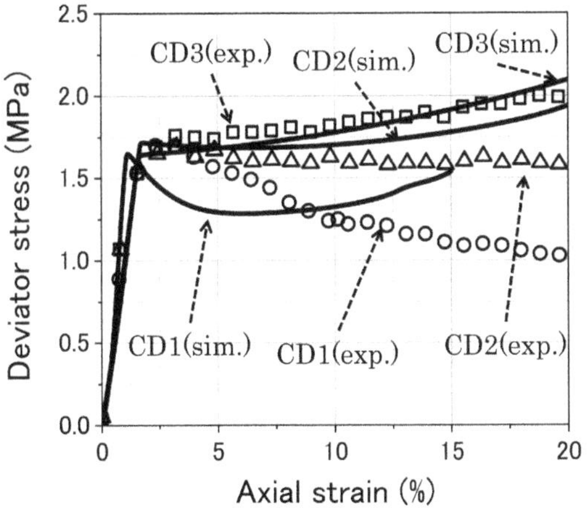

Figure 4.10a Comparison for the stress–strain relations between the experimental and the simulation results with the strain dependency of viscoplastic parameter C_2. (Case CD1, CD2, CD3) (After Oka, F., Kimoto, S., Higo, Y., Ohta, H., Sanagawa, T. and Kodaka, T. 2011. An elasto-viscoplastic model for diatomaceous mudstone and a numerical simulation of compaction bands. *Int. J. Numer. Anal. Meth. Geomech.* 35(2): 244–263.)

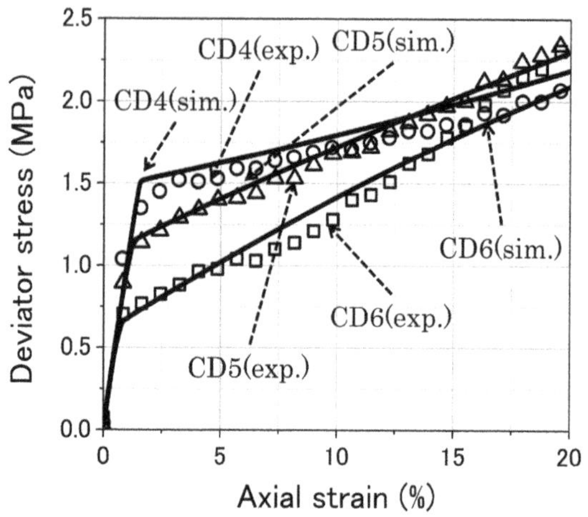

Figure 4.10b Comparison for the stress–strain relations between the experimental and the simulation results with the strain dependency of viscoplastic parameter C_2. (Case CD4, CD5, CD6). (After Oka, F., Kimoto, S., Higo, Y., Ohta, H., Sanagawa, T. and Kodaka, T. 2011. An elasto-viscoplastic model for diatomaceous mudstone and a numerical simulation of compaction bands. *Int. J. Numer. Anal. Meth. Geomech.* 35(2): 244–263.)

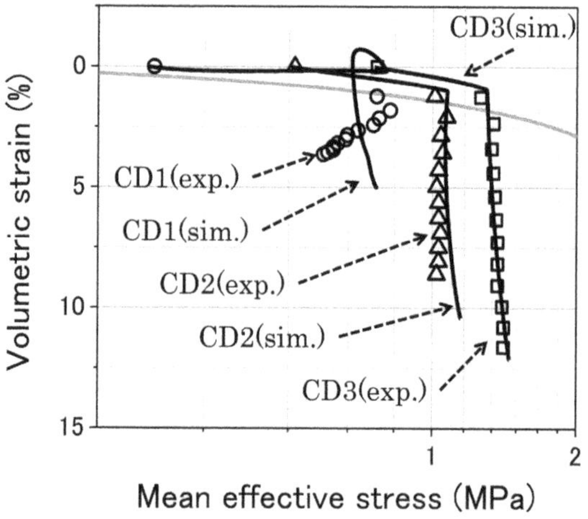

Figure 4.10c Comparison for the volumetric strain-mean effective stress relations between the experimental and the simulation results with the strain dependency of viscoplastic parameter C_2. (Case CD1, CD2, CD3) (After Oka, F., Kimoto, S., Higo, Y., Ohta, H., Sanagawa, T. and Kodaka, T. 2011. An elasto-viscoplastic model for diatomaceous mudstone and a numerical simulation of compaction bands. *Int. J. Numer. Anal. Meth. Geomech.* 35(2): 244–263.)

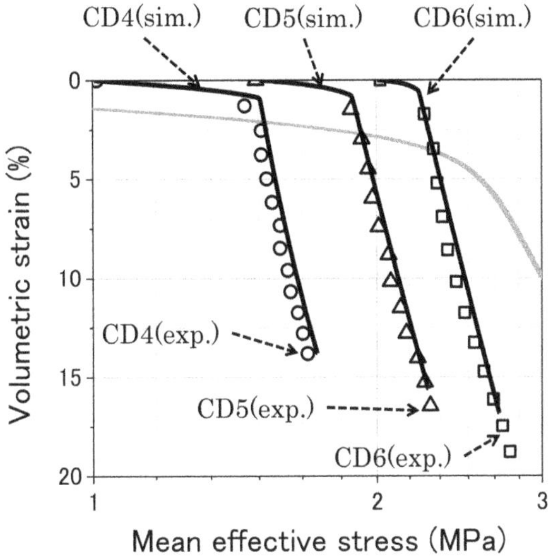

Figure 4.10d Comparison for the volumetric strain-mean effective stress relations between the experimental and the simulation results with the strain dependency of viscoplastic parameter C_2. (Case CD4, CD5, CD6) (After Oka, F., Kimoto, S., Higo, Y., Ohta, H., Sanagawa, T. and Kodaka, T. 2011. An elasto-viscoplastic model for diatomaceous mudstone and a numerical simulation of compaction bands. *Int. J. Numer. Anal. Meth. Geomech.* 35(2): 244–263.)

4.2.5 Summary

The present study involved the numerical simulation of the strain localization behavior of Noto diatomaceous mudstone, in particular the capture of compaction bands. By way of an experimental investigation into diatomaceous mudstone, the formation of compaction bands was put into evidence using PTV image analysis. In the experiments, it was observed that porous rock undergoes large volumetric compression under relatively high effective confining pressure. Strain localization occurs in compaction bands that are perpendicular or very slightly inclined deformation bands with respect to the direction with the most compressive stress.

Using an elasto-viscoplastic constitutive model that describes the structural degradation of materials, we numerically analyzed drained triaxial compression tests on diatomaceous mudstone. This was achieved by solving coupled soil deformation–water flow equations using finite elements formulated within an updated Lagrangian description for large deformations. Comparing the simulated results with the experiments, it was found that the elasto-viscoplastic model can very well reproduce the compaction band type of localized deformation mode and the stress–strain responses of diatomaceous mudstone.

4.3 NUMERICAL ANALYSIS OF DYNAMIC STRAIN LOCALIZATION OF SATURATED AND UNSATURATED SOILS

4.3.1 Introduction

Many researchers have numerically studied the static and dynamic strain localization problems of geomaterials using rate-independent constitutive models (e.g., Schrefler et al. 1995, 2006; Borja 2004; Loret and Prevost 1991; Schrefler and Gawin 1996; Ehlers and Volk 1998; Borja and Aydin 2004; Andrade and Borja 2006; Zhang and Schrefler 2000; Zhang et al. 2001). Viscoplasticity can be introduced as a general procedure to regularize the elastic-plastic porous solids (Loret and Prevost 1991). For geomaterials, in particular clay, the inelastic response is also inevitably rate dependent, and this rate dependency is essential to consider in their constitutive models. Using an elasto-viscoplastic constitutive model, Oka et al. (1994, 1995, 2000), and Oka and Kimoto (2005) studied the development of shear bands in water-saturated clay under quasi-static deformations. They used the Biot type of two-phase mixture theory in their formulation in order to describe the flow and deformation behavior of soil. They demonstrated that strain localization is closely linked to material instability, and that it can be effectively simulated through a finite element analysis using the elasto-viscoplastic model for both normally consolidated and overconsolidated clays. Oka et al. (2002) examined the effects of dilatancy and permeability on static strain localization in water-saturated elasto-viscoplastic soils. Shahbodagh et al. (2014) proposed a computational framework for the large deformation dynamic analysis of strain localization, and numerically studied the dynamic strain localization in water-saturated elasto-viscoplastic clay. Lazari et al. (2015) developed a local and non-local elasto-viscoplastic model for the strain localization analysis of multiphase geomaterials under quasi-static loading.

Oka et al. (2019) proposed a computational model for the dynamic analysis of strain localization in unsaturated soils. The contents of Section 4.3 are mainly based on the work by Oka et al. (2019). The u-p finite element formulation based on the multiphase mixture theory is derived in the finite deformation regime. The updated Lagrangian approach is employed along with the Jaumann rate of Cauchy's stress tensor for the weak form of the equation of motion. The cyclic elasto-viscoplastic model developed by Kimoto et al. (2015) (see Sections 2.2 and 2.4) for saturated clay is extended to include the effect of suction on the behavior of partially saturated clay (Oka et al. 2006), and is employed in the formulation. The structural degradation of the soil skeleton is considered in the constitutive model as the strain softening with respect to the viscoplastic strain. The mesh-size dependency is studied to provide stable and convergent solutions, and the inception and progression of strain localization in fully saturated and unsaturated clay under dynamic compressive loading are investigated.

4.3.2 Dynamic strain localization analysis of soil

4.3.2.1 *Discretization of governing equations and constitutive equations*

The discretization of the governing equations for the analysis is presented in Chapter 3. The cyclic elasto-viscoplastic model for partially saturated soils (Section 2.4) is used, which is an extension of the model in Section 2.2. Namely, the cyclic elasto-viscoplastic constitutive model proposed by Shahbodagh et al. (2014) and Kimoto et al. (2015) is extended to take the unsaturation effects into account. The extended elasto-viscoplastic model includes the effect of suction in a similar manner to that taken by Kato et al. (2009) and Oka and Kimoto (2012) for partially saturated sand. Consideration of the effect of suction on the constitutive model has been successfully applied to unsaturated sandy soil behaviors (e.g., Kato et al. 2009, 2014; Akaki et al. 2016; Higo et al. 2015). In addition, we use the soil-water characteristic curve by Van Gunuchten (1980) shown in Section 2.4.

4.3.2.2 *Finite element mesh, loading, and boundary conditions*

The strain localization of a homogeneous elasto-viscoplastic specimen is simulated for both the saturated and unsaturated cases under dynamic compressive loading. The finite element mesh and the boundary conditions for this problem are shown in Figure 4.11.

The simulation is performed using a three-dimensional mesh system under plane strain conditions, for which the deformation in the Y direction is constrained. The specimen is assumed to be 10 m in width and 20 m in height. A 10×20 mesh pattern (200 elements) is considered as the default mesh configuration in the analysis. For the finite element analysis, a 20-node hexahedron element is used for the displacement of the solid skeleton with a reduced Gaussian integration to eliminate shear locking and to reduce the appearance of a spurious hourglass mode. For the coupled formulation, the restriction imposed by the Babuska–Brezzi condition does not allow the use of equal-order interpolation for all field variables (Babuska 1973; Brezzi 1974). Hence, an 8-node hexahedron element is used for pore fluid pressures.

In this analysis, the displacement boundary conditions are adjusted so that symmetric conditions can be provided. The horizontal displacement at the corners of the specimen is fixed as a trigger for strain localization. All the boundaries are assumed to be impermeable to both water and gas, while pore fluids are allowed to flow within the specimen. The specimen is subjected to an axial compressive acceleration distributed uniformly on the upper surface. The applied acceleration rises from zero to a maximum value of 3.5 gal within 0.1 s, and then remains constant until 15.0 s, as illustrated in Figure 4.11. The applied acceleration results in an overall axial strain of about 20% within 15 s.

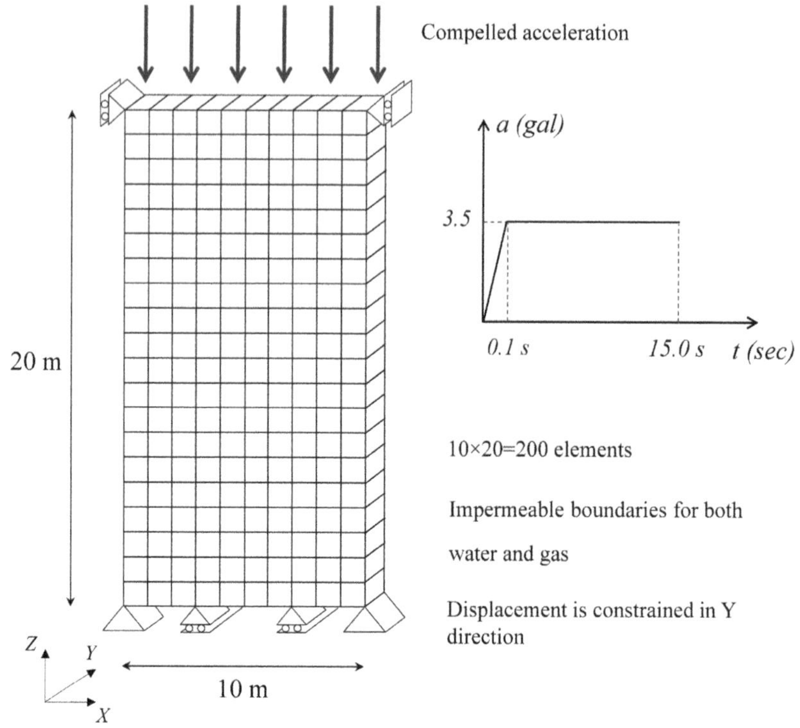

Figure 4.11 Finite element mesh, boundary conditions, and applied acceleration profile for dynamic strain localization analysis. (After Oka, F., Shahbodagh, B. and Kimoto, S. 2019. A computational model for dynamic strain localization in unsaturated elasto-viscoplastic soils. *Int. J. Numer. Anal. Meth. Geomech.* 43(1): 138–165.)

4.3.2.3 Material and numerical parameters

In this analysis, the material parameters of a clay sample from Torishima, Osaka, Japan, are used (as listed in Table 4.2). Some of the parameters, such as density, the swelling index, the initial void ratio, and the compression index, were determined by mechanical laboratory tests. For other parameters, a one-element simulation was carried out to reproduce the results of laboratory triaxial tests under boundary and initial conditions. According to the plastic limit of Torishima clay (PL = 30%), the soil-water characteristic curve parameters for low-plasticity clay, α and n', were chosen based on the study by Lu and Likos (2004). These curve-adjusting parameters and the parameters related to the hydraulic properties of the unsaturated soil are given as:

water permeability coefficient under saturated conditions (m/s) $k_s^W = 5.77 \times 10^{-10}$; gas permeability coefficient (m/s) $k_s^G = 1.00 \times 10^{-3}$; van Genuchten parameter (1/kPa) $\alpha = 0.033$; van Genuchten parameter

Table 4.2 Material parameters and initial conditions of Torishima clay

Initial void ratio	e_0	1.250
Density (t/m³)	ρ_{sat}	1.66
Compression index	λ	0.341
Swelling index	κ	0.019
Initial elastic shear modulus (kPa)	G_0	15040
Initial mean effective stress (kPa)*	σ'_{m0}	200
Stress ratio at compression*	M^*_m	1.24
Viscoplastic parameter (kPa)	σ'_{mk}	1.0
Viscoplastic parameter	m'	24.68
Viscoplastic parameter (1/s)	C_1	1.00×10^{-5}
Viscoplastic parameter (1/s)	C_2	3.83×10^{-6}
Softening parameter (kPa)	σ'_{mai}	200
Structural parameter (kPa)	σ'_{maf}	60
Structural parameter	β	3.6
Kinematic hardening parameter	B^*_0	100
Kinematic hardening parameter	B^*_1	40
Kinematic hardening parameter	C_f	10
Reference value of plastic strain (%)	$\gamma^{vp*}_{(n)r}$	1.25
Strain-dependent modulus parameter	A	10
Strain-dependent modulus parameter	R	0.4
Hardening parameter	A^*_2	5.9
Hardening parameter	B^*_2	1.8
Hardening parameter	B^*_3	0.1
Suction-dependent parameter	S_I	0.5
Suction-dependent parameter	s_d	0.25

Source: Data from Oka, F., Shahbodagh, B. and Kimoto, S. 2019. A computational model for dynamic strain localization in unsaturated elasto-viscoplastic soils. *Int. J. Numer. Anal. Meth. Geomech.* 43(1): 138–165.

Note: *Not a material parameter but the initial value of the skeleton stress.

$n' = 1.038$; minimum saturation $S_{r\min} = 0.0$; maximum saturation $S_{r\max} = 0.99$; shape parameter of water permeability $a = 3.0$; and shape parameter of gas permeability $b = 2.3$.

The methods for determining the numerical parameters are described as follows. In the time integration, the parameters of the Newmark-β method ($\beta = 0.3025$ and $\gamma = 0.6$) are used to satisfy the numerical stability (Oka and Kimoto 2012). It should be noted that the damping of the system is considered using Rayleigh damping which is proportional to the velocity. The initial stiffness-dependent type of Rayleigh damping, with a constant of $\alpha_1 = 0.0023$, is applied in the present study. Further details on the formulation of the weak form of the equation of motion are given in Shahbodagh (2011). For the time increment, the Courant–Friedrichs–Lewy condition is considered and Δt 0.001 s is adopted, confirming the stability and accuracy considerations (Oka and Kimoto 2012).

4.3.3 Numerical results of strain localization

In this section, the results of the numerical simulation are presented to illustrate the progress of the shear strain localization in both fully saturated and partially saturated clay. The results of the saturated and unsaturated analyses are compared at various strain levels.

The initial pore pressures considered for this analysis are for the unsaturated case: $P^W(\text{kPa}) = -1240$, $P^G(\text{kPa}) = 20$, $P^C(\text{kPa}) = 1260$, $Sr(\%) = 72$; and for the saturated case, $P^W(\text{kPa}) = 100$, $Sr(\%) = 100$.

Figure 4.12 shows the distributions of the accumulated viscoplastic shear strain, $\gamma^{vp} = \int (D'^{vp}_{ij} D'^{vp}_{ij})^{1/2} dt$ (D'^{vp}_{ij}: deviatoric viscoplastic stretching), at overall axial strains of 5%, 10%, 15%, and 20%. It is observed that shear bands start to develop from the trigger points at the corners of the specimen. With the increase in the overall axial strain, the bands continue to propagate toward the center of the opposing walls and eventually narrow to four distinct symmetric diagonal bands. Similar patterns of shear banding are observed in both saturated and unsaturated cases. The thickness of the localized zone is, however, narrower in the unsaturated case compared to the saturated case. In the unsaturated soil, a breakdown in the interfacial effects due to shearing particularly reduces strength and causes pronounced shear banding. This can be observed from the large level of accumulated viscoplastic shear strain, γ^{vp}, developed inside the shear bands in the unsaturated case. Larger accumulated shear strain in a material can be interpreted as more instability in that material. Following this point of view, the unsaturated clay is more unstable than the saturated clay.

For saturated clay, Oka et al. (1995) conducted an instability analysis and demonstrated that the preferred orientation of the shear bands is 45 degrees under a plane strain locally undrained condition, i.e., $k^W = 0$. In this example, the water permeability of the soil considered is very small, and the load is applied in a very short time duration. This makes the water drainage condition very similar to the locally undrained condition, resulting in the formation of shear bands with the same inclination angle as the one analytically obtained by Oka et al. (1995) for saturated clay. The inclination angle is smaller (almost 43 degrees) in the unsaturated case.

As shown by Oka et al. (1994), in undrained triaxial compression tests on fully saturated normally consolidated clay, the level of water content inside the shear bands reduces, i.e., contraction occurs inside the bands.

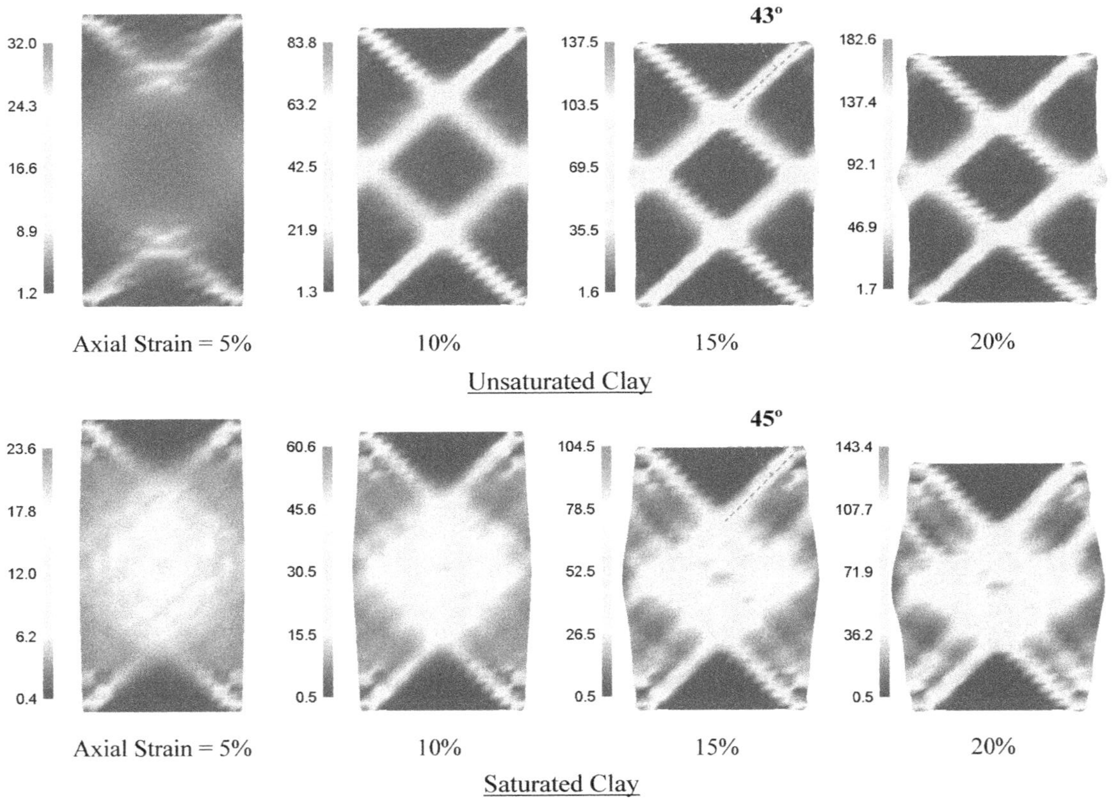

Figure 4.12 Distributions of accumulated viscoplastic shear strain for saturated and unsaturated clays (Legend unit: %). (After Oka, F., Shahbodagh, B. and Kimoto, S. 2019. A computational model for dynamic strain localization in unsaturated elasto-viscoplastic soils. *Int. J. Numer. Anal. Meth. Geomech.* 43(1): 138–165.)

This analysis shows that the volumetric strain, $\varepsilon_v = \int D_{kk}\, dt$, is almost zero throughout the whole specimen under transient loading since water does not have enough time to move through the pores due to both the very low water permeability of clay and the short duration of loading. According to the assumption of the incompressibility of water and soil particles, the volume change equal to zero in the saturated case is completely logical. In the unsaturated case, the volume contraction occurs in the whole specimen due to the compaction of the air voids. This is more intense inside the localized zone, where shearing results in a further reduction in volume. Such contractant shear banding is a primary mode of localized deformation in normally consolidated and lightly overconsolidated unsaturated soils under compression by Peric et al. (2014).

The distributions of pore water pressure are shown in Figure 4.13. In the unsaturated case, the negative pore water pressure indicates a meniscus with a suction force that behaves as a capillary force between the soil particles. In the partially saturated case, it is observed that pore water pressure starts to increase in the entire specimen. This development is strongly affected by the formation of shear bands, which makes the pore water pressure distribution inhomogeneous. It is seen that the water pressure increases dramatically inside the shear bands. This can be interpreted as the collapse of the water meniscus which is caused by shearing. In the saturated clay, the pore water pressure distributions have the same patterns as those in the unsaturated clay. Comparing the two cases, the pressure is more localized in the unsaturated case.

Figure 4.14 shows the distributions of the pore gas pressure in the unsaturated clay. Similar to the pore water pressure, the gas pressure increases with the progress of the overall axial strain. This increase is larger along the shear bands than in the rest of the specimen. The formation of the contractant shear bands in the unsaturated clay results in an increase in both the pore water and the pore gas pressures in the bands, producing both water and gas pressure gradient fields in the specimen. As a result, both fluids are squeezed out of the bands as the domain is compressed. Due to the very low water permeability of the unsaturated clay, a high water pressure gradient is observed in the specimen, while the distribution of the gas pressure is relatively uniform due to the high gas permeability of the soil. The magnitude of the increase in gas pressure is low for the present unsaturated simulated case with high suction. This is consistent with the experimental results by Kimoto et al. (2011) in which the gas pressure was lower for soil with higher suction while the experiments were performed under a lower strain rate than the present analysis.

As a result of the contraction inside the shear bands in normally consolidated or lightly overconsolidated clay subjected to triaxial compression loading, the degree of saturation increases and, consequently, the suction

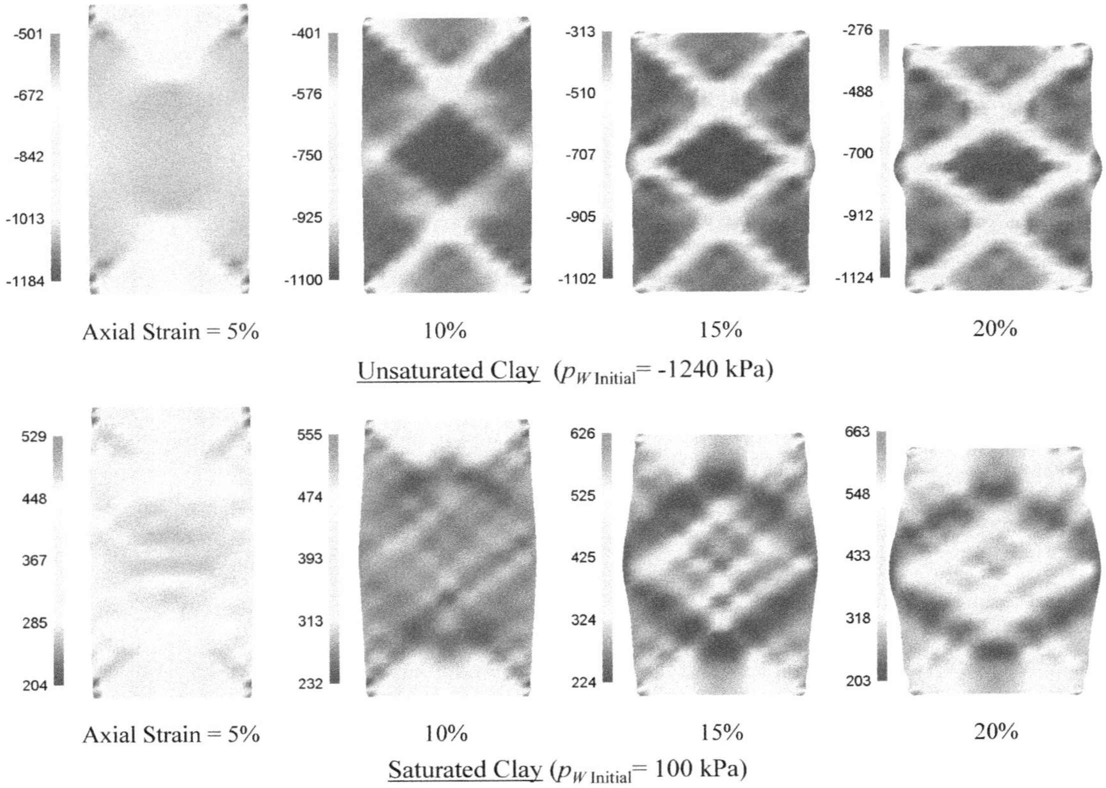

Figure 4.13 Distributions of pore water pressure for saturated and unsaturated clays (Unit: kPa). (After Oka, F., Shahbodagh, B. and Kimoto, S. 2019. A computational model for dynamic strain localization in unsaturated elasto-viscoplastic soils. *Int. J. Numer. Anal. Meth. Geomech.* 43(1): 138–165.)

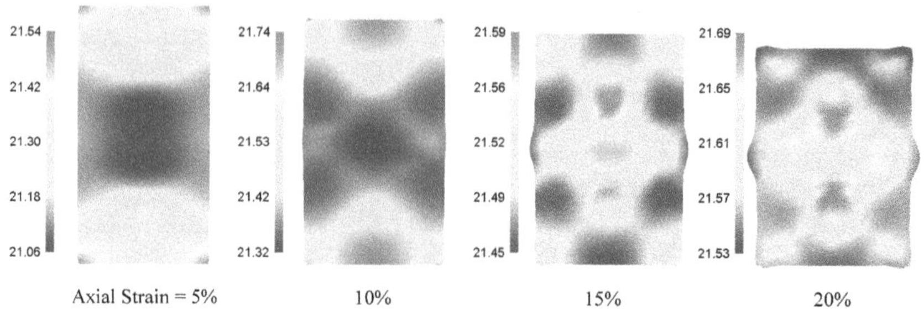

Figure 4.14 Distributions of pore gas pressure for unsaturated clay (Unit: kPa). (After Oka, F., Shahbodagh, B. and Kimoto, S. 2019. A computational model for dynamic strain localization in unsaturated elasto-viscoplastic soils. *Int. J. Numer. Anal. Meth. Geomech.* 43(1): 138–165.)

drops along the shear bands. This behavior can be observed in Figure 4.15. In this figure, the contours for the degree of saturation and for the suction are shown at different levels of overall axial strain.

The decrease in suction can be especially dangerous in loose soils where the suction force provides stability to the soil particles. It is important to note that even though suction generally improves the mechanical properties of soil, i.e., increasing the stiffness and shear strength, the hypothesis for saturated material, which is widely used in the analyses of geotechnical problems, is not at all on the side of safety since partially saturated soil can experience a different type of failure, i.e., shear banding, due to the drastic loss in soil strength associated with a collapse of suction causing a strong strain localization.

The stress–strain relations for both fully saturated and partially saturated clay are shown in Figure 4.16. It is observed that the peak deviator stress in the partially saturated case is higher than that in the saturated case. In general, partially saturated soil can show higher peak strength than air-dried or fully saturated soil due to the suction (capillary) force among the soil particles. Comparing the levels of overall axial strain at the peak strengths in the two cases, it is observed that the unsaturated clay reaches the peak strength at smaller values of overall strain than the saturated clay. After reaching the peak strength, the deviator stress falls abruptly as the overall strain increases. This strain-softening behavior, i.e., brittleness, controls the progressive failure and post-failure phenomena. From Figure 4.16, it is seen that the unsaturated clay exhibits a more brittle failure than the fully saturated clay. This is again due to the drastic loss of suction force inside the shear bands in the unsaturated clay.

Passing the peak stress level, the strength of the soil reaches a reasonably stable value after large deformations along the shear bands, i.e., residual strength. Higo et al. (2011) experimentally showed that the residual stress levels are almost the same for partially saturated sand, air-dried sand,

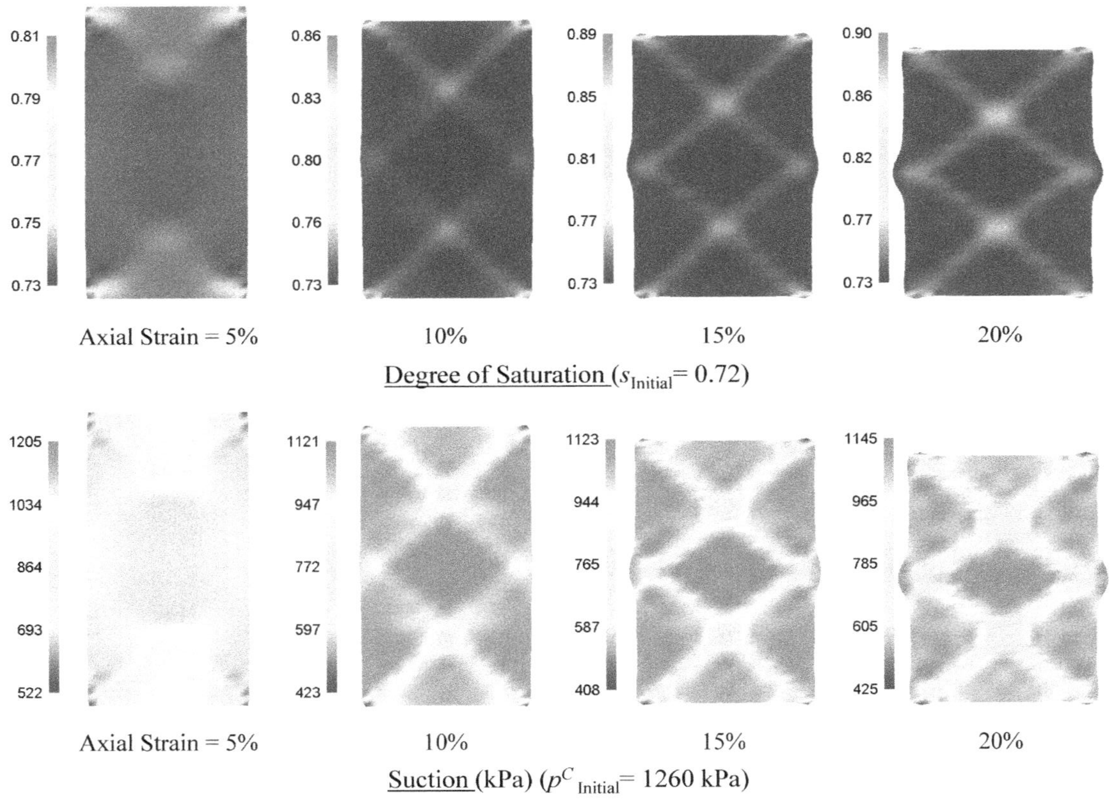

Figure 4.15 Distributions of the degree of saturation and suction for unsaturated clay. (After Oka, F., Shahbodagh, B. and Kimoto, S. 2019. A computationalmodelfordynamicstrainlocalizationinunsaturatedelasto-viscoplasticsoils.*Int.J.Numer.Anal.Meth.Geomech.*43(1):138–165.)

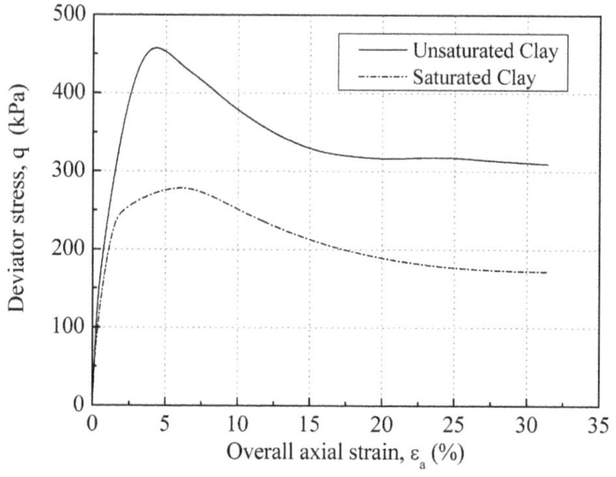

Figure 4.16 Stress–strain relations for saturated and unsaturated clays. (After Oka, F., Shahbodagh, B. and Kimoto, S. 2019. A computational model for dynamic strain localization in unsaturated elasto-viscoplastic soils. *Int. J. Numer. Anal. Meth. Geomech.* 43(1): 138–165.)

and fully saturated sand. Cunningham et al. (2003) also observed that in unsaturated silty clay under a condition of constant suction, the post-rupture failure envelope is independent of the suction applied to the sample. It should be noted that the post-rupture strength, which takes place just after the peak, must be clearly distinguished from the residual strength, which requires much larger relative displacements to develop. However, based on the present results, it is seen that both the post-rupture strength and the residual strength for the unsaturated clay are higher than those for the saturated clay. The reason may be that even though the suction drops dramatically along the shear bands, there is still a large value of suction force inside the shear bands, which is shown in Figure 4.15.

Burland (1990) demonstrated that in unconsolidated undrained triaxial tests on clay, localization essentially coincides with peak strength. The present results are consistent with those of Burland's study. These analyses indicate the distributions of the accumulated viscoplastic shear strains: just before the peak strength ($\varepsilon^a = 3.1\%$), at the peak strength ($\varepsilon^a = 4.2\%$), and immediately after the peak strength ($\varepsilon^a = 5.5\%$). It is observed that shear bands start to appear at the same time as the strength reaches its peak value.

Cunningham et al. (2003) observed that unsaturated silty clay contracted during triaxial tests, but that once the peak strength had been reached, further changes in volume ceased abruptly. The same behavior is seen in the volumetric strain–overall axial strain relations obtained in the present analysis. This indicates that after the peak strength, further straining results from rigid blocks sliding along the shear bands and not from changes in volume.

Many researchers have discussed that the material rate dependence substantially eliminates the pathological mesh sensitivity under quasi-static and/or dynamic loading conditions (e.g., see Loret and Prevost 1991). By using the viscoplastic model for the strain localization analysis of saturated clay under quasi-static loading, Oka et al. (1995) reported a slightly greater softening response with the finer mesh configurations. In later work by Oka et al. (2002), the effects of the mesh size on the stress–strain relations were found to be insignificant, whereas all the cases with different mesh configurations showed identical shear band formation. Oka et al. (2019) confirmed that the numerical results converge with an increase in the number of finite element meshes.

4.3.4 Summary

The dynamic strain localization phenomenon in saturated and unsaturated clay was discussed within the finite deformation regime using the multiphase mixture theory and the cyclic elasto-viscoplastic constitutive model. The non-linear kinematic hardening rule, softening due to the structural degradation of soil particles, and the effect of suction on the behavior of partially saturated soil were taken into account in the model. The updated Lagrangian approach was employed along with the Jaumann rate of Cauchy's stress tensor for the weak form of the equation of motion. The inception and progression of localization in fully saturated and unsaturated clay under dynamic compressive loading were investigated. From the results, the following conclusions can be made:

1. In both saturated and unsaturated cases, similar patterns of shear bands were observed. However, in the case of unsaturated clay, the strain localized more prominently, and much clearer shear bands were formed.
2. Comparing the stress–strain relationships, it was found that the peak strength was higher in the unsaturated case than in the saturated case due to the suction force. It was also seen that the unsaturated soil reached the peak stress level at a smaller axial strain than the saturated soil.
3. In the unsaturated case, more brittleness, i.e., strain softening, was observed due to the collapse of the suction force caused by shearing. In addition, the rate of shear strength reduction, from peak to residual, was higher in the unsaturated case. This rate controls the post-failure deformation in geo-structures, where the higher rate of strength reduction can result in more devastating consequences following failure. It demonstrates that even though suction improves the mechanical properties of soil, i.e., increasing the stiffness and shear strength, the hypothesis of full saturation, which is widely used in the analysis of geotechnical problems, is not at all on the side of safety since the drastic loss of strength associated with the collapse of suction causes a strong strain localization.

APPENDIX A4.1 THREE-DIMENSIONAL MOHR STRESS CIRCLE (MOHR 1882; MALVERN 1969)

Let a stress vector acting on a plane, t_i, as

$$t_i = \sigma_{ji} n_j \tag{A4.1}$$

where $n_i\,(i = 1,2,3)$ are the components of the unit normal vector to the plane.

As shown in Figure A4.1, the normal component is

$$\sigma_n = \mathbf{t} \cdot \mathbf{n} = \sigma_{ji} n_j n_i \tag{A4.2}$$

and the tangential component is

$$\tau_n = t_i t_i - \sigma_n^2 \tag{A4.3}$$

The principal components are given by

$$t_1 = \sigma_1 n_1,\ t_2 = \sigma_2 n_2,\ t_3 = \sigma_3 n_3,\ \ \sigma_1 \geq \sigma_2 \geq \sigma_3 \tag{A4.4}$$

Then,

$$\sigma_n = \sigma_1 n_1^2 + \sigma_2 n_2^2 + \sigma_3 n_3^2 \tag{A4.5}$$

$$t_i t_i = \tau_n^2 + \sigma_n^2 = \sigma_1^2 n_1^2 + \sigma_2^2 n_2^2 + \sigma_3^2 n_3^2 \tag{A4.6}$$

where

$$n_1^2 + n_2^2 + n_3^2 = 1 \tag{A4.7}$$

Mohr's circles for three-dimensional stress conditions are shown in Figure A4.2.

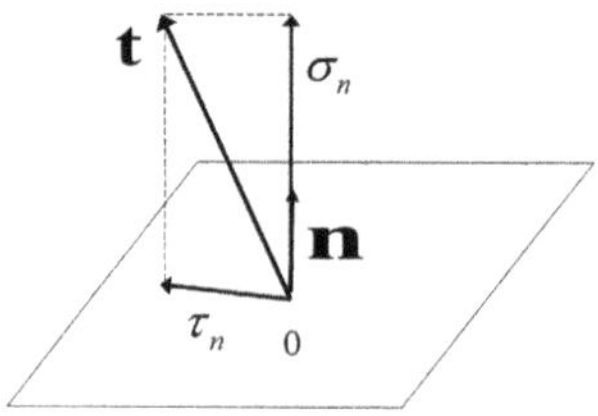

Figure A4.1 Normal and tangential components of stress vector acting on the plane.

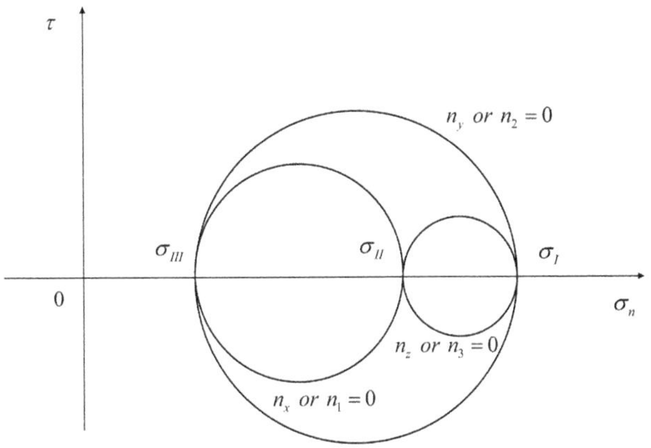

Figure A4.2 Mohr's circles for three-dimensional stress condition.

From Equations A4.5–A4.7, we have

$$n_1^2 = \frac{\sigma_n^2 + \tau_n^2 - \sigma_n(\sigma_2 + \sigma_3) + \sigma_2\sigma_3}{(\sigma_2 - \sigma_1)(\sigma_3 - \sigma_1)}$$

$$n_2^2 = \frac{\sigma_n^2 + \tau_n^2 - \sigma_n(\sigma_3 + \sigma_1) + \sigma_3\sigma_1}{(\sigma_3 - \sigma_2)(\sigma_1 - \sigma_2)} \tag{A4.8}$$

$$n_3^2 = \frac{\sigma_n^2 + \tau_n^2 - \sigma_n(\sigma_1 + \sigma_2) + \sigma_1\sigma_2}{(\sigma_1 - \sigma_3)(\sigma_2 - \sigma_3)}$$

Rewriting them, we have

$$\left[\sigma_n - \frac{1}{2}(\sigma_2 + \sigma_3)\right]^2 + \tau_n^2 = R_1^2$$

$$\left[\sigma_n - \frac{1}{2}(\sigma_3 + \sigma_1)\right]^2 + \tau_n^2 = R_2^2 \tag{A4.9}$$

$$\left[\sigma_n - \frac{1}{2}(\sigma_1 + \sigma_2)\right]^2 + \tau_n^2 = R_3^2$$

$$R_1^2 = \frac{1}{4}(\sigma_2 + \sigma_3)^2 + (\sigma_1 - \sigma_2)(\sigma_1 - \sigma_3)n_1^2 - \sigma_2\sigma_3$$

$$R_2^2 = \frac{1}{4}(\sigma_3 + \sigma_1)^2 - (\sigma_1 - \sigma_2)(\sigma_2 - \sigma_3)n_2^2 - \sigma_3\sigma_1 \tag{A4.10}$$

$$R_3^2 = \frac{1}{4}(\sigma_1 + \sigma_2)^2 + (\sigma_1 - \sigma_3)(\sigma_2 - \sigma_3)n_3^2 - \sigma_1\sigma_2$$

Hence, the possible range of R_i is

$$\frac{1}{2}(\sigma_2-\sigma_3)\le R_1\le\left[\sigma_1-\frac{1}{2}(\sigma_2+\sigma_3)\right]$$
$$\left[\sigma_2-\frac{1}{2}(\sigma_1+\sigma_3)\right]\le R_2\le\frac{1}{2}(\sigma_1-\sigma_3) \tag{A4.11}$$
$$\frac{1}{2}(\sigma_1-\sigma_2)\le R_3\le\left[\sigma_3-\frac{1}{2}(\sigma_1+\sigma_2)\right]$$

REFERENCES

Adachi, T. and Oka, F. 1982. Constitutive equations for normally consolidated clay based on elasto-viscoplasticity. *Soils Found.* 4(4): 57–70.

Akaki, T., Kimoto, S. and Oka, F. 2016. Chemo-thermo-mechanically coupled seismic analysis of methane hydrate-bearing sediments during a predicted Nankai through earthquake. *Int. J. Numer. Anal. Meth. Geomech.* 40(16): 2207–2237.

Andrade, J.E. and Borja, R.I. 2006. Capturing strain localization in dense sands with random density. *Int. J. Numer. Meth. Eng.* 67(11): 1531–1564.

Arroyo, M., Castellanza, R. and Nova, R. 2005. Compaction bands and oedometric testing in cemented soils. *Soils Found.* 45(2): 181–194.

Babuška, I. 1973. The finite element method with Lagrangian multipliers. *Numer. Math.* 20(3): 179–192.

Borja, R.I. 2004. Computational modeling of deformation bands in granular media, II: Numerical simulations. *Com. Meth. Appl. Mech. Eng.* 193: 2699–2718.

Borja, R.I. and Aydin, A. 2004. Computational modeling of deformation bands in granular media, I: Geological and mathematical framework. *Com. Meth. Appl. Mech. Engng.* 193: 2667–2698.

Brezzi, F. 1974. On the existence, uniqueness and approximation of saddle-point problems arising from Lagrangian multipliers. *RAIRD* 8(R2): 129–151.

Burland, J.B. 1990. On the compressibility and shear strength of natural clays. *Geotechnique* 40(3): 329–378.

Cunningham, M.R., Ridley, A.M., Dineen, K. and Burland, J.B. 2003. The mechanical behaviour of a reconstituted unsaturated silty clay. *Geotechnique* 53(2): 183–194.

Ehlers, W. and Volk, W. 1998. On theoretical and numerical methods in the theory of porous media based on polar and non-polar elasto-plastic solid materials. *Int. J. Solids Struct.* 34–35(34–35): 4597–4617.

Garbrecht, D. 1973. *Kornbruch in Irregularen Haufwerken aus Elastischsproden Kugeln*, Dissertation. University Karlsruhe, Veroffentlichungen IBF, Heft Nr. 56.

Higo, Y., Oka, F., Kodaka, T. and Kimoto, S. 2006. Three-dimensional strain localization of water-saturated clay and numerical simulation using an elasto-viscoplastic model. *Philos. Mag. Struct. Prop. Condens. Matter* 86(21–22): 3205–3240.

Higo, Y., Oka, F., Kimoto, S., Sanagawa, T. and Matsushima, Y. 2011. Observation of microstructural changes and strain localization of unsaturated sands using microfocus X-ray CT. In: *Proceedings of 9th International Workshop on Bifurcation and Degradation in Geomaterials*, ed. S. Bonelli, C. Dascalu and F. Niot, 37–43. Springer, Berlin.

Higo, Y., Lee, C.-W., Doi, T., Kinugawa, T., Kimura, M., Sayuri Kimoto, S. and Oka, F. 2015. Study of dynamic stability of unsaturated embankments with different water contents by centrifugal model tests. *Soils Found.* 55(1): 112–126.

Issen, K.A. and Rudnicki, J.W. 2000. Conditions for compaction bands in porous rock. *J. Geophys. Res.* 105(B9): 529–536.

Issen, K.A. and Rudnicki, J.W. 2001. Theory of compaction bands in porous rock. *Phys. Chem. Earth* 26(1–2): 95–100.

Issen, K.A. 2002. The influence of constitutive models on localization conditions for porous rock. *Eng. Fract. Mech.* 69(17): 1891–1906.

Johnson, G.C. and Bammann, D. 1984. A discussion of stress rates in finite deformation problems. *Int. J. Solids Struct.* 20(8): 725–737.

Kato, R., Oka, F., Kimoto, S., Kodaka, T. and Sunami, S. 2009. A method of seepage-deformation coupled analysis of unsaturated ground and its application to river embankment. *Trans. JSCE* C65(1): 226–240 (in Japanese).

Kato, R., Oka, F. and Kimoto, S. 2014. A numerical simulation of seismic behavior of highway embankments considering seepage flow. In: *Computer Methods and Recent Advances in Geomechanics*, ed. F. Oka, A. Murakami, R. Uzuoka and S. Kimoto (*Proceedings of the 14th International Conference of International Association for Computer Methods and Recent Advances in Geomechanics*), 755–760. CRC/Taylor & Francis, Boca Raton, FL.

Kimoto, S. 2002. *Constitutive Models for Geomaterials Considering Structural Changes and Anisotropy*, PhD thesis. Graduate School of Engineering, Kyoto University, Japan.

Kimoto, S., Oka, F. and Higo, Y. 2004. Strain localization analysis of elasto-viscoplastic soil considering structural degradation. *Int. J. Comp. Meth. App. Mech. Engng.* 193(27–29): 2845–2866.

Kimoto, S. and Oka, F. 2005. An elasto-viscoplastic model for clay considering destructuralization and consolidation analysis of unstable behavior. *Soils Found.* 45(2): 29–42.

Kimoto, S., Oka, F., Fukutani, J., Yabuki, T. and Nakashima, K. 2011. Monotonic and cyclic behavior of unsaturated sandy soil under drained and fully undrained conditions. *Soils Found.* 51(4): 663–681.

Kimoto, S., Shahbodagh, B., Mirjalili, M. and Oka, F. 2015. Cyclic elastoviscoplastic constitutive model for clay considering nonlinear kinematic hardening rules and structural degradation. *Int. J. Geomech.* 15(5): A4014005-1–A4014014.

Kodaka, T., Higo, Y. and Takyu, T. 2001. Deformation and failure characteristics of rectangular clay specimens under three-dimensional conditions. In: *Proceedings of the 15th International Conference on Soil Mechanics and Geotechnical Engineering*, Istanbul, 1: 167–170. CRC Press, Lisse.

Kodaka, T., Oka, F., Kitahara, H., Ohta, H. and Otani, J. 2006. Observation of compaction bands under triaxial conditions for diatomaceous mudstone. In: *Proceedings of the International Symposium on Geomechanics and Geotechnics of Particulate Media*, ed. M. Hyodo, H. Murata and Y. Nakata, 69–75. CRC Press/Taylor & Francis, Leiden.

Lazari, M., Sanavia, L. and Schrefler, B.A. 2015. Local and non-local elasto-viscoplasticity in strain localization analysis of multiphase geomaterials. *Int. J. Numer. Anal. Meth. Geomech.* 39(14): 1570–1592.

Loret, B. and Prevost, J.H. 1991. Dynamic strain localization in fluid-saturated porous media. *J. Engng. Mech. ASCE* 117(4): 907–922.

Lu, N. and Likos, W.J. 2004. *Unsaturated Soil Mechanics*. John Wiley & Sons, New York.

Maekawa, H. and Miyakita, K. 1983. Mechanical properties of diatomaceous soft rock. *J. JSCE* 334: 135–143 (in Japanese).

Maekawa, H. 1992. *Study on the Mechanical Characteristics of Soft Mud Rock and Its Application*, PhD thesis. Kyoto University (in Japanese).

Malvern, L.E. 1969. *Introduction to the Mechanics of a Continuous Medium*. Prentice Hall, Englewood Cliffs, NJ.

Mohr, O. 1882. Über die Darstellung des Spannungazustandes und des Deformation-Zustandes eines Körper-elements. *Zivilingenieur*: 113.

Mollema, P.N. and Antonellini, M.A. 1996. Compaction bands: A structural analog for anti-mode I cracks in aeolian sandstone. *Tectonophysics* 267(1–4): 209–228.

Oka, F. 1981. Prediction of time-dependent behaviour of clay. In: *Proceedings of 10th International Conference on Soil Mechanics and Foundation Engineering*, Stockholm, 1: 215–218. Balkema, Rotterdam.

Oka, F., Adachi, T. and Yashima, A. 1994. Instability of an elasto-viscoplastic constitutive model for clay and strain localization. *Mech. Mater.* 18(2): 119–129.

Oka, F., Adachi, T. and Yashima, A. 1995. A strain localization analysis of clay using a strain softening viscoplastic model. *Int. J. Plast.* 11(5): 523–545.

Oka, F., Yashima, A., Sawada, K. and Aifantis, E.C. 2000. Instability of gradient-dependent elasto-viscoplastic model for clay and strain localization analysis. *Com. Meth. Appl. Mech. Engng.* 183: 67–86.

Oka, F., Higo, Y. and Kimoto, S. 2002. Effect of dilatancy on the strain localization of water-saturated elasto-viscoplastic soil. *Int. J. Solids Struct.* 39(13–14): 3625–3647.

Oka, F. and Kimoto, S. 2005. An elasto-viscoplastic model for clay considering destructuralization and prediction of compaction bands. In: *Geomechanics, Testing, Modeling and Simulation*, *Geotechnical Special Publication*, ed. J.A. Yamamuro and J. Koseki, 143, 71–80. ASCE, Reston, VI.

Oka, F., Kodaka, T., Kimoto, S., Kim, Y.-S. and Yamasaki, N. 2006. An elasto-viscoplastic model and multiphase coupled FE analysis for unsaturated soil. In: *Proceedings* of *Unsaturated Soils Conference*, Phoenix, Geotechnical Special Publication, ASCE 147(2): 2039–2050.

Oka, F., Kimoto, S., Higo, Y., Ohta, H., Sanagawa, T. and Kodaka, T. 2011. An elasto-viscoplastic model for diatomaceous mudstone and a numerical simulation of compaction bands. *Int. J. Numer. Anal. Meth. Geomech.* 35(2): 244–263.

Oka, F. and Kimoto, S. 2012. *Computational Modeling of Multiphase Geomaterials*. CRC Press/Taylor & Francis, Boca Raton, FL.

Oka, F., Shahbodagh, B. and Kimoto, S. 2019. A computational model for dynamic strain localization in unsaturated elasto-viscoplastic soils. *Int. J. Numer. Anal. Meth. Geomech.* 43(1): 138–165.

Olsson, W.A. 1999. Theoretical and experimental investigation of compaction bands in porous rock. *J. Geophys. Res.* 104(B4): 7219–7228.
Papamichos, E., Vardoulakis, I. and Quadfel, H. 1993. Permeability reduction due to grain crushing around a perforation. *Int. J. Rock Mech. Min. Sci. Geomech. Abstr.* 30: 1223–1229.
Perić, D., Zhao, G. and Khalili, N. 2014. Strain localization in unsaturated elastic-plastic materials subjected to plane strain compression. *J. Eng. Mech.* 140(7): 04014050.
Perrin, G. and Leblond, J.B. 1993. Rudnicki and Rice's analysis of strain localization revisited. *J. Appl. Mech. ASME* 60(4): 842–846.
Perzyna, P. 1963. The constitutive equations for work-hardening and rate sensitive plastic materials. *Proc. Vibr. Probl.* 4(3): 281–290.
Rudnicki, W. and Rice, R. 1975. Conditions for the localization of deformation in pressure-sensitive dilatant materials. *J. Mech. Phys. Solids* 23(6): 371–394.
Schrefler, B.A., Majorana, C.E. and Sanavia, L. 1995. Shear band localization in saturated porous media. *Arch. Mech.* 47(3): 577–599.
Schrefler, B.A. and Gawin, D. 1996. The effective stress principle: Incremental or finite form? *Int. J. Numer. Anal. Meth. Geomech.* 20(11): 785–814.
Schrefler, B.A., Zhang, H.W. and Sanavia, L. 2006. Interaction between different internal length scales in strain localization analysis of fully and partially saturated porous media: The 1-D case. *Int. J. Numer. Anal. Meth. Geomech.* 30(1): 45–70.
Shahbodagh, B. 2011. *Large Deformation Dynamic Analysis Method for Partially Saturated Elasto-Viscoplastic Soils*, PhD thesis. Kyoto University.
Shahbodagh, B., Mirjalili, M., Kimoto, S. and Oka, F. 2014. Dynamic analysis of strain localization in water-saturated clay using a cyclic elasto-viscoplastic model. *Int. J. Numer. Anal. Meth. Geomech.* 38(8): 771–793.
van Genuchten, M.Th. 1980. A closed-form equation for predicting the hydraulic conductivity of unsaturated soils. *Soil Sci. Soc. Am. J.* 44(5): 892–898.
Vardoulakis, I. and Sulem, J. 1995. *Bifurcation Analysis in Geomechanics*. Blackie Academic & Professional, an imprint of Chapman & Hall, Glasgow.
Wong, T.-f., David, C. and Zhu, W. 1997. The transition from brittle faulting to cataclastic flow in porous sandstones mechanical deformation. *J. Geophys. Res.* 102(B2): 3009–3025.
Zhang, H.W. and Schrefler, B.A. 2000. Gradient-dependent plasticity model and dynamic strain localisation analysis partially saturated porous media: One dimensional model. *Eur. J. Mech. A Solids* 19(3): 503–524.
Zhang, H.W., Sanavia, L. and Schrefler, B.A. 2001. Numerical analysis of dynamic strain localisation in initially water saturated dense sand with a modified generalized plasticity model. *Comput. Struct.* 79(4): 441–459.

Chapter 5

Instability analysis of water infiltration into an unsaturated elasto-viscoplastic material

5.1 INTRODUCTION

The behavior of unsaturated soil subjected to water infiltration plays an important role in geomechanics because the failure of natural slopes, embankments, and artificial soil structures is most often due to water infiltration. It is known that the failure of soil structures can be triggered by a wetting process from an unsaturated stage resulting from an increase in moisture content and a decrease in suction.

The stability of a water infiltration problem in an unsaturated soil and its numerical simulation are presented in this chapter. We begin with a linear stability analysis of elasto-viscoplastic unsaturated material in order to investigate which variables have a significant effect on the onset of the instability subjected to water infiltration. Then, the parametric study of the numerical simulation of the problem with special reference to the suction–saturation relation and the numerical simulation results are shown.

In this section, the material parameters and the conditions that contribute to the growth rate of the fluctuation are examined via a linear instability analysis. From this analysis, it is found that the onset of the instability of the material depends on the specific moisture capacity (slope of the soil-water characteristic curve), the suction, and the hardening parameter. Then, the results of the numerical simulations of the one-dimensional water infiltration problem are presented to show the effect of the specific moisture capacity and the initial suction on the development of volumetric strain.

5.2 ONE-DIMENSIONAL INSTABILITY ANALYSIS OF WATER INFILTRATION INTO UNSATURATED VISCOPLASTIC POROUS MEDIA

Many researchers have studied the behavior of unsaturated soils (e.g., Alonso et al. 2003; Cunningham et al. 2003; Ehlers et al. 2004; Khalili et al. 2004; Kimoto et al. 2007; Oka et al. 2010). Buscarnera and Nova (2009, 2011) presented a theoretical approach to cope with possible soil

DOI: 10.1201/9781003200031-5

instabilities in unsaturated conditions. Their results show that unsaturated soil specimens are prone to instability when they are subjected to water infiltration.

In order to study the effects of the constitutive parameters on the deformation of an unsaturated material, an instability analysis and numerical simulations were performed by Garcia et al. (2010). Following the study by Garcia et al. (2010), a linear instability analysis conducted on an unsaturated material in a viscoplastic state is presented in this section.

Due to the high non-linearity of the hydraulic and the constitutive equations involved in the unsaturated coupled seepage-deformation analysis method, only simplified one-dimensional analytical solutions for the infiltration problem can be obtained for elastic materials (e.g., Wu and Zhang 2009). Therefore, the linear instability analysis is applied to a one-dimensional viscoplastic unsaturated material based on the multiphase coupled seepage-deformation framework presented in Chapter 3. The results of the instability analysis obtained here are discussed with the numerical simulation results in Section 5.3.

5.2.1 Governing equations

Following Chapter 1, one-dimensional total and partial stress values can be rewritten as

$$\sigma^W = -n^W P^W \tag{5.1}$$

$$\sigma^G = -n^G P^G \tag{5.2}$$

$$\sigma^S = \sigma' - n^S P^F \tag{5.3}$$

$$\sigma = \sigma^S + \sigma^W + \sigma^G \tag{5.4}$$

$$\sigma' = \sigma + P^F \tag{5.5}$$

The average fluid pressure acting on the solid phase is given by Equation 1.67 as

$$P^F = S_r P^W + (1 - S_r) P^G \tag{5.6}$$

where S_r is the saturation, which is a function of suction P^C.

A simplified viscoplastic constitutive model is used in this analysis. The stress–strain relation can be written as

$$\sigma' = H\varepsilon + \mu\dot{\varepsilon} \tag{5.7}$$

where ε is the strain, the superimposed dot denotes the differentiation with respect to time t; H is the strain hardening–softening parameter, which is a function of suction P^C; and μ is the viscoplastic parameter.

Suction P^C is included in the constitutive model and is defined by

$$P^C = P^G - P^W \tag{5.8}$$

The one-dimensional equilibrium equation for the whole mixture can be written from Equation 3.20 as follows:

$$\frac{\partial \sigma}{\partial x} + \rho \bar{F} = \frac{\partial \sigma'}{\partial x} - \frac{\partial P^F}{\partial x} + \rho \bar{F} = 0 \tag{5.9}$$

Darcy's laws for the flow of water and air can be obtained from Equation 3.17, namely,

$$w^W = -\frac{k^W}{\gamma_W}\left\{\frac{\partial P^W}{\partial x} - \rho_W \bar{F}\right\} \tag{5.10}$$

$$w^G = -\frac{k^G}{\gamma_G}\left\{\frac{\partial P^G}{\partial x} - \rho_G \bar{F}\right\} \tag{5.11}$$

where w^W and w^G are the relative velocities of water and gas to the solid, respectively.

Considering Equations 3.8 and 3.11, the continuity equations for the water and air phases in the one-dimensional analysis can be written as

$$S_r \dot{\varepsilon} + n\dot{S}_r = -\frac{\partial w^W}{\partial x} \tag{5.12}$$

$$(1 - S_r)\dot{\varepsilon} - n\dot{S}_r + n(1 - S_r)\frac{\dot{\rho}_G}{\rho_G} = -\frac{\partial w^G}{\partial x} \tag{5.13}$$

5.2.2 Perturbed governing equations

In the following, the perturbation of the pore water pressure, P^W, the pore air pressure, P^G, and strain, ε, in one-dimensional form is considered for the governing equations.

The perturbation of the equilibrium equation (Equation 5.9) is given by

$$\frac{\partial \tilde{\sigma}}{\partial x} = \frac{\partial \tilde{\sigma}'}{\partial x} - \frac{\partial \tilde{P}^F}{\partial x} = 0 \tag{5.14}$$

where the perturbed variables are indicated by a tilde. The perturbation of the skeleton stress σ' can be written from Equation 5.7 as

$$\tilde{\sigma}' = \tilde{H}\varepsilon + H\tilde{\varepsilon} + \mu\tilde{\dot{\varepsilon}} \tag{5.15}$$

In Equation 5.7, the strain hardening–softening parameter H is a function of the suction; hence, the perturbation of H is given as

$$\tilde{H} = \frac{\partial H}{\partial P^C}\tilde{P}^C = A\tilde{P}^C \tag{5.16}$$

where A ($=\partial H/\partial P^C$) indicates the slope of curve H–P^C.

The perturbation of the terms in periodic form for Equations 5.8 and 5.15 can be written as

$$\tilde{P}^C = \tilde{P}^G - \tilde{P}^W \tag{5.17}$$

$$\tilde{P}^G = P^{G^*}\exp(iqx + \omega t) \tag{5.18}$$

$$\tilde{P}^W = P^{W^*}\exp(iqx + \omega t) \tag{5.19}$$

$$\tilde{\varepsilon} = \varepsilon^*\exp(iqx + \omega t) \tag{5.20}$$

$$\tilde{\dot{\varepsilon}} = \omega\varepsilon^*\exp(iqx + \omega t) \tag{5.21}$$

where q is the wave number ($=2\pi/l$, l: wave length); ω is the growth rate of the fluctuation; and superscript * indicates the amplitude of each variable.

Using Equations 5.17–5.21 in Equation 5.15, the space differentiation of the perturbed skeleton stress is

$$\frac{\partial\tilde{\sigma}'}{\partial x} = (A\varepsilon P^{G^*} - A\varepsilon P^{W^*} + H\varepsilon^* + \mu\omega\varepsilon^*)iq\exp(iqx + \omega t) \tag{5.22}$$

Similarly, by means of Equation 5.6, the perturbation of the average pore pressure, P^F, can be written as

$$\tilde{P}^F = \tilde{S}_r P^W + S_r\tilde{P}^W + (1 - S_r)\tilde{P}^G - \tilde{S}_r P^G \tag{5.23}$$

where the degree of saturation, S_r, is a function of the suction. Then, the perturbation of the degree of saturation is given as

$$\tilde{S}_r = \frac{\partial S_r}{\partial P^C}\tilde{P}^C = B\tilde{P}^C \tag{5.24}$$

where B ($=\partial S_r/\partial P^C$) indicates the slope of curve S_r–P^C; B is called the specific moisture capacity.

Using Equations 5.17–5.19 and Equations 5.23 and 5.24, the gradient of the perturbed average pore pressure is

$$\frac{\partial \tilde{P}^F}{\partial x} = \left\{(-BP^W + S_r + BP^G)P^{W*} + (BP^W + 1 - s - BP^G)P^{G*}\right\} iq \exp(iqx + \omega t) \quad (5.25)$$

Substituting Equations 5.22 and 5.25 into Equation 5.14 and rearranging the terms, we obtain

$$-(A\varepsilon + BP^C + S_r)P^{W*} + (A\varepsilon + BP^C + S_r - 1)P^{G*} + (H + \mu\omega)\varepsilon^* = 0 \quad (5.26)$$

The perturbation of the continuity equation for the water phase (Equation 5.12) is given by

$$\tilde{S}_r\dot{\varepsilon} + S_r\tilde{\dot{\varepsilon}} + n\tilde{\dot{S}}_r = -\frac{\partial \tilde{w}^W}{\partial x} \quad (5.27)$$

where $\dot{\varepsilon}$ is the current strain rate.

The perturbation of the rate of the degree of saturation is

$$\tilde{\dot{S}}_r = \omega B\tilde{P}^C \quad (5.28)$$

The perturbation of the spatial differentiation of the Darcy equation (Equation 5.10) for water can be written as

$$\frac{\partial \tilde{w}^W}{\partial x} = -\frac{\partial}{\partial x}\left[\frac{k^W}{\gamma_W}\left\{\frac{\partial \tilde{P}^W}{\partial x} - \rho_W\bar{F}\right\}\right] = -\frac{k^W}{\gamma_W}\frac{\partial^2 \tilde{P}^W}{\partial x^2} \quad (5.29)$$

Substituting Equations 5.17–5.19, 5.21, 5.24, 5.28, and 5.29 into Equation 5.27 and rearranging the terms, we obtain

$$-\left(B\dot{\varepsilon} + n\omega B - q^2\frac{k^W}{\gamma_W}\right)P^{W*} + (B\dot{\varepsilon} + n\omega B)P^{G*} + S_r\omega\varepsilon^* = 0 \quad (5.30)$$

For the sake of simplicity, it is considered that the time rate of the air density is equal to zero in the present state; as a result, the perturbation of the continuity equation for the air phase (Equation 5.13) is given by

$$(1 - S_r)\tilde{\dot{\varepsilon}} - \tilde{S}_r\dot{\varepsilon} - n\tilde{\dot{S}}_r = -\frac{\partial \tilde{w}^G}{\partial x} \quad (5.31)$$

The perturbation of the spatial differentiation of the Darcy-type equation (Equation 5.11) for air can be written as

$$\frac{\partial \tilde{w}^G}{\partial x} = \frac{\partial}{\partial x}\left[-\frac{k^G}{\gamma_G}\left\{ \frac{\partial \tilde{P}^G}{\partial x} \right\} - \rho_G \bar{F} \right] = -\frac{k^G}{\gamma_G}\frac{\partial^2 \tilde{P}^G}{\partial x^2} \tag{5.32}$$

Substituting Equations 5.17–5.19, 5.21, 5.24, 5.28, and 5.32 into Equation 5.31 and rearranging the terms, we obtain

$$(B\dot{\varepsilon} + n\omega B)P^{W^*} - \left(B\dot{\varepsilon} + n\omega B - q^2 \frac{k^G}{\gamma_G} \right) P^{G^*} + (1 - S_r)\omega\varepsilon^* = 0 \tag{5.33}$$

We can rewrite the perturbation of the equilibrium and the continuity equations (Equation 5.26, 5.30, and 5.33) in matrix form as

$$\begin{bmatrix} -(A\varepsilon + BP^C + S_r) & (A\varepsilon + BP^C + S_r - 1) & (H + \mu\omega) \\ -\left(B\dot{\varepsilon} + Bn\omega - q^2 \dfrac{k^W}{\gamma_W} \right) & (B\dot{\varepsilon} + Bn\omega) & (S_r\omega) \\ (B\dot{\varepsilon} + Bn\omega) & -\left(B\dot{\varepsilon} + Bn\omega - q^2 \dfrac{k^G}{\gamma_G} \right) & (1 - S_r)\omega \end{bmatrix} \begin{Bmatrix} P^{W^*} \\ P^{G^*} \\ \varepsilon^* \end{Bmatrix} \tag{5.34}$$

$$= [A]\{y\} = 0$$

For non-zero values of P^{W^*}, P^{G^*}, and ε^*, the determinant of matrix $[A]$ must be equal to zero. From $\det[A] = 0$, we have a polynomial function of ω as

$$\omega^2 + \alpha_1\omega + \alpha_2 = 0 \tag{5.35}$$

in which

$$\begin{aligned} \alpha_1 = & \frac{1}{-Bn\left\{ \gamma_G\gamma_W + (\gamma_W k^G + \gamma_G k^W)q^2\mu \right\}} \\ & \Big[q^2\{\gamma_W k^G S_r^2 + (A\varepsilon + BP^C)(\gamma_W k^G S_r - \gamma_G k^W(1 - S_r)) \\ & + \gamma_G k^W(S_r^2 - 2S_r + 1) - B(\gamma_W k^G + \gamma_G k^W)(nH + \dot{\varepsilon}\mu) + k^W k^G \mu q^2\} - B\dot{\varepsilon}\gamma_W\gamma_G \Big] \end{aligned} \tag{5.36}$$

$$\begin{aligned} \alpha_2 = & \frac{1}{-Bn\left\{ \gamma_G\gamma_W + (\gamma_W k^G + \gamma_G k^W)q^2\mu \right\}} \\ & \{-B\dot{\varepsilon}(\gamma_W k^G + \gamma_G k^W) + k^W k^G \mu q^2\}q^2 H \end{aligned} \tag{5.37}$$

5.2.3 Instability conditions

In the following, we discuss the onset of the instability of the material system. If the growth rate of perturbation ω, which is the root of Equation 5.35, is positive for the real part, the material system is unstable. On the contrary, if ω is negative, the material system is stable. In order to estimate whether ω is negative or positive, we adopt the Routh–Hurwitz criterion. The roots of Equation 5.35 have negative real parts when the coefficients of the characteristic polynomial satisfy

$$\alpha_1 > 0, \ \alpha_2 > 0 \tag{5.38}$$

The first factor in Equations 5.36 and 5.37 is positive because B is negative. Thus, it is sufficient to consider the sign of the variables in the second factor for the analysis, namely,

$$\alpha_1^* = \Big[q^2\{\gamma_W k^G S_r^2 + (A\varepsilon + BP^C)(\gamma_W k^G S_r - \gamma_G k^W (1 - S_r)) \\ + \gamma_G k^W (S_r^2 - 2S_r + 1) - B(\gamma_W k^G + \gamma_G k^W)(nH + \dot{\varepsilon}\mu) + k^W k^G \mu q^2\} - B\dot{\varepsilon}\gamma_W \gamma_G\Big] > 0 \tag{5.39}$$

$$\alpha_2^* = \{-B\dot{\varepsilon}(\gamma_W k^G + \gamma_G k^W) + k^W k^G q^2\} q^2 H > 0 \tag{5.40}$$

where all the terms q, s, n, μ, k^W, k^G, γ_W, and γ_G are positive, as well as the term $(S_r^2 - 2S_r + 1)$, because $S_r < 1$. In the term $(\gamma_W k^G S_r - \gamma_G k^W (1 - S_r))$, γ_G is small, i.e., $\gamma_G / \gamma_W \approx 0.001$, which makes the term positive for the common permeability values of the soils.

The possibility for α_1^* and α_2^* to be negative depends on the strain hardening–softening parameter H, the values of A and B, strain ε, and strain rate $\dot{\varepsilon}$. Firstly, the hardening parameter decreases when suction decreases; consequently, the slope of curve H–P^C is positive, i.e., $A > 0$. Secondly, saturation increases when suction decreases, this means that the slope of curve S_r–P^C is negative, i.e., $B < 0$. Thirdly, strain ε is positive in tension and negative in compression, and the strain rate $\dot{\varepsilon}$ can be positive or negative.

Let us consider the model when parameter H is positive, i.e., the viscoplastic hardening case. In this case, it is possible that α_1^* is negative. Thus, the onset of the instability of the material system appears. Both α_1^* and α_2^* can be negative in the following cases:

1. Large B
   B is negative, hence when B is large, term BP^C in α_1^* is more negative. In addition, if strain rate $\dot{\varepsilon}$ is negative, term $-B\dot{\varepsilon}$ included in α_1^* and α_2^* becomes negative, i.e., the possibility of instability is more likely.

2. Large suction
   If P^C increases, the term BP^C becomes more negative; consequently, α_1^* can be more easily negative.
3. Large A ($=\partial H/\partial P^C$) and negative strain $\varepsilon < 0$
   In this case, when A becomes larger while the strain is negative, $\varepsilon < 0$ (compression), term $A\varepsilon$ is more negative. Consequently, it is possible that α_1^* becomes more negative.
4. Viscoplastic parameter μ and negative strain rate $\dot{\varepsilon} < 0$
   If the strain rate is negative, namely, compressive or contractant, term $\dot{\varepsilon}\mu$ in α_1^* becomes negative. In addition, if the viscoplastic parameter μ is larger, term α_1^* becomes more negative.

Similarly, for the case in which the strain hardening–softening parameter H is negative, i.e., viscoplastic softening, α_1^* and α_2^* can also be negative. Then, similar conditions exist for the onset of the instability of the material system.

Until now, the conditions for the onset of the instability of an unsaturated material system have been shown by an analysis using a viscoplastic model. From the analysis, it can be said that in both the hardening and softening ranges, the onset of the instability of a material in a viscoplastic state mainly depends on terms BP^C and $A\varepsilon$, as well as strain rate $\dot{\varepsilon}$. In Section 5.3, the results of numerical simulations of the one-dimensional infiltration problem are presented in order to study the material instability by the model proposed in Section 5.2. The numerical analyses are based on the effect of the variation in parameters α and n', which controls the soil-water characteristic curve as well as the specific moisture capacity B (slope of the $SWCC$) and the initial suction $P^C{}_i$. The effect of these parameters on the generation of volumetric strain ε is presented.

5.3 NUMERICAL SIMULATION OF ONE-DIMENSIONAL INFILTRATION PROBLEM

The multiphase coupled elasto-viscoplastic formulation proposed by Oka et al. (2006) and presented in Chapter 3, is used to describe the water infiltration into a one-dimensional column. The numerical analyses presented here are based on the fundamental concept of the theory of porous media. The materials are assumed to be composed of solid, water, and air, which are thought to be continuously distributed throughout space at a macroscopic level. An elasto-viscoplastic constitutive model is adopted for the soil skeleton (Kimoto and Oka 2005; Oka et al. 2006) with the extension to unsaturated soils, which is described in Section 2.4. The skeleton stress, which is determined from the difference between the total stress and the average pore fluid pressure, is used for the stress variable in the governing equations. In addition, the constitutive parameters are functions of the matric suction, by which the shrinkage or expansion of the overconsolidation boundary surface and the static yield surface can be described.

In this section, numerical simulations of the one-dimensional water infiltration problem are presented to discuss the effect of the specific moisture capacity and the initial suction on the development of volumetric strain. For the numerical simulation, an updated Lagrangian method with the objective Jaumman rate of Cauchy's stress is adopted (Kimoto et al. 2004; Oka et al. 2006; Chapter 3 of this book). It is shown that the instability obtained by the numerical analyses is consistent with the theoretical results obtained by the linear instability analysis.

In the numerical simulation, a finite element formulation based on the finite deformation theory has been used in which the strain rate tensor, $\dot{\varepsilon}_{ij}$, is replaced by the stretching tensor, D_{ij}, for the constitutive model (Kimoto et al. 2004; Oka et al. 2006).

5.3.1 Simulation results of the one-dimensional infiltration problem

Weak forms of the continuity equations for water and air and the rate type of the conservation of momentum are discretized in space and solved by the finite element method. In the finite element formulation, the independent variables are the pore water pressure, the pore air pressure, and the nodal velocity. An eight-node quadrilateral element with a reduced Gaussian integration is used for the displacement, and four nodes are used for the pore water and the pore air pressures. The backward finite difference method is used for the time discretization.

The finite element mesh and the boundary conditions for the simulations are shown in Figure 5.1. A homogeneous soil column with a depth of 1 m is employed in the simulations. An undrained boundary for water is assigned at the bottom and on the lateral sides of the column. Air flux is allowed both at the bottom and at the top of the column. The top of the column is subjected to a pore water pressure equal to 4.9 kPa. The simulations of the infiltration process start from an unsaturated condition where the initial suction P^C_i is the same along the column. At $t = 0$, water starts to infiltrate due to the pore water pressure applied at the top until a hydrostatic condition is attained in the column. Different values for the van Genuchten parameters (1980) – parameter α, from $\alpha = 0.1$ to $\alpha = 10$ (1/kPa), and parameter n', from $n' = 1.01$ to $n' = 9.0$, as well as two different levels of initial suction, $P^C_i = 25.5$ (Case 1) and 100 kPa (Case 2) – were considered to study the instability of the unsaturated material system. Table 5.1 lists the material parameters required by the constitutive model. Figure 5.2 shows an example of the results for the time history of the pore water pressure distribution obtained by the simulations. Figure 5.2 shows the transition of

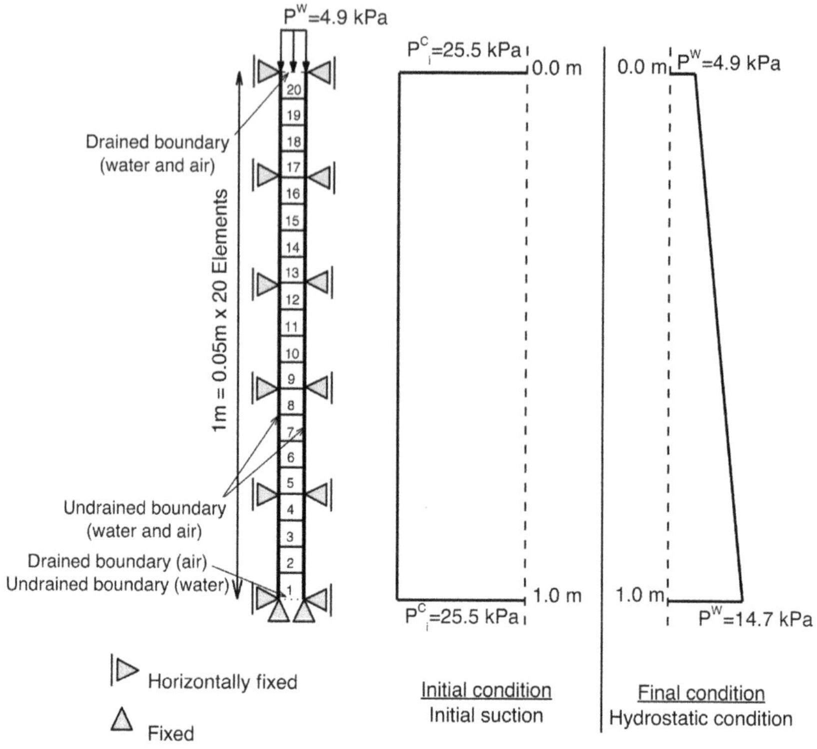

Figure 5.1 Finite element mesh and boundary conditions.

the soil from an initial unsaturated state ($P^C{}_i = 25.5$ kPa) to a saturated state corresponding to the hydrostatic condition.

In this section, the discussion shows a trend in the deformation behavior of unsaturated soil as well as the consistency between the numerical results and the theoretical results obtained in Section 5.2. It is worth noting that the onset of the instability of the unsaturated viscoplastic material subjected to a wetting process can be interpreted as the sudden increase in compressive volumetric deformation during the numerical analysis shown in Figure 5.3. This compressive behavior (collapse behavior) can be attained if the soil presents an open potentially unstable unsaturated structure which can also be attained when relatively high suction or applied stress exists (Barden et al. 1973; Lloret and Alonso 1980; Wheeler and Sivakumar 1995). Finally, in order to show the potentially stable and unstable regions of the one-dimensional infiltration problem, the results of the simulation will be plotted in the α–n' space, where the effect of the initial suction is also included.

Garcia et al. (2010) investigated the significance of the values for BP^C and $A\varepsilon$ on the onset of the instability of the unsaturated material. Equation 5.39

Table 5.1 Material parameters for the simulations

Viscoplastic parameter	m'	23.0
Viscoplastic parameter (1/s)	C_1	1.0×10^{-08}
Viscoplastic parameter (1/s)	C_2	1.0×10^{-08}
Stress ratio at critical state	M^*_m	0.947
Parameter of tangent line rigid method	θ	0.5
Coefficient of water permeability at s = 1.0 (m/s)	k^W_s	1.83×10^{-05}
Coefficient of air permeability at s = 0.0 (m/s)	K^G_s	1.00×10^{-03}
Compression index	λ	0.136
Swelling index	κ	0.0175
Initial shear elastic modulus (kPa)	G_0	4000
Initial void ratio	e_0	0.5983
Structural parameter	β	0.0
Suction parameter	S_I	0.2
Suction parameter	S_d	0.25
Minimum saturation	s_{min}	0.0
Maximum saturation	s_{max}	0.99
Parameter of coefficient of water permeability	a	3.0
Parameter of coefficient of air permeability	b	2.3

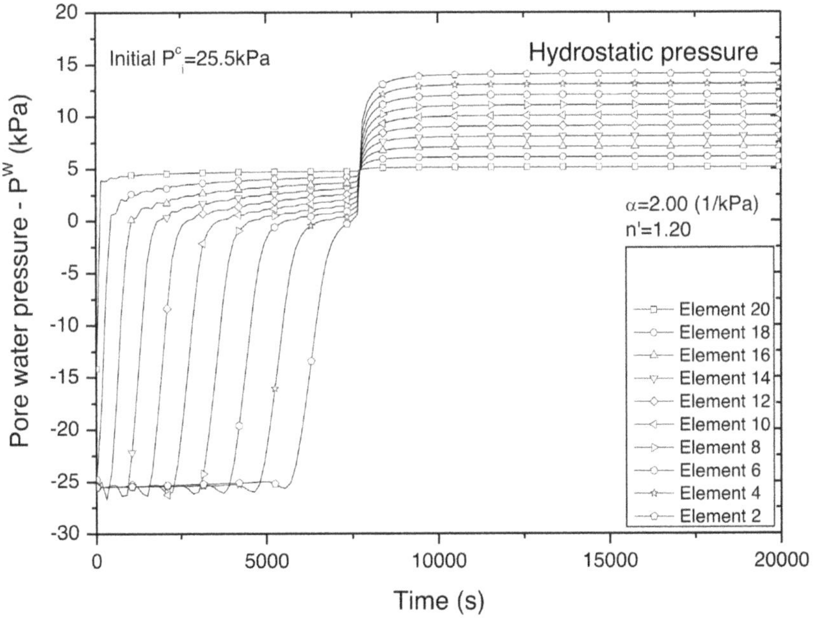

Figure 5.2 Time history of pore water pressure (α = 2.00 1/kPa, n' = 1.20, P^C_i = 25.5 kPa).

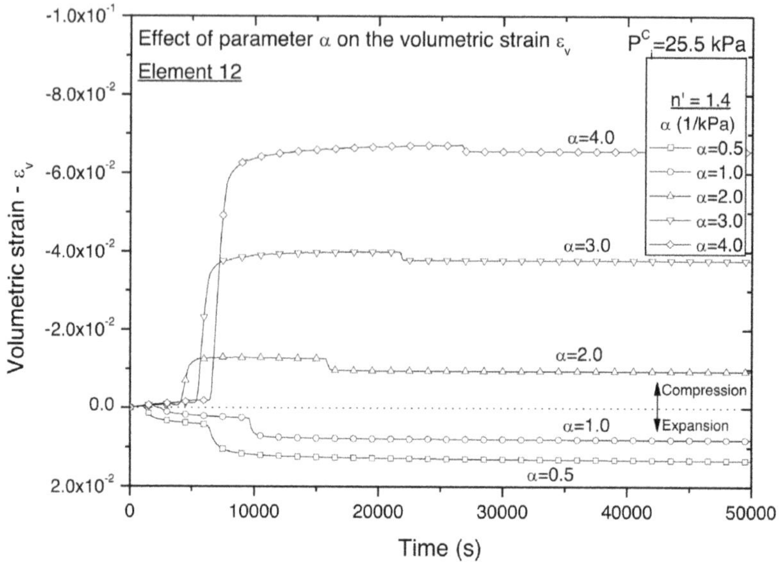

Figure 5.3 Development of volumetric strain for various values of the parameter α (P^C_i = 25.5 kPa).

shows that the onset of the instability of the material system mainly depends on the values BP^C and $A\varepsilon$ and the strain rate $\dot{\varepsilon}$. Among these terms, $\dot{\varepsilon}$ depends on the deformation pattern of the soil during the infiltration process, and its effect on the onset of the instability of the unsaturated material is addressed. To estimate which of the other two terms has a more significant effect on the onset of the instability of the material, the values corresponding to $A\varepsilon$ and BP^C, included in Equation 5.39, are investigated during the infiltration process. The strength degradation due to the decrease in suction is shown schematically in Figure 2.19 and the relation between the strain-hardening parameter and suction is given by Equations 2.108 and 2.109. From these equations, A is obtained as

$$A = \frac{\partial \sigma'_{mb}}{\partial P^C} = \sigma'_{ma} \exp\left(\frac{1+e_0}{\lambda-\kappa}\varepsilon_{kk}^{vp}\right)\left[S_I \exp\left(-S_d\left(\frac{P_i^C}{P^C}-1\right)\right)\right]\left[S_d \frac{P_i^C}{P^{C^2}}\right] \tag{5.41}$$

In the same manner, the slope of the $SWCC$ can be obtained from Equations 2.124 and 2.125 as

$$B = \frac{\partial s}{\partial P^C} = -\alpha m n'\left(s_{\max} - s_{\min}\right)\left(\alpha P^C\right)^{n'-1}\left\{1+\left(\alpha P^C\right)^{n'}\right\}^{-m-1} \tag{5.42}$$

In the analysis of the one-dimensional soil column, strain ε in Equation 5.7 corresponds to both the axial strain and the volumetric strain ($\varepsilon = \varepsilon_x = \varepsilon_v$). The results of the simulations are summarized in a stability chart in order

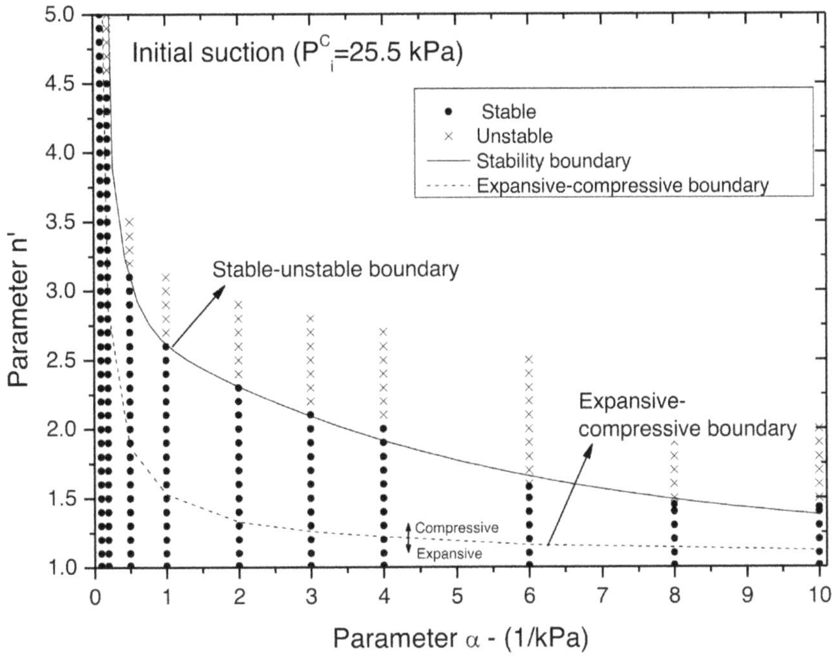

Figure 5.4 Stable and unstable regions for parameters α and n' during the infiltration process (P^C_i = 25.5 kPa).

to observe the potentially stable region for the one-dimensional water infiltration problem. Figures 5.4 and 5.5 show the results of the simulation for different values of parameters α and n' and for the initial levels of suction P^C_i = 25.5 kPa and 100 kPa, respectively. In these figures, the solid circle indicates the stable simulation results, while "x" indicates the unstable simulation results (large increase in the compressive volumetric deformation). On the graphs, the boundaries between the material-stable and material-unstable regions (continuous line) and the boundary between expansive and compressive behaviors (dashed line) are shown. From Figures 5.4 and 5.5, the following characteristics can be understood:

1. The expansive behavior is presented for materials with smaller parameters α and n', but the compressive behavior is obtained when these parameters increase.
2. For the same parameter α, the potential for instability increases when parameter n' is large. A larger parameter n' leads to an increase in B and an increase in the deformation.
3. For the same parameter n', the instability potential increases when parameter α becomes large. This larger parameter α leads to an increase in B and, consequently, to an increase in the deformation.

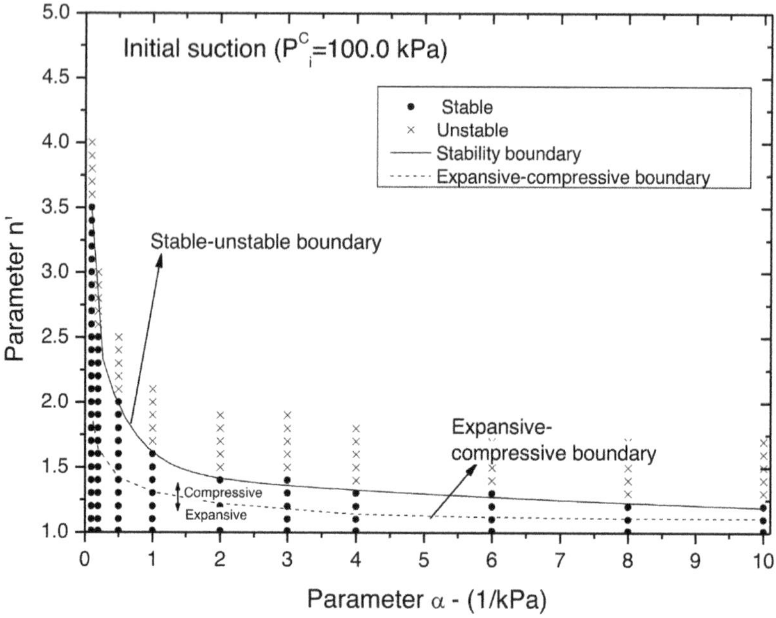

Figure 5.5 Stable and unstable regions for parameters α and n' during the infiltration process (P^C_i = 100 kPa).

4. It is possible to see that the material stable–unstable boundaries and the expansive–compressive boundaries shrink as suction increases. It occurs when P^C_i is changed from 25.5 kPa to 100 kPa.
5. The instability obtained by the numerical analyses is consistent with the theoretical results obtained by the linear instability analysis presented in Section 5.2.

According to the instability results, it can be said that unsaturated soils that are prone to instability during a wetting process are those with higher levels of initial suction, e.g., clays, as well as soils that are represented by a steeper soil-water characteristic curve, e.g., sands.

5.3.2 Discussions on stability

An instability analysis was conducted to study the instability of an unsaturated material system during a wetting process. It was found that in both the hardening and softening ranges, the onset of instability of an unsaturated material system grows if suction, P^C, and specific moisture capacity, $B = \partial s/\partial P^C$, increase. Moreover, the onset of instability can be increased if the strain rate is negative ($\dot{\varepsilon} < 0$) and the behavior of the material is compressive ($\varepsilon_v < 0$), while the slope of curve $\sigma'_{mb} - P^C (A = \partial \sigma'_{mb}/\partial P^C)$ increases.

It was also found that the effect of the specific moisture capacity and the suction is more significant than the effect of the slope of curve $\sigma'_{mb} - P^C$.

From the numerical study of the one-dimensional infiltration problem, it was found that the elasto-viscoplastic material system is more unstable when parameters α and n' are large, namely, for large values of the slope of the soil-water characteristic curve (specific moisture capacity). For the effect of suction, it was observed that the instability significantly increased when the initial suction of the unsaturated material was increased ($P^C{}_i$=25.5 kPa to 100 kPa). These trends are consistent with the theoretical results obtained by the linear instability analysis.

The simulated settlements show that for smaller parameters α and n', the material behavior is expansive. Nevertheless, when these parameters become large, the expansive behavior of the soils changes to a compressive one. Parameters α and n' indicate a great effect on the development of the deformation. The greater the parameters α and n', the larger the settlements obtained. Larger rates of settlement were also obtained for higher levels of initial suction. This suggests that rapid transitions from unsaturated to saturated states and higher levels of suction lead to compressive behavior and instability, e.g., a wetting-induced collapse.

5.4 SIMULATION OF THE EXPERIMENTAL RESULTS

Experimental results of infiltration into unsaturated soils are scarce and costly, but are valuable and useful input data for simulations. They allow the calibration and verification of numerical models used to investigate the behavior of unsaturated soils. Yang et al. (2004, 2006) performed interesting experimental work and provided measurements of rainfall intensity, the pore water pressure, the volumetric water content, and the drainage rate instantaneously and continuously during rainfall infiltration. The results showed that rainfall intensity and its duration had an important effect on the rainfall infiltration in the finer soil layer. In addition, from the measurement of the negative pore water pressures and water contents, different paths were shown for the wetting and drying processes during the rainfall infiltration, i.e., the soil exhibited hysteretic behavior.

Some of the laboratory test results of vertical infiltration into a layered soil were compared with the numerical results obtained by the present multiphase coupled finite element analysis method. The experimental results were obtained by performing column tests on finer-over-coarser soil subjected to simulated rainfall under conditions of no-ponding at the surface and constant head at the bottom. The laboratory tests were started from a hydrostatic condition where the water level was located at the bottom of the soil column; constant rainfall intensities lower than the permeability of a clayey sand were applied to the top of the column for 24 h. Garcia et al. (2011) performed numerical simulations by Yang et al. (2006), which

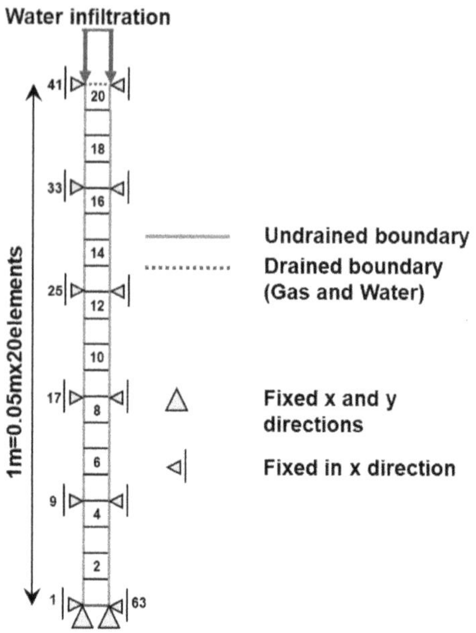

Figure 5.6 Model of the soil column. (After Garcia, E., Oka, F., Kimoto, S. 2011. Numerical analysis of a one-dimensional infiltration problem in unsaturated soil by a seepage-deformation coupled method. *Int. J. Numer. Anal. Meth. Geomech.*, 35(5): 544–568.)

are presented in this section. The same boundary conditions as those used in the experiments were used for the simulations. A 20-element column mesh with two soil layers and a drained boundary at the top were used for the space discretization. The horizontal dimensions of the column are irrelevant because the flow pattern is one-dimensional. The finite element mesh of the soil column is shown in Figure 5.6. The main parameters for the material properties of the soil particles, water, and gas, required by the constitutive model and used for the simulation, are listed in Table 5.2. The saturated permeability coefficients measured in the experiment (k for clayey sand = 8.8×10^{-7} m/s, k for fine sand = 2.7×10^{-4} m/s) and two soil-water characteristic curves similar to those measured experimentally by Yang et al. (2006) were used in the simulation (clayey sand: $\alpha = 2.0$ and $n' = 1.14$; fine sand: $\alpha = 1.0$ and $n' = 5.0$). The results of the experiments for two different rainfall intensities (CF-R1: $q = 1.6 \times 10^{-7}$ m/s and CF-R2: $q = 3.3 \times 10^{-7}$ m/s) are shown in Figures 5.7 and 5.8.

Figures 5.7 and 5.8 also show the calculated pore water pressure profiles obtained in the simulations at similar elevations to those shown in the experimental results. Both figures show the water front advancing with time from the initial pressure head profile at $t = 0.0$ h. When infiltration starts, the pore water pressure increases for the clayey sand, notably in depths relatively close to the soil surface, and the suction becomes small.

Table 5.2 Material parameters

Material		*Clayey sand*	*Fine sand*
Compression index	λ	0.03	0.03
Swelling index	κ	0.002	0.002
Initial shear elastic modulus (kPa)	G_0	28700	28700
Initial void ratio	e_0	0.762	0.762
Permeability of water at s = 1 (m/s)	k_s^W	8.8×10^{-7}	2.7×10^{-4}
Minimum saturation	s_{min}	0.0	0.0
Maximum saturation	s_{max}	0.99	0.99
Parameter of coefficient of water permeability	α	2.0	1.0
Parameter of coefficient of air permeability	n'	1.14	5.00

Source: Data from Garcia, E., Oka, F., Kimoto, S., 2011. Numerical analysis of a one-dimensional infiltration problem in unsaturated soil by a seepage-deformation coupled method. *Int. J. Numer. Anal. Meth. Geomech.*, 35(5): 544–568.

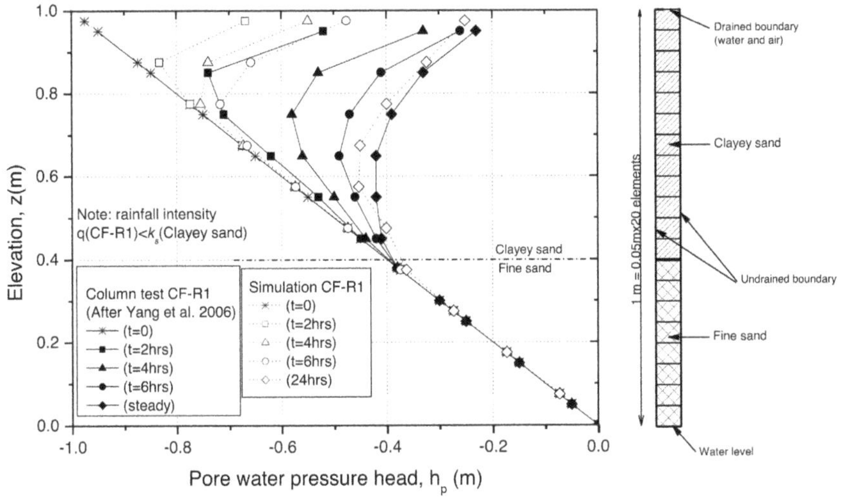

Figure 5.7 Pore-water head profile CF-R1. (After Garcia, E., Oka, F., Kimoto, S. 2011. Numerical analysis of a one-dimensional infiltration problem in unsaturated soil by a seepage-deformation coupled method. *Int. J. Numer. Anal. Meth. Geomech.*, 35(5): 544–568.)

This trend gradually progresses downward toward the fine sand layer as the wetting front advances. When the results of the experiments are compared with the results of the simulations, it is seen that the following features observed in the experiments are captured by the numerical analysis. (a) The pore water pressure increases initially at the surface of the clayey sand and gradually increases down the soil column while the water infiltrates. (b) The pore water pressure is higher and the infiltration is faster when the intensity of the rainfall is higher (CF-R2). (c) The pore water pressure is always

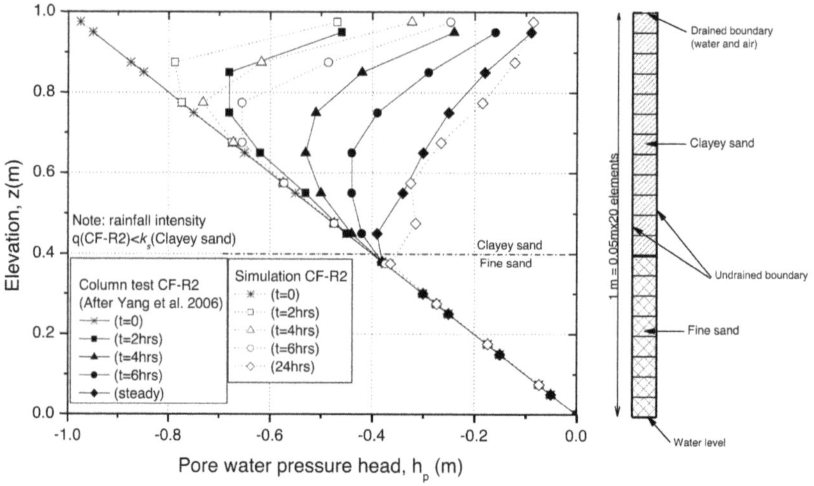

Figure 5.8 Pore-water head profile CF-R2. (After Garcia, E., Oka, F., Kimoto, S. 2011. Numerical analysis of a one-dimensional infiltration problem in unsaturated soil by a seepage-deformation coupled method. *Int. J. Numer. Anal. Meth. Geomech.*, 35(5): 544–568.)

negative even after 24 h of rainfall infiltration, and it does not develop in the fine sand layer. (d) Although the infiltration during the simulation was slower compared with the experiments, the final results were very close to the steady state measured.

Despite the same rainfall intensities, permeabilities and soil-water characteristic curves similar to those reported in the experiments were used for the numerical analysis, a time delay in the simulated infiltration is observed. This time delay can be explained mainly by the uncertainty in the soil-water characteristic curve and permeabilities which are highly dependent on the void ratio. More accurate information about the soil-water characteristic curve and permeabilities is necessary to improve the simulation. It emphasizes the significance of the experimental data related to the infiltration and deformation of unsaturated soils in the improvement of the numerical models in order to understand the complex response of unsaturated soils.

REFERENCES

Alonso, E.E., Gens, A. and Delahaye, C.H. 2003. Influence of rainfall on the deformation and stability of a slope in overconsolidated clays: A case study. *Hydrogeol. J.* 11(1): 174–192.

Barden, L., McGown, A. and Collins, K. 1973. The collapse mechanism in partly saturated soil. *Eng. Geol.* 7(1): 49–60.

Buscarnera, G. and Nova, R. 2009. Modelling the onset of instability in oedometric tests on unsaturated bounded soils. In: *Computational Geomechanics*, ed. S. Pietruszczak, G.N. Pande, C. Tamagnini, and R. Wan, First International

Symposium on Computational Geomechanics, Juan-les-Pins, France, Int. Centre for Computational Engineering, Rhodes, Greece: 226–238.

Buscarnera, G. and Nova, R. 2011. Modelling instabilities in triaxial testing on unsaturated soil specimens. *Int. J. Numer. Anal. Meth. Geomech.* 35(2): 179–200.

Cunningham, M.R., Ridley, A.M., Dineen, K. and Burland, J.B. 2003. The mechanical behaviour of a reconstituted unsaturated silty clay. *Géotechnique* 53(2): 183–194.

Ehlers, W., Graf, T. and Ammann, M. 2004. Deformation and localization analysis of partially saturated soil. *Comput. Methods Appl. Mech. Eng.* 193(27–29): 2885–2910.

Garcia, E., Oka, F. and Kimoto, S. 2010. Instability analysis and simulation of water infiltration into an unsaturated elasto-viscoplastic material. *Int. J. Solids Struct.* 47(25–26): 3519–3536.

Garcia, E., Oka, F. and Kimoto, S. 2011. Numerical analysis of a one-dimensional infiltration problem in unsaturated soil by a seepage-deformation coupled method. *Int. J. Numer. Anal. Meth. Geomech.* 35(5): 544–568.

Khalili, N., Geiser, F. and Blight, G.E. 2004. Effective stress in unsaturated soils: Review with new evidence. *Int. J. Geomech.* 4(2): 115–126.

Kimoto, S. and Oka, F. 2005. An elasto-viscoplastic model for clay considering destructuralization and consolidation analysis of unstable behavior. *Soils Found.* 45(2): 29–42.

Kimoto, S., Oka, F., Fushita, T. and Fujiwaki, M. 2007. A chemo-thermo-mechanically coupled numerical simulation of the subsurface ground deformations due to methane hydrate dissociation. *Comput. Geotech.* 34(4): 216–228.

Kimoto, S., Oka, F. and Higo, Y. 2004. Strain localization analysis of elasto-viscoplastic soil considering structural degradation. *Comput. Methods Appl. Mech. Eng.* 193(27–29): 2845–2866.

Lloret, A. and Alonso, E.E. 1980. Consolidation of unsaturated soils including swelling and collapse behaviour. *Géotechnique* 30(4): 449–477.

Oka, F., Kodaka, T., Kimoto, S., Kim, Y.S. and Yamasaki, N. 2006. An elasto-viscoplastic model and multiphase coupled FE analysis for unsaturated soil. In: *Fourth International Conference on Unsaturated Soils*, eds. G.A. Miller, C.E. Zapata, S.L. Houston and D.G. Fredlund. Geotechnical Special Publication, 147(2): 2039–2050. ASCE, Carefree, AZ.

Oka, F., Kodaka, T., Suzuki, H., Kim, Y.S., Nishimatsu, N. and Kimoto, S. 2010. Experimental study on the behavior of unsaturated compacted silt under triaxial compression. *Soils Found.* 50(1): 27–44.

Van Genuchten, M.Th. 1980. A closed-form equation for predicting the hydraulic conductivity of unsaturated soils. *Soil Sci. Soc. Am. J.* 44(5): 892–898.

Wheeler, S.J. and Sivakumar, V. 1995. An elasto-plastic critical state framework for unsaturated soil. *Géotechnique* 45(1): 35–53.

Wu, L.Z. and Zhang, L.M. 2009. Analytical solution to 1D coupled water infiltration and deformation in unsaturated soils. *Int. J. Numer. Anal. Meth. Geomech.* 33(6): 773–790.

Yang, H., Rahardjo, H., Wibawa, B. and Leong, E.C. 2004. A soil column apparatus for laboratory infiltration study. *Geotech. Test. J.* 27(4): 347–355.

Yang, H., Rahardjo, H., Leong, E.C. 2006. Behavior of unsaturated layered soil columns during infiltration. *J. Hydrol. Eng.* 11(4): 329–337.

Chapter 6

Numerical simulation of rainfall infiltration on unsaturated soil slope

6.1 INTRODUCTION

River embankments frequently fail due to heavy rains and typhoons associated with climate warming. Heavy rainfall causes a rise in the water level of rivers as well as a rise in the groundwater level within river embankments. As a result, river embankments have failed due to rainfall infiltration and the generation of seepage flow. In addition, sometimes the upper parts of river levees are connected to surrounding mountains and are exposed to continuous seepage flow from hilly areas. This, in turn, increases the pore water pressure within a river embankment leading to the degradation of the material (e.g., seepage-induced erosion). This deformation could trigger the progressive failure of the river embankment endangering nearby structures.

Many researchers have reported on embankment and slope failure due to rainfall infiltration (e.g., Yoshida et al. 1991; Au 1998; Alonso et al. 2003; Yamagishi et al. 2004, 2005; Matsushi et al. 2006; Chen et al. 2006; Nakata et al. 2010; Kimoto et al. 2013); similarly, studies have addressed the effect of rainfall infiltration on slope stability using a statistical approach (e.g., Au 1993; Okada and Sugiyama 1994).

The failure of embankments has also been studied using the numerical approach. Several researchers have implemented numerical solutions to analyze the effect of hydraulic characteristics on the instability of unsaturated slopes (e.g., Cai and Ugai 2004; Rahardjo et al. 2007; Oka et al. 2011a,b). In these formulations, the effects of rainfall infiltration on the generation of pore water pressure and the instability of a slope are generally evaluated by a seepage analysis using the finite element method followed by a slope stability analysis; thus, the study of the coupling of the deformation and the transient flow is disregarded. However, it is also possible to use numerical methods that can simultaneously consider the unsaturated seepage flow and the deformation of soil structures to study the infiltration processes (Cho and Lee 2001; Alonso et al. 2003; Ehlers et al. 2004; Ye et al. 2005; Oka et al. 2009, 2010, 2011a, b; Kimoto et al. 2013; Oka and Kimoto 2012).

This chapter presents a study to clarify the effect of rainfall infiltration considering seepage flow on the deformation of unsaturated river

DOI: 10.1201/9781003200031-6

embankments based on the work by Garcia (2010), Oka et al. (2010), and Kimoto et al. (2013). Two-dimensional numerical analyses have been performed on an embankment. The case study, namely Seta River, corresponds to a three-layered river embankment which is subjected to the effect of both rainfall infiltration and seepage flow from a mountain side. In the simulation, the rainfall record measured at the Seta River in Shiga prefecture, Japan, and the variation in the water level measured at the right side of the embankment are used for the analyses, and the water level of the river is considered constant. In the Seta River case, the effects of different water permeabilities for the upper layer on the seepage flow velocity and on the surface deformation are emphasized. Furthermore, the effects of horizontal drains and the degree of compaction on the generation of deformation were investigated by Kimoto et al. (2013).

The numerical analyses were carried out employing a seepage-deformation coupled method for unsaturated soil using the method presented in Section 2.4 and Chapter 3, neglecting the acceleration term. Constitutive and hydraulic parameters that represent the soils found in the Seta River embankment were employed in the simulations. From the numerical results, it was seen that the deformation of the embankment significantly depends on the water permeability of the soil and it is localized on the slope surface at the river side. The larger the saturated water permeability of the soil, the larger the velocity of the seepage flow and the larger the deformation on the surface of the river embankment.

6.2 CASE STUDY OF SLOPE STABILITY

6.2.1 Measurement data and numerical analysis

The river embankment used in the analysis is located in Otsu city, Shiga prefecture, Japan. Figure 6.1 is a sketch of the geomorphology of the study area of the Seta River embankment. As observed in Figure 6.1, the back of the embankment is surrounded by a hilly area. The embankment is located in a flat surface which is divided by small rivers that flow toward the Seta River. The study area was farmland but nowadays is a residential area. The soil investigation and the observation of the behavior of the embankment at the Seta River show that the water infiltration from the surrounding mountains is permanent and has significant impact, damaging and degrading the soil close to the embankment surface at the river side. Figure 6.2 depicts the mechanism of the water flow and the degradation of the soil surface at the river side of the embankment. The fluctuation in the water level of the river at the Seta River site is 1 m per year.

Figure 6.3 shows the rainfall record of the Seta River and the groundwater level observations at the right side of the embankment (mountain side) from January 2009 to February 2010. Figure 6.3 shows that the groundwater

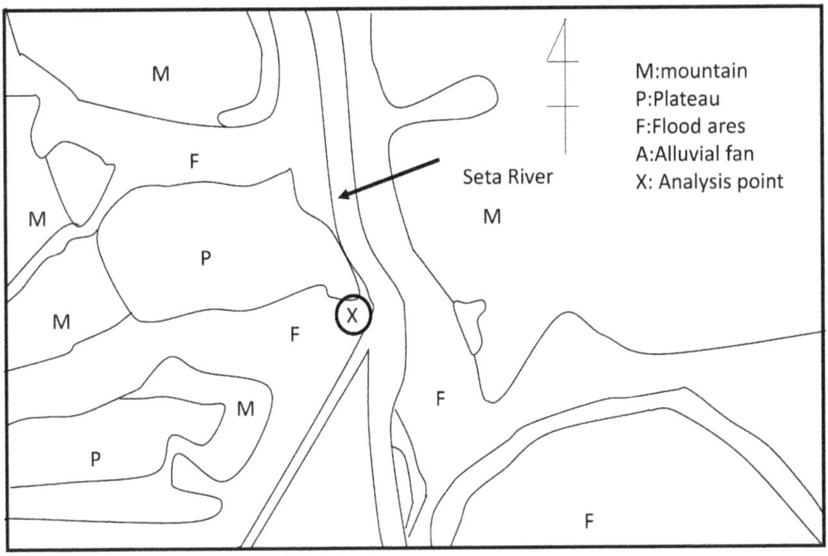

Figure 6.1 Geomorphology and location of the degradation of Seta River embankment.

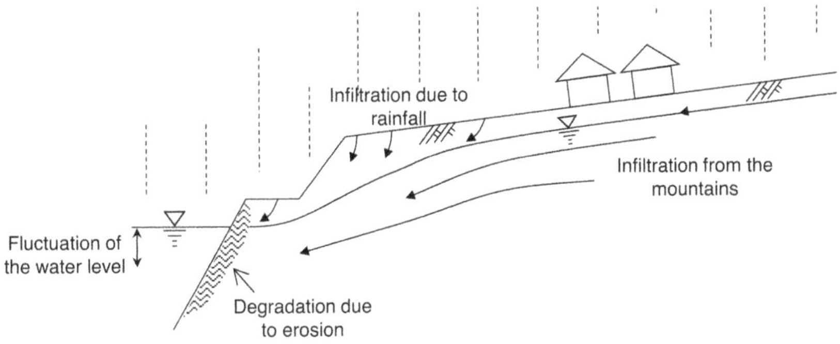

Figure 6.2 Mechanism of the water flow and degradation of the slope surface. (After Kimoto, S., Oka, F. and Garcia, E. 2013. Numerical simulation of the rainfall infiltration on unsaturated soil slope considering a seepage flow. *J. SAGS & AGSEA* 44(3): 1–13.)

level changes over time and increases during rainfall. This is an indication of seepage flow induced by rainfall infiltration.

The cross section and the boundary conditions for the simulation of the rainfall infiltration and seepage flow into the unsaturated river embankment are shown in Figure 6.4. The top surface of the river embankment has an inclination of 0.95°. The slopes of the embankment have gradients of 1V:1H at the upper part and 1V:3H at the middle of the embankment. For displacement, the embankment is fixed at the bottom in both the horizontal

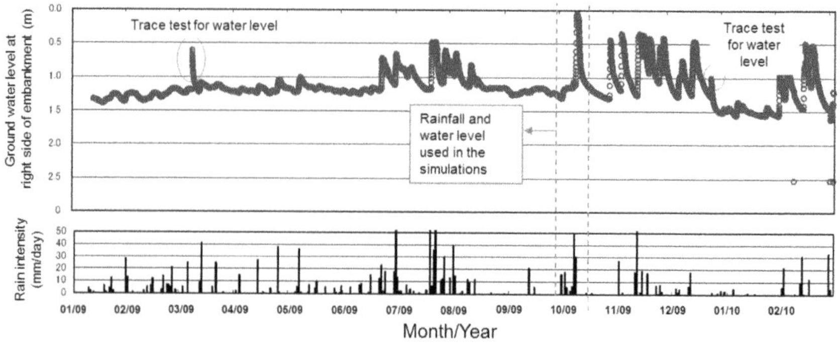

Figure 6.3 Rainfall and water level data from January 2009 to February 2010. (After Kimoto, S., Oka, F. and Garcia, E. 2013. Numerical simulation of the rainfall infiltration on unsaturated soil slope considering a seepage flow. *J. SAGS & AGSEA* 44(3): 1–13.)

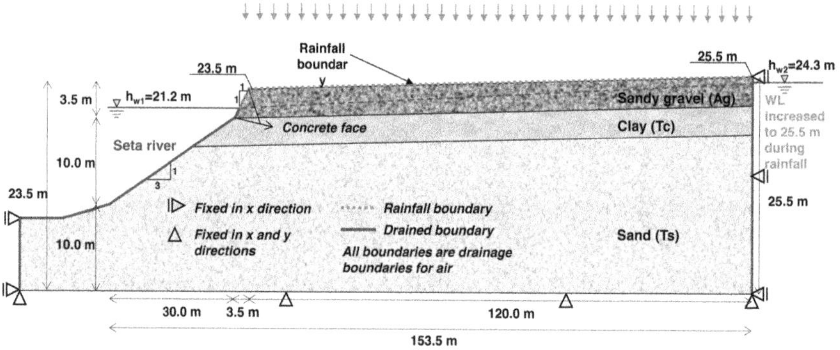

Figure 6.4 Cross section of the river embankment and its boundary conditions. (After Kimoto, S., Oka, F. and Garcia, E. 2013. Numerical simulation of the rainfall infiltration on unsaturated soil slope considering a seepage flow. *J. SAGS & AGSEA* 44(3): 1–13.)

and vertical directions and the lateral boundaries are fixed only in the horizontal direction. The initial negative pore water pressure distribution, i.e., suction, in the top sandy gravel layer is considered linear. The water level at the boundaries of the embankment is located at 21.2 m on the river side and at 24.3 m on the mountain side. The water level inside the embankment is linearly distributed with the water levels at the river and mountain sides, and its inclination is about 1.5°. Air flux is allowed for the entire boundaries and the initial air pressure, P_i^G, is assumed to be zero. The boundary conditions for water flux are described as follows: an impermeable boundary is assigned to the bottom or the soil foundation; for the lateral sides of the embankment below the water levels, the boundary is considered permeable;

above the water levels the boundaries are initially impermeable, but change to permeable if the pore water pressure turns positive. In the case of rainfall, if the soil is unsaturated, a flow boundary is used on the slope surface and the whole water infiltrates into the soil; however, once the surface is saturated, a prescribed pore water pressure equal to zero is used on the surface; in this case, the incoming water is controlled by the gradient of the matric suction inside the embankment. Thus, it is possible to control the volume of water that enters the soil and excess water is dissipated as runoff.

The material parameters required by the constitutive model used in the analysis are listed in Table 6.1. These parameters can be determined from boring tests such as *N* values and the parameters of materials similar to those of soils. Details of the parametric calculation can be found in Kimoto and Oka (2005), and also in section 5.7 of Oka and Kimoto (2012). The parameters in Table 6.1 represent the three different soils found in the river embankment, namely, sandy gravel (Ag), clay (Tc), and sand (Ts). The upper

Table 6.1 Material parameters for the Seta River simulations

Material parameter	*Sandy gravel (Ag)*	*Clay (Tc)*	*Sand (Ts)*
Viscoplastic parameter m'	40	27	40
Viscoplastic parameter (1/s) C_1	1×10^{-15}	2×10^{-14}	1×10^{-20}
Viscoplastic parameter (1/s) C_2	2×10^{-15}	2×10^{-13}	2×10^{-20}
Stress ratio critical state M^*_m	1.27	1.25	1.27
Compression index λ	0.0804	0.4910	0.0804
Swelling index κ	0.0090	0.0760	0.0090
Elastic shear modulus (kPa) G_0	3000	23000	20000
Initial void ratio e_0	0.344	1.23	0.535
Structural parameter β	5	15	0
Structural parameter $\sigma'_{maf}/\sigma'_{mai}$	0.60	0.579	0.60
Ver. water permeab. (m/s) k^W_{sv}	**	1×10^{-8}	1×10^{-6}
Hor. water permeab. (m/s) k^W_{sh}	**	1×10^{-7}	1×10^{-5}
Dry gas permeability (m/s) k^G_s	1×10^{-3}	1×10^{-3}	1×10^{-3}
van Genuchten par. (1/kPa) α	0.1	0.13	2
van Genuchten par. n'	4	1.65	1.2
Suction parameter S_I	0.2	0.2	0.2
Suction parameter S_d	0.2	5	0.2
Minimum saturation s_{min}	0	0	0
Maximum saturation s_{max}	0.97	0.99	0.99
Shape parameter a	3	3	3
Shape parameter b	1	1	1

Source: Data from Kimoto, S., Oka, F. and Garcia, E. 2013. Numerical simulation of the rainfall infiltration on unsaturated soil slope considering a seepage flow. *J. SAGS & AGSEA* 44(3): 1–13.

Note: **The permeability of the sandy gravel layer depends on the simulation case.

sandy gravel layer is a loose soil with an N value of 2–4, it is approximately 3.0 m in height and it overlies on a clay layer approximately 4.0 m in height. The bottom of the embankment is composed of a stiff silty sand layer with an N value of 50–60.

In order to study the effect of rainfall infiltration and seepage flow on the unsaturated river embankment, the rainfall record and the water level measured from October 2 to 8, 2009, corresponding to the Seta River site were used in the simulation. As shown in Figure 6.3, the water level measured during this time on the mountain side was maximum, reaching almost the top of the embankment. Enlargements of the rainfall record and the water measured at that time are shown in Figure 6.5. The total precipitation is 106 mm, with a maximum hourly rainfall of 14 mm. According to the precipitation record, the rainfall is concentrated between noon of October 7 to the morning of October 8 with a total rainfall of 97 mm (t = 132 h to t = 152 h). During the simulation, the non-uniform rainfall was applied on the top and the slope of the river embankment, while the water level increased on the right side (mountain side). The initial degree of saturation for the unsaturated soil (sandy gravel layer) was considered to be about 60%.

At the site, it was observed that the deformation of the river embankment is localized on the river side, just below the concrete face that covers

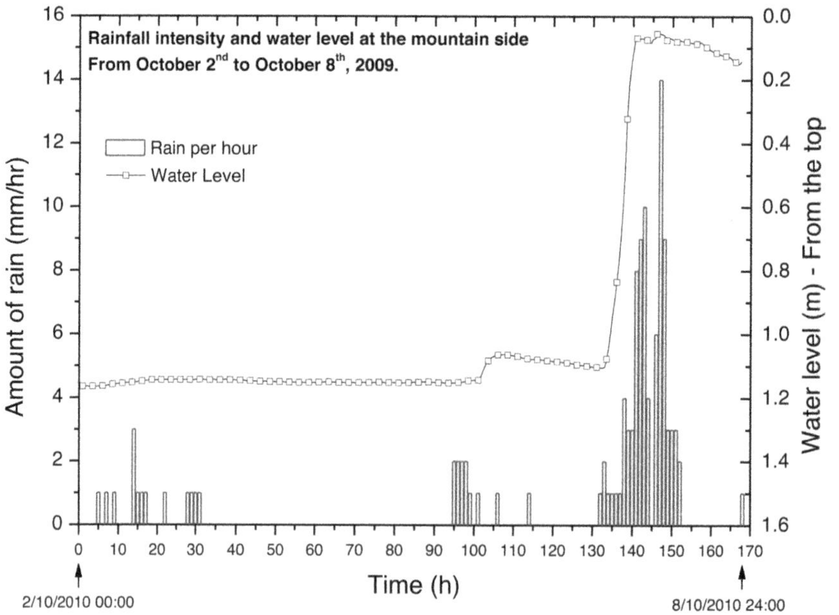

Figure 6.5 Rainfall intensity and water level at the mountain side from October 2 to October 8, 2009. (After Kimoto, S., Oka, F. and Garcia, E. 2013. Numerical simulation of the rainfall infiltration on unsaturated soil slope considering a seepage flow. *J. SAGS & AGSEA* 44(3): 1–13.)

the surface of the slope. It is thought that this localized accumulation of deformation is triggered by the combined effect of the seepage flow due to rainfall infiltration over the embankment surface, the seepage flow from the mountain side, and the difference in the permeabilities between the sandy gravel soil and the slope surface.

6.2.2 Effect of permeability

Different cases were analyzed in order to see the effect of the permeabilities on the seepage flow velocity and the development of deformation in the unsaturated river embankment. The simulation cases are listed in Table 6.2. The analyzed cases consist of different combinations of saturated permeabilities for the upper sandy gravel layer and the slope surface. In Cases 1–4, both the saturated permeability of the sandy gravel and the saturated permeability of the slope surface are equal, and increase simultaneously, namely, $k^{W(G)}_{sv} = 1 \times 10^{-6}$, 3×10^{-6}, 6×10^{-6}, and 1×10^{-5} m/s. For Cases 5–8, a concrete face with a smaller permeability than the top soil layer is assumed at the slope surface. This condition is similar to the current case at the Seta River site. The saturated permeability for the sandy gravel is increased, i.e., $k^{W(G)}_{sv} = 1 \times 10^{-6}$, 3×10^{-6}, 6×10^{-6}, and 1×10^{-5} m/s, respectively, while the permeability of the slope surface is kept constant, i.e., $k^{W(S)}_{sv} = 1 \times 10^{-7}$ m/s. In addition, the material of the concrete at the slope surface is assumed to be an elastic material in order to simulate a concrete face over the slope of the embankment. For the elastic material, the elastic Young's modulus is considered to be $E = 1$ GPa and Poisson's ratio is $\nu = 0.1$.

Table 6.2 Saturated vertical permeabilities considered for the analysis

	Vertical permeability	
Case no.	*$k^{W(G)}_{sv}$ sandy gravel (m/s)*	*$k^{W(S)}_{sv}$ slope surface (m/s)*
1	1×10^{-6}	1×10^{-6}
2	3×10^{-6}	3×10^{-6}
3	6×10^{-6}	6×10^{-6}
4	1×10^{-5}	1×10^{-5}
5	1×10^{-6}	1×10^{-7}
6	3×10^{-6}	1×10^{-7}
7	6×10^{-6}	1×10^{-7}
8	1×10^{-5}	1×10^{-7}

Source: Data from Kimoto, S., Oka, F. and Garcia, E. 2013. Numerical simulation of the rainfall infiltration on unsaturated soil slope considering a seepage flow. *J. SAGS & AGSEA* 44(3): 1–13.

Note: *The horizontal permeability is assumed to be 10 times the vertical permeability $k^{W}_{sh} = 10\, k^{W}_{sv}$.

One of the greatest uncertainties concerning the study of rainfall infiltration and seepage flow lies in the field values of the saturated water permeability. Because the accumulation of deformation due to the seepage flow observed at the site is localized in the slope of the embankment composed of a sandy gravel layer, it is thought that this phenomenon is due to water infiltration. The results presented herein are based on the variation of the permeabilities for the upper layer, namely, the sandy gravel layer, as well as the permeabilities of the slope surface. The permeabilities of the clay layer and the sand layer are kept constant. Different combinations of permeabilities for the sandy gravel layer and for the slope of the embankment are used in order to find the permeability combinations that may lead to the localization of deformation at the slope of the embankment.

Cases 1–4 intend to evaluate the effect of the saturated water permeability on a homogeneous soil. A comparison of the horizontal hydraulic gradient distributions near the slope of the embankment at time t = 151 h for these cases is presented in Figure 6.6. The figure shows that the horizontal hydraulic gradient is very similar for the four cases, regardless of the permeability. Maximum values are obtained at the middle of the slope surface, immediately above the river water table. A comparison of the distributions

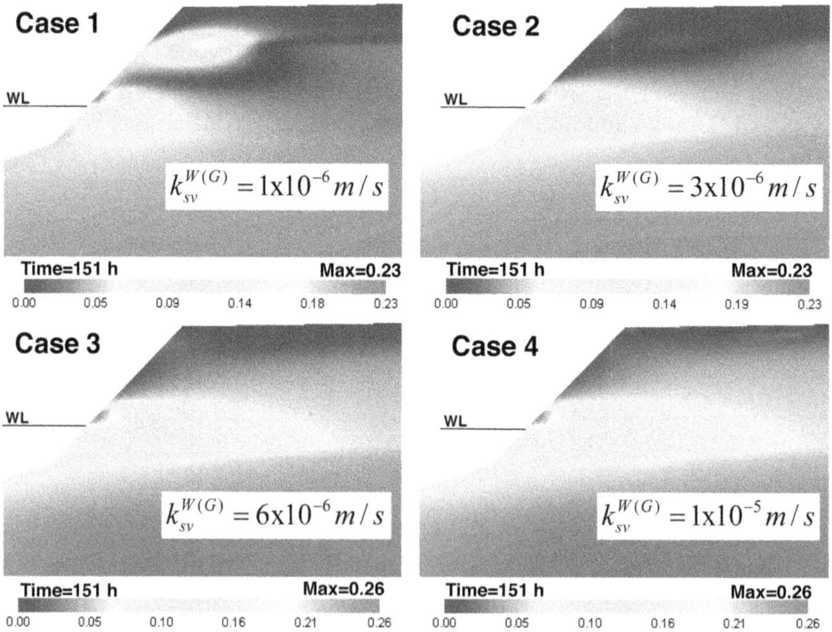

Figure 6.6 Distribution of the horizontal hydraulic gradient: Cases 1–4 (t = 151 h). (After Kimoto, S., Oka, F. and Garcia, E. 2013. Numerical simulation of the rainfall infiltration on unsaturated soil slope considering a seepage flow. *J. SAGS & AGSEA* 44(3): 1–13.)

of the accumulated viscoplastic shear strain for these cases at time t = 151 h is shown in Figure 6.7; $\gamma^{vp} = \int_0^t (D'^{vp}_{ij} D'^{vp}_{ij})^{1/2} dt$; D'^{vp}_{ij}: deviatoric viscoplastic stretching. It is shown that the accumulation of viscoplastic shear strain is generated on the slope of the embankment immediately above the water level of the river. None of the cases shows a large development of deformation on the slope of the embankment; however, a small accumulation of viscoplastic shear strain is evident above the water level near the slope surface. This can be attributed to the large hydraulic gradients present on the surface of the slope above the water level. It can be said that due to the homogeneity of the upper layer, the water can flow easily toward the river side avoiding an increase in the pore water pressure inside the embankment.

In Cases 5–8, a concrete face with small permeability is assumed to be installed at the slope surface. This is similar to the real case found at the Seta River site. Figure 6.8 shows a comparison of the distribution of the horizontal hydraulic gradient after major rainfall at time t = 151 h for Cases 5–8, respectively. Larger gradients are obtained for the cases with the concrete face; this suggests a large accumulation of water and the generation of pore water pressure at the back of the concrete face.

Figure 6.9 presents a comparison of the distribution of viscoplastic shear strain for Cases 5–8 at time t = 151 h. Figure 6.9 shows that the

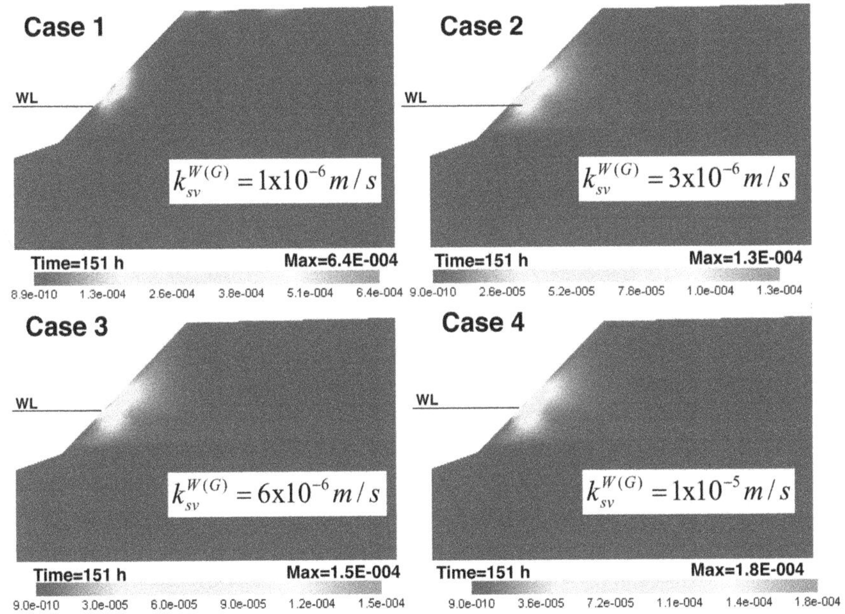

Figure 6.7 Distribution of the viscoplastic shear strain: Cases 1–4 (t = 151 h). (After Kimoto, S., Oka, F. and Garcia, E. 2013. Numerical simulation of the rainfall infiltration on unsaturated soil slope considering a seepage flow. *J. SAGS & AGSEA* 44(3): 1–13.)

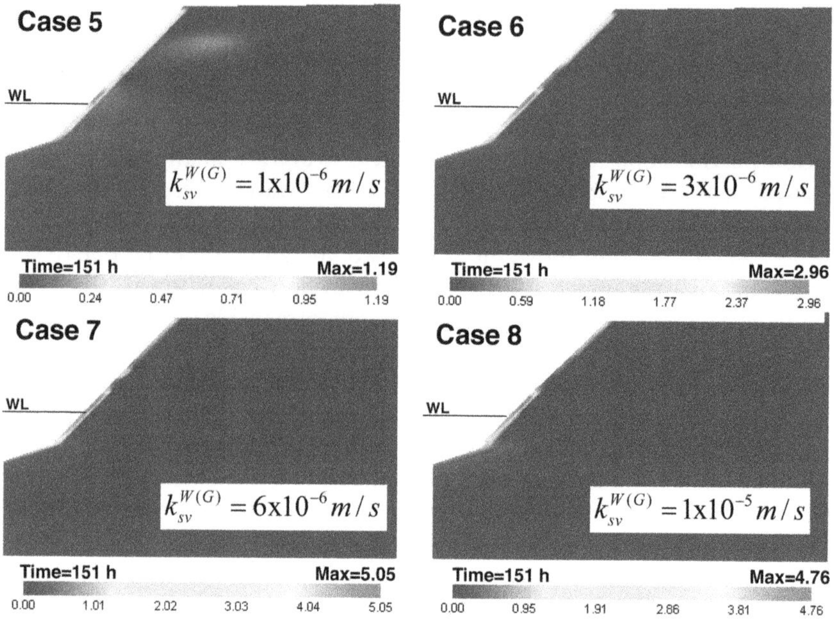

Figure 6.8 Distribution of the horizontal hydraulic gradient: Cases 5–8 (t = 151 h). (After Kimoto, S., Oka, F. and Garcia, E. 2013. Numerical simulation of the rainfall infiltration on unsaturated soil slope considering a seepage flow. *J. SAGS & AGSEA* 44(3): 1–13.)

accumulation of viscoplastic shear strain is localized below the concrete face. This result is very similar to the field observation, where a large deformation is present below the concrete face in the embankment. Cases 5–7 show that the accumulated viscoplastic strains, after major rainfall, are relatively small; however, in Case 8, a large amount of accumulated viscoplastic strain (up to 49%) is generated. The water accumulated inside the embankment results in the accumulation of deformation of the soil at the back of the concrete face. A comparison of Figures 6.8 and 6.9 shows that the localization of deformation below the concrete face is related to the hydraulic gradient and the permeabilities of the soil layers.

By comparing the results of Cases 1–8 in Figures 6.6–6.9, it is possible to see that the effect of the permeability of the slope surface of a river embankment on its local instability is very significant. Larger hydraulic gradients and accumulated viscoplastic shear strains are obtained in the cases where the permeability of the slope surface is smaller compared to the permeability of the soil layer. This could be explained by the accumulation of water behind the slope surface owing to the impediment of the water flow; as a result, larger pore water pressures and hydraulic gradients are generated at the back of the slope that may induce soil erosion. This suggests that some

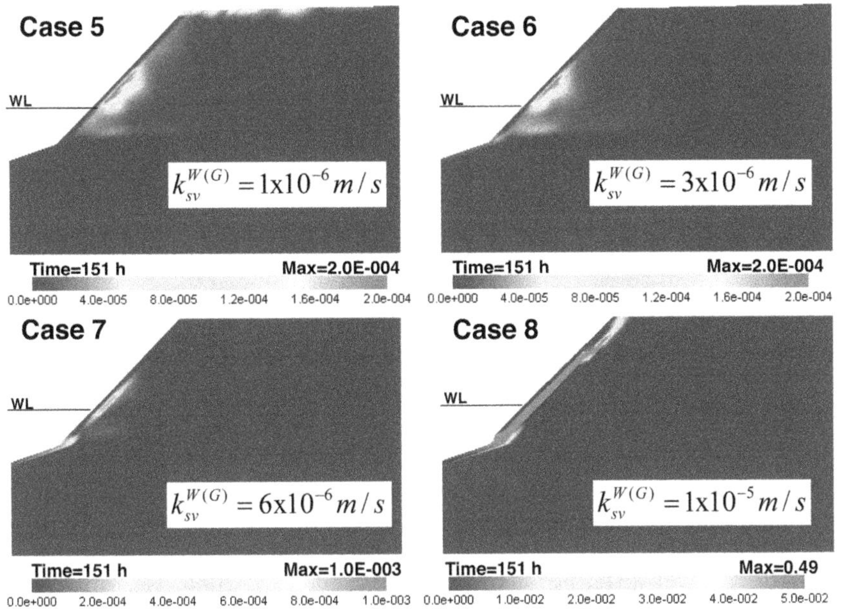

Figure 6.9 Distribution of the viscoplastic shear strain: Cases 5–8 (t = 151 h). (After Kimoto, S., Oka, F. and Garcia, E. 2013. Numerical simulation of the rainfall infiltration on unsaturated soil slope considering a seepage flow. *J. SAGS & AGSEA* 44(3): 1–13.)

geotechnical measures to avoid rainfall infiltration into the slopes, such as concrete faces and cement–soil mixtures, can be harmful to the local stability of the embankments if they are not accompanied by additional measures to reduce the water levels generated by the accumulation of water during rainfall infiltration. This supports the importance of subsurface drainage of the soil structures to improve their local and general stability.

Garcia (2010) and Kimoto et al. (2013) proposed an effective countermeasure to the degradation of an embankment, i.e., horizontal drains in the embankment (Hritzuk et al. 2003; Rahardjo et al. 2003) and compaction of the embankment using the same analysis method mentioned in this chapter.

REFERENCES

Alonso, E.E., Gens, A. and Delahaye, C.H. 2003. Influence of rainfall on the deformation and stability of a slope in overconsolidated clays: A case study. *Hydrogeol. J.* 11(1): 174–192.

Au, S.W.C. 1993. Rainfall and slope failure in Hong Kong. *Eng. Geol.* 36(1–2): 141–147.

Au, S.W.C. 1998. Rain-induced slope instability in Hong Kong. *Eng. Geol.* 51(1): 1–36.
Cai, F. and Ugai, K. 2004. Numerical analysis of rainfall effects on slope stability. *Int. J. Geomech.* 4(2): 69–78.
Chen, H., Dadson, S. and Chi, Y.G. 2006. Recent rainfall-induced landslides and debris flow in Northern Taiwan. *Geomorphology* 77(1–2): 112–125.
Cho, S.E. and Lee, S.R. 2001. Instability of unsaturated soil slopes due to infiltration. *Comput. Geotech.* 28(3): 185–208.
Ehlers, W., Graf, T. and Amman, M. 2004. Deformation and localization analysis of partially saturated soil. *Com. Meth. Appl. Mech. Engng.* 193: 2885–2910.
Garcia, E. 2010. *Numerical Analysis of the Rainfall Infiltration Problem in Unsaturated Soil*, PhD thesis. Kyoto University.
Hritzuk, K.J., Leong, E.C., Rezaur, R.B. and Rezaur, R.B. 2003. Effectiveness of horizontal drains for slope stability. *Eng. Geol.* 69(3–4): 295–308.
Kimoto, S. and Oka, F. 2005. An elasto-viscoplastic model for clay considering destructuralization and consolidation analysis of unstable behavior. *Soils Found.* 45(2): 29–42.
Kimoto, S., Oka, F. and Garcia, E. 2013. Numerical simulation of the rainfall infiltration on unsaturated soil slope considering a seepage flow. *J. SAGS & AGSEA* 44(3): 1–13.
Matsushi, Y., Ayalew, L., Hattanji, T. and Matsukura, Y. 2006. Mechanisms of shallow landslides on soil-mantled hillslopes with permeable and impermeable bedrocks in the Boso Peninsula, Japan. *Geomorphology* 76(1–2): 92–108.
Nakata, Y., Liu, D., Hyodo, M., Yoshimoto, N. and Kato, Y. 2010. Numerical simulation of an expressway embankment slope failure. In: *Unsaturated Soils, Theoretical and Numerical Advances in Unsaturated Soil Mechanics*, Proc. 4th Asian Pacific Conf. on Unsaturated Soils, ed. O. Buzzi, S. Fityus and D. Sheng, 719–724. CRC Press/Taylor & Francis, Boca Raton, FL.
Oka, F., Kimoto, S., Takada, N. and Higo, Y. 2009. A multi-phase elasto-viscoplastic analysis of an unsaturated river embankment associated with seepage flow. In: *Proc. Int. Symp. on Prediction and Simulation Methods for Geohazard Mitigation*, ed. F. Oka, A. Murakami and S. Kimoto, 127–132. CRC Press/Balkema Taylor & Francis Group, Leiden.
Oka, F., Kimoto, S., Takada, N., Gotoh, H. and Higo, Y. 2010. A seepage-deformation coupled analysis of an unsaturated river embankment using a multiphase elasto-viscoplastic theory. *Soils Found.* 50(4): 483–494.
Oka, F., Kimoto, S., Garcia, E. and Yamamoto, H. 2011a. Effect of the rainfall infiltration on an unsaturated levee during seepage flow, computer methods for geomechanics: Frontiers and new applications. In: *Proc. 13th Int. Conf. on IACMAC*, ed. N. Khalili and M. Oeser, 614–619. Center for Infrastructures and Safety, Sydney.
Oka, F., Garcia, E. and Kimoto, S. 2011b. Instability of unsaturated soil during the water infiltration. In: *Advances in Bifurcation and Degradation in Geomaterials*, Proc. 9th Int. Workshop on Bifurcation and Degradation in Geomaterials, ed. S. Bonelli, C. Dascalu and F. Nicot, 293–297. Springer, New York.
Oka, F. and Kimoto, S. 2012. *Computational Modeling of Multiphase Geomaterials.* CRC Press/Taylor & Francis, Boca Raton, FL.

Okada, K. and Sugiyama, T. 1994. A risk estimation method of railway embankment collapse due to heavy rainfall. *Struct. Saf.* 14(1–2): 131–150.

Rahardjo, H., Ong, T.H., Rezaur, R.B. and Leong, E.C. 2007. Factors controlling instability of homogeneous soil slope under rainfall. *J. Geotech. Geoenviron. Eng.* 133(12): 1532–1543.

Rahardjo, H., Hritzuk, K.J., Leong, E.C. and Rezaur, R.B. 2003. Effectiveness of horizontal drains for slope stability. *Eng. Geol.* 69(3–4): 295–308.

Yamagishi, H., Watanabe, N. and Ayalew, L. 2005. Heavy-rainfall induced landslides on July 13, 2004 in Niigata Region, Japan, Monitoring. In *Prediction and Mitigation of Water Related Disasters*, eds. Takara, K., Tachikawa, Y. and Bandara, N.M.N.S, DPRI, Kyoto Univ, Kyoto:6.

Ye, G., Zhang, F., Yashima, A., Sumi, T. and Ikemura, T. 2005. Numerical analyses on progressive failure of slope due to heavy rain with 2D and 3D FEM. *Soils Found.* 45(2): 1–15.

Yoshida, Y., Kuwano, J. and Kuwano, R. 1991. Rain-induced slope failures caused by reduction in soil strength. *Soils Found.* 31(4): 187–193.

Chapter 7

Dynamic analysis of a levee during earthquakes

7.1 INTRODUCTION

Many researchers have reported their observations of damage patterns to river embankments due to earthquakes. Sasaki et al. (1994) and Kaneko et al. (1995) reported various failure modes in dikes and road embankments due to the 1993 Kushiro-Oki earthquake which hit the eastern part of Hokkaido, Japan. Typical damage to river facilities in Hokkaido consisted of dike failure, revetment failure, and sluice damage. Their main aim was, however, to report a new failure mode followed by examples of severely damaged embankments along the Kushiro and Tokachi rivers. In this mode of failure, the earthquake motion triggered cracking in the crest and liquefied the saturated zone at the bottom of the embankment. The slopes slid toward the river on the liquefied material, and the crest largely settled. Such failure mode was not circular, and the shear bands faced the inside or the center of the embankment. Sasaki et al. (1994) explained that this failure mechanism was due to the water trapped inside the embankment body, which was higher than the ground level, and due to the subsidence of the embankment in its foundations below the water table prior to the earthquake. The high-water level in embankments can be due to rainfall infiltration, while pre-subsidence mainly occurs in embankments founded on compressible or soft ground such as loose sand and soft peat foundations.

Harder et al. (2011) and Sasaki et al. (2012) also reported that in the 2011 off the Pacific Coast of Tohoku Earthquake, the most severe damage to the embankments was observed in sections with continuous liquefiable layers in the foundations or sections with compressible soft foundations which allowed the lower portion of the liquefiable sandy embankment fills to settle into the ground below a shallow water table and to become saturated at the time of the earthquake. Finn et al. (1998) studied the failure mechanism suggested by Sasaki et al. (1994) and proposed geometric criteria for the prediction of settlement in dikes against seismic liquefaction. Their simple criteria included the dike's height, the side slopes, and the thickness of both the liquefiable layer and the overlaying foundation layer. Oka et al. (2012) reviewed the main causes and patterns of river

DOI: 10.1201/9781003200031-7

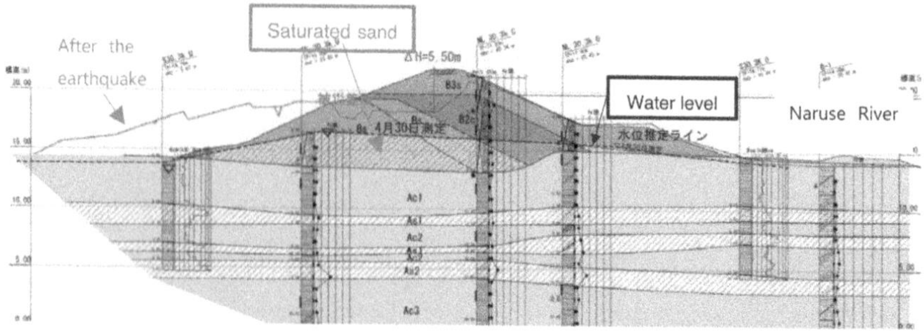

Figure 7.1 Cross section of the embankment at Shinonakanome-jyoryu of the Naruse River. (After Tohoku Regional Development Bureau of MLIT 2011.)

embankment damage in the 2011 Tohoku earthquake with a complete report and examples of river dike failures in many regions based on in situ research carried out by the authors, the Ministry of Land, Infrastructure, Transport and Tourism (MLIT), and the Japan Institute of Construction Engineering (JICE). An example of a severely damaged dike reported by Oka et al. (2012) is the failure of the right bank of the Naruse River, shown in Figure 7.1 (Tohoku Regional Development Bureau of MLIT 2011).

The river dikes of the Naruse River were severely damaged and failed at seven sites. The settled crest and the interior inclined cracks resembled the failure mechanism reported by Sasaki et al. (1994) and Kaneko et al. (1995). Oka and Uzuoka (2018) also reported observations of damage to the surface road and cracks in the levees along Kizu River, Oji River, Akuta River, and Yodo River caused by the earthquake that hit Osaka, Japan, in June 2018. The magnitude 6.1 earthquake caused liquefaction at the toe of the embankment which led to the lateral displacement of the foundations and cracks in the embankment body.

Along with their reported observations, Oka et al. (2012) also used an infinitesimal strain analysis method to analyze the different damage patterns of river embankments subjected to seismic loading. They demonstrated that the impact of long-duration earthquakes can be significantly different from short-duration earthquakes in terms of damage patterns and total settlement. Although the results show good qualitative agreement with what has occurred in the field, the infinitesimal strain analysis method does not seem sufficiently reliable for a post-liquefaction and post-localization response which includes large deformations. Moreover, some modifications to the dimensions of the assumed sublayers seem necessary. Improving the horizontal boundary conditions would yield more realistic simulation results. Subsequently, Sadeghi et al. (2014) and Shahbodagh et al. (2020) showed the importance of the subsoil clay's stiffness on the failure mode of an embankment. They showed how easily the mechanical properties of the foundations can affect the failure mode of embankments with identical dimensions subjected to the same input accelerations.

The damage pattern that has been discussed thus far, although seen in many fields, has not yet been rigorously studied. This is mainly due to the complex behavior of soils and earth structures during earthquakes and especially after liquefaction. In this section, the large deformation dynamic response of river embankments subjected to seismic excitations is presented using a $u - p$ formulation developed in the context of the mixture theory. The spatial discretization of the governing equations is achieved using the finite element method (FEM) in a large deformation regime with the updated Lagrangian description, whereas the time integration is conducted using the Newmark technique. A cyclic elasto-viscoplastic constitutive model is used to simulate the rate-dependent behavior of clay under dynamic loading conditions, while a cyclic elasto-plastic model is adopted for sand. Efforts are made to model the failure modes and damage patterns of river embankments observed in the 2011 Tohoku earthquake. The effects of the ground profile, water table, and earthquake motion on the seismic response and damage pattern of river embankments are particularly emphasized.

7.2 PATTERNS OF FAILURE IN RIVER EMBANKMENTS SUBJECTED TO SEISMIC LOADING

The major seismic failure modes observed in embankments are shown in Figure 7.2. Pattern 1 represents shallow circular sliding, a sort of slight and recoverable damage that happens when an earthquake is not strong enough and the foundations do not settle too much. Shallow sliding is minor damage observed in many embankments subjected to seismic loading and can be easily repaired. Pattern 2 indicates a deep circular sliding failure mode. In this mechanism, sliding slices are long enough to reach the

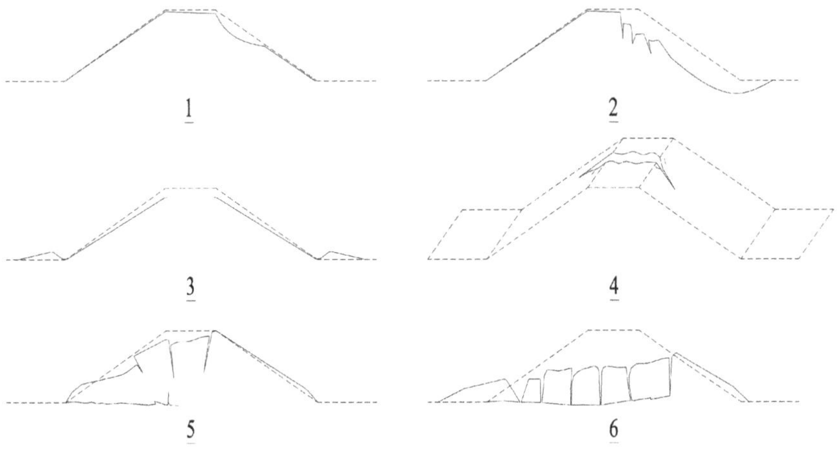

Figure 7.2 Different damage patterns of river embankment.

foundations. Patterns 1 and 2 are the usual modes of failure considered in common geotechnical engineering practice. Pattern 3 shows the settlement of an embankment with the heaving of the toes. This generally occurs in, but is not limited to, embankments on stiff or incompressible foundations. This can also happen to embankments on loose or compressible foundations, provided that short-duration earthquakes do not cause liquefaction. No apparent or shallow tension cracks can be observed in this mode of failure since the embankment body is in compression. This mode of failure was observed in the right embankment at the Kimazuka section of the Naruse River in Miyagi after the 2011 Tohoku earthquake (Oka et al. 2012). Pattern 4 depicts a transversal crack and fissures running the length of an embankment. This mode can be brought about by transverse asynchronous excitations along an embankment. This type of damage pattern was observed on the right bank of the Eai River, in Fukunuma, Miyagi (Oka et al. 2012), and the traverse crack occurred on the left bank of the Kushiro Retarding Basin dike during the 1993 Kushiro-oki earthquake (Sasaki et al. 1995).

Pattern 5 happens when the soil foundations lose their strength due to liquefaction and allow for excessive settlement of the embankment crest and shoulders. The lateral expansion of the embankment toe, which leads to tension fields in the body, creates longitudinal cracks in the embankment body. Although liquefaction of the foundations is a major reason for this behavior, a highly compressible and soft foundation can also lead to the same state even without the occurrence of liquefaction. This type of damage occurred in the left embankment of the Abukuma River (28.6–30.0 km) in Noda district, the left embankment of the Eai River in Uetani district (14.0 km–14.6 km) (Tohoku Regional Development Bureau of MLIT 2011), and the right bank of the Edo River at Nishisekiyado (57.5 km) during the 2011 Tohoku earthquake (Oka et al. 2012).

Pattern 6 shows one of the most severe types of damage that can occur to embankments due to earthquake-induced liquefaction. Lateral expansion of the toe and lateral movement of the embankment slopes caused by the loosened liquefied or excessively settled soft foundations create fissured blocks in the embankment body with large settlements. This mode of failure was observed in Fuchishiri-jyoryu (27.5–27.8 km) and Fukunuma (26.6–26.8 km) along the Eai River, in Edano (30.6–31.4 km) along the Abukuma River, and in Shimonakanome (30.0–30.5 km) along the Naruse River after the 2011 Tohoku earthquake.

7.3 DYNAMIC ANALYSIS OF EMBANKMENT

7.3.1 Governing equations

The governing equations for the dynamic analysis of multiphase geomaterials, i.e., air–water–soil three-phase materials, are described within the framework of the theory of porous media. In the theory, the governing equations are described for the multiphase mixture, i.e., an immiscible mixture of solid and fluids (Oka and Kimoto 2012). The $u - p$ formulation based on the updated Lagrangian method has been employed along with the Jaumann rate of Cauchy's stress tensor for the weak form of the equilibrium equation to deal with large deformation. An elasto-plastic constitutive model and an elasto-viscoplastic constitutive model considering non-linear kinematic hardening were used to describe the plastic and viscoplastic behaviors of the soil skeleton during cyclic loading, respectively. In addition, a soil-water characteristic curve was used as a constitutive equation.

The formulation of the finite deformation analysis by Oka et al. (2018, 2019) and the three-phase porous theory for unsaturated soil by Kato et al. (2014) and Kimoto et al. (2019) were combined to derive the governing equations in the present study. Detailed descriptions of the dynamic finite element formulations are shown in Chapter 3, Shahbodagh (2011), and Oka et al. (2018).

7.3.2 Constitutive equations

The elasto-plastic constitutive model by Oka et al. (1999) and the elasto-viscoplastic constitutive model by Kimoto et al. (2015) are used for the saturated sandy soil and the saturated clayey soil, respectively. The cyclic viscoplastic model was presented in Section 2.2 and the cyclic elasto-plastic model by Oka et al. (1999) was presented in section 9.3 of Oka and Kimoto (2012). An extended elasto-plastic model considering the effect of suction is used for the partially saturated sand (Section 2.4 and Kato et al. 2014).

The soil-water characteristic curve describes the relationship between the volumetric water content or the degree of saturation and the suction of the soil. The curve can be obtained as a measure of the water-holding capacity of the soil because the water content changes when it is subjected to various levels of suction. In the model, the equation proposed by van Genuchten (1980) is adopted.

7.3.3 Conservation equations and discretization

The mass conservation equations for the three phases, that is, the solid, water, and gas phases, are considered. In addition, the conservation equations of the linear momentum for the three phases are used. The sum of the conservation laws of linear momentum for the three phases gives the equation of motion for the whole multiphase mixture. The governing equations presented in Chapter 3 are used.

The weak forms of the continuity equations for water and gas, and the equation of motion are discretized in space using the finite element method. Assuming a two-dimensional plane strain condition, an 8-node isoparametric element is used for the displacement, velocity, and acceleration of the solid skeleton, and a 4-node isoparametric element is used for the pore water pressure and the pore gas pressure. For the time discretization, Newmark's method is adopted. In the following calculations, $\beta = 0.3025$, $\gamma = 0.6$, and the calculated time increment, Δt, is 0.001 s. It should be mentioned that the pore air pressure is assumed to be zero in the following simulations since the effect of the pore air pressure is negligible under drained conditions at the surface boundary.

The numerical program titled "COMVI2D-DY", developed in the geomechanics laboratory at Kyoto University, was used to perform the two-dimensional plane strain finite element analysis under seismic loading.

7.3.4 Simulation model of river embankment and input earthquake

Figure 7.3 shows the finite element mesh with the boundary conditions used for the simulations. A typical river embankment on a clay foundation is modeled. The height of the embankment from the ground surface is 6.0 m, and the inclination of the slope is set to be 1V:2H. To reduce the side boundary effects, semi-infinite elements with equi-displacement conditions are used on the left and right ends. A rigid base is used so that the bottom boundary of the model is fixed. Simulation cases are shown in Table 7.1 and the geometry of the embankment model is depicted in Figure 7.4. In total, seven cases with different ground profiles were simulated.

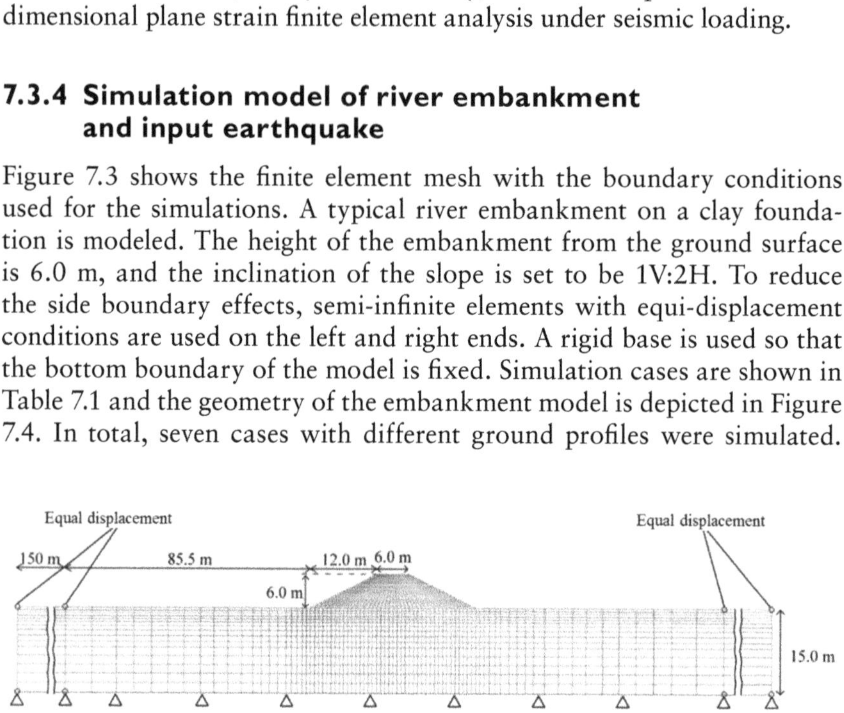

Figure 7.3 Finite element mesh.

Table 7.1 Simulation cases

Case	*Water table*	*Top of water table*	*Settlement depth (m)*	*Thickness of saturated sand (m)*
1-1	Flat	GL −1.5 m	–	–
1-2	Flat	GL −1.5 m	1.0	1.0
2-1	Parabola	GL +0.0 m	1.0	2.5
2-2	Parabola	GL +0.8 m	1.0	3.3
2-3	Parabola	GL +0.0 m	0.5	2.0
2-4	Parabola	GL +0.5 m	0.5	2.5
2-5	Parabola	GL +1.0 m	–	2.5

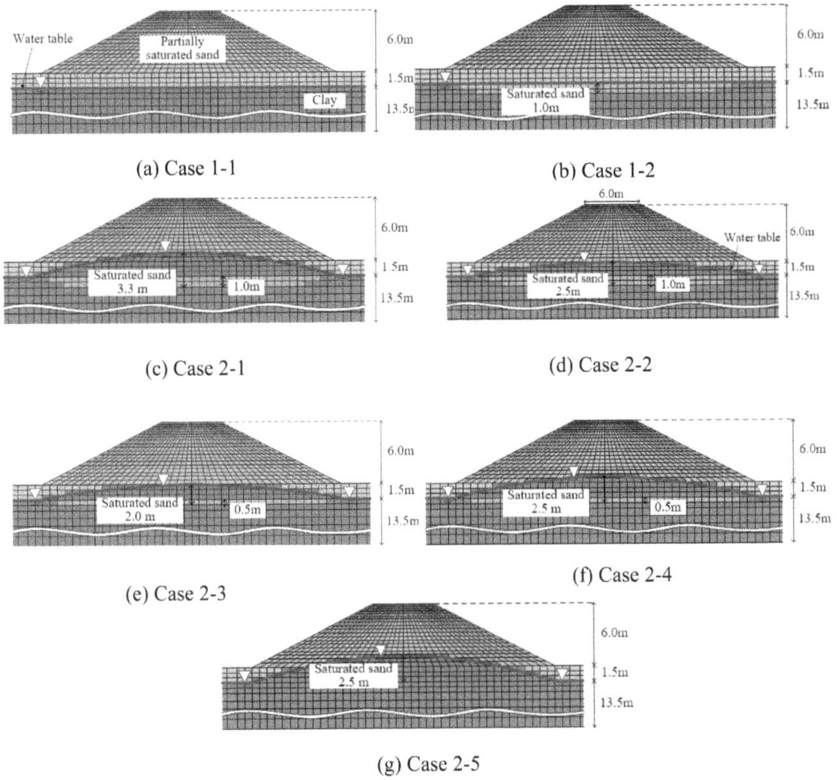

Figure 7.4 Geometry of the embankment model.

Cases 2-1, 2-2, 2-3, and 2-4 correspond to the case in which the bottom of the embankment has settled into the clayey subsoil layer, and the water table exists inside the embankment body in an upward convex shape. The maximum thickness of the saturated zone inside the embankment is from 1.0 m to 3.3 m. Case 2-5 corresponds to the case in which the embankment has not settled and the water table exists inside the embankment body in an upward convex shape.

The embankment body over the saturated zone is assumed to be composed of the partially saturated sandy soil with a maximum saturation of 60%. The initial pore water pressure below the water table is given by the hydrostatic pressure. The pore water pressure and the pore air pressure on the boundaries of the embankment are assumed to be zero (see Figure 7.5). Figure 7.6 shows the input earthquake motion record used in this analysis. The acceleration profile was obtained during the 2011 Tohoku earthquake off the Pacific Coast and it was recorded at a depth of 80 m at Tajiri in Miyagi prefecture (MYGH06, KiK-net) with a maximum acceleration of 152.2 gal.

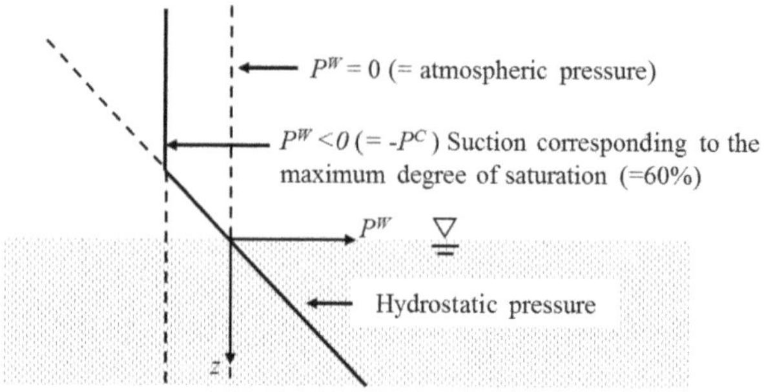

Figure 7.5 Initial water pressure.

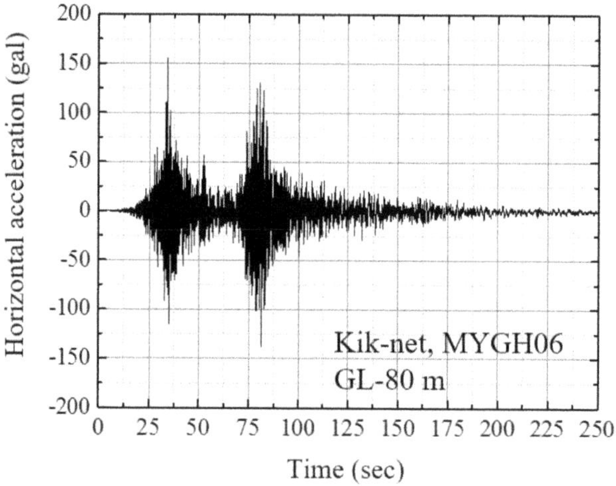

Figure 7.6 Input acceleration (MYGH06, KiK-net).

The elasto-plastic model and the elasto-viscoplastic model are used for the embankment soil and the clayey foundation, respectively. The material parameters of the sandy and clayey soils are listed in Tables 7.2 and 7.3. They were determined based on the experimental data of loose sand (e.g., at Akita Port damaged due to liquefaction [Iai and Kameoka 1993]), and the parameters of the soft clay were determined based on the data for alluvial clay at Torishima along the Yodo River (Mirjalili 2010). The cyclic strength curves for each soil are shown in Figure 7.7, and the soil-water characteristic curve by the van Genuchten–type model for the sand layer is shown in Figure 7.8.

7.3.5 Simulation results

Although Kimoto et al. (2016) and Matsuoka (2016) studied all the simulation cases, herein, we will discuss Case 2-2 in detail. First, the results of

Table 7.2 Material parameters for the constitutive models

Parameter	*Sand*	*Clay*
Density ρ (t/m^3)	1.8	1.66
Initial void ratio e_0	0.8	1.25
Compression index λ	0.025	0.341
Swelling index κ	0.0003	0.019
Quasi-overconsolidation ratio $\sigma'_{mbi}/\sigma'_{m0}$	1.2	1.0
Initial elastic shear modulus ratio G_0/σ'_{m0}	761	400
Stress ratio at phase transformation (sand) M_m^*	0.909	1.24
Stress ratio at failure M_f^*	1.229	1.24
Hardening parameters B_0^*, B_1^*, C_f	5000, 300, 1000	100, 40, 10
Structural parameters n, b	0.50, 50	0.30, 3.6
Dilatancy parameters D_0^*, n	1.0, 4.0	–
Reference value of plastic strain γ_r^{p*}	0.001	–
Reference value of plastic strain γ_r^{e*}	0.003	–
Control parameter of anisotropy C_d	2000	–
Viscoplastic parameter m'	–	24.68
Viscoplastic parameter C_1 (1/s)	–	1.00×10^{-6}
Viscoplastic parameter C_2 (1/s)	–	3.83×10^{-7}
Scalar hardening parameters A_2^*, B_2^*	–	5.9, 1.8
Strain-dependent parameters a', r	–	10, 0.4
Suction parameter S_I	0.2	0.50
Suction parameter s_d	0.1	0.25
Suction parameter S_{IB}	0.2	–
Suction parameter s_{dB}	0.1	–

Table 7.3 Material parameters for the hydraulic property

Parameter	*Sand*	*Clay*
Water coefficient of permeability k_{ws} (m/s)	2.25×10^{-4}	5.87×10^{-10}
Gas coefficient of permeability k_{gs} (m/s)	2.25×10^{-2}	5.87×10^{-8}
Van Genuchten parameter a (1/kPa)	0.4	0.033
Van Genuchten parameter n'	3.0	1.083
Minimum saturation s_{min}	0.99	0.99
Maximum saturation s_{max}	0.00	0.00
Shape parameter of water permeability a	3.0	3.0
Shape parameter of gas permeability b	1.0	2.3

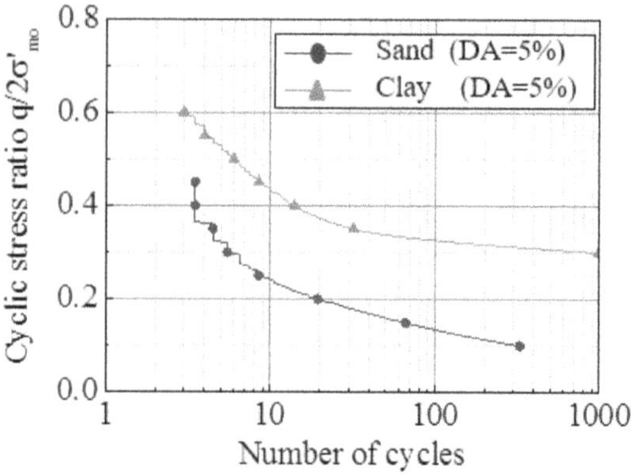

Figure 7.7 Cyclic strength curve of materials.

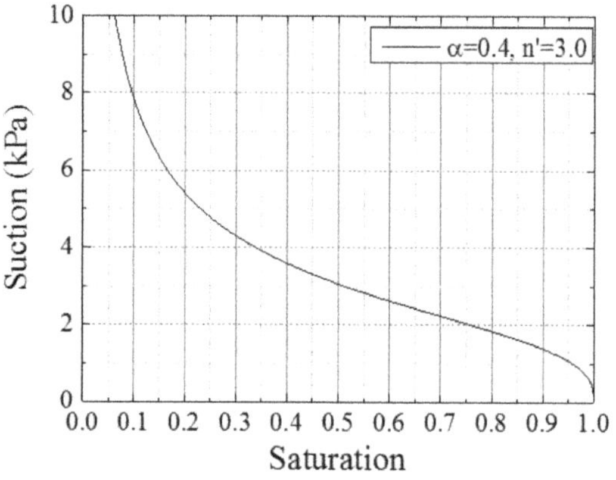

Figure 7.8 Soil-water characteristic curve for sand.

Case 2-2, in which the thickness of the saturated sand is 3.3 m, are shown to discuss the typical behavior of an embankment with liquefaction of the soils inside the levee. The distributions of the skeleton stress decreasing ratio (SSDR), which is defined by $\left(\sigma'_{m0}-\sigma'_{m}\right)/\sigma'_{m0}$ and gives the degree of liquefaction, are shown in Figure 7.9. Liquefaction of the water-saturated region in the embankment can be seen at 66 s after the earthquake. Figure 7.10 shows the distribution of the degree of saturation. The degree of saturation increases inside the embankment in the region where the suction decreases.

The associated distribution of the matric suction is illustrated in Figure 7.11. The decrease in suction leads to the degradation of the embankment. The distributions of the accumulated plastic deviatoric strains are presented in Figure 7.12. Localization of the accumulated plastic shear

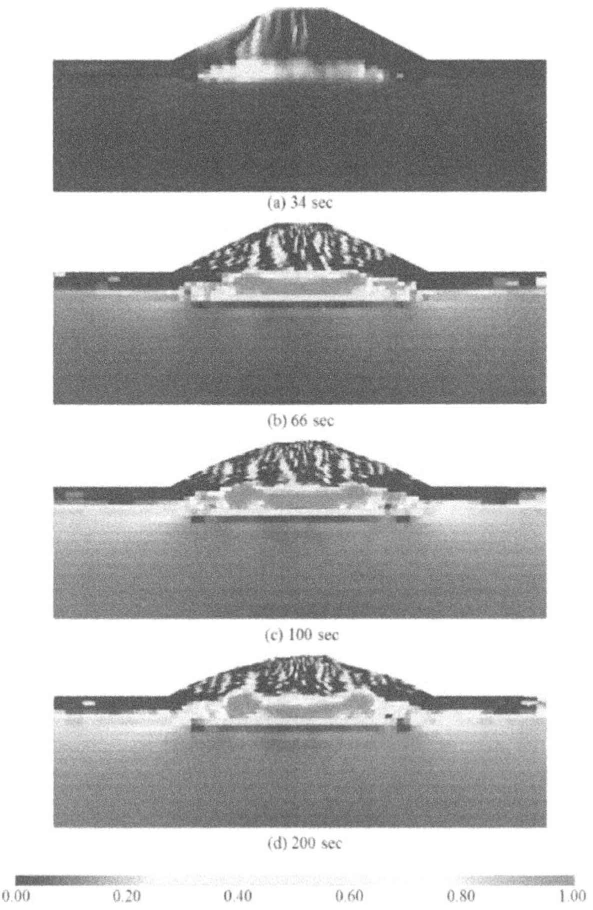

Figure 7.9 Distribution of skeleton stress decreasing ratio (Case 2-2).

strain from the top of the slope, that is, slip failure, can be seen at 66 s after the earthquake. In addition, large strain is obtained in the liquefied region beneath the water table (see Figure 7.12). At 200 s after the earthquake, shear localization can be seen in several regions and complex deformation occurs in the embankment. The distribution of the displacement vectors shown in Figure 7.13 is consistent with the accumulated plastic shear strain. The displacement profiles at P-1, P-2, and P-3 (Figure 7.14) are further discussed here. The vertical and horizontal displacement-time profiles at the top are compared for Cases 2-1, 2-2, and 2-3 in Figure 7.15. The larger settlement occurred for the case with larger thickness of the saturated sand layer. The vertical and horizontal displacement-time profile at the toe of the embankment is shown in Figure 7.16. The largest lateral spreading is

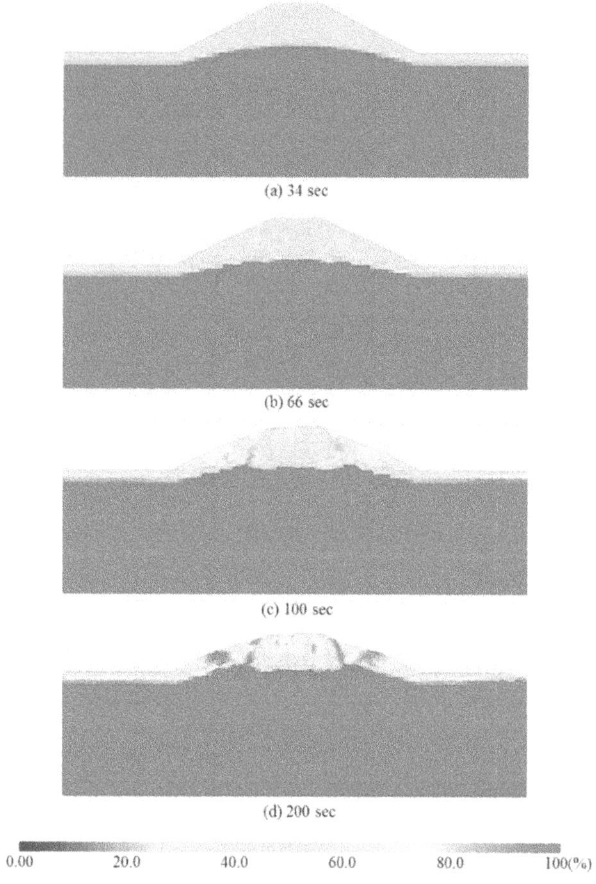

Figure 7.10 Distribution of saturation (Case 2-2).

seen for Case 2-2. The settlement rate, that is, the ratio of settlement to the height of the embankment, is summarized for all the calculation cases in Figure 7.17. Linear relations can be seen between the thickness of the saturated sand layer and the settlement rate (S/H; S: settlement, H: thickness of the sand layer) irrespective of the position of the water table.

7.4 SUMMARY

The non-linear dynamic response of river embankments subjected to seismic excitations was studied using a multiphase-coupled FEM analysis method. Localization of the shear strain was observed in several regions and complex deformation occurred in the embankment due to the liquefaction

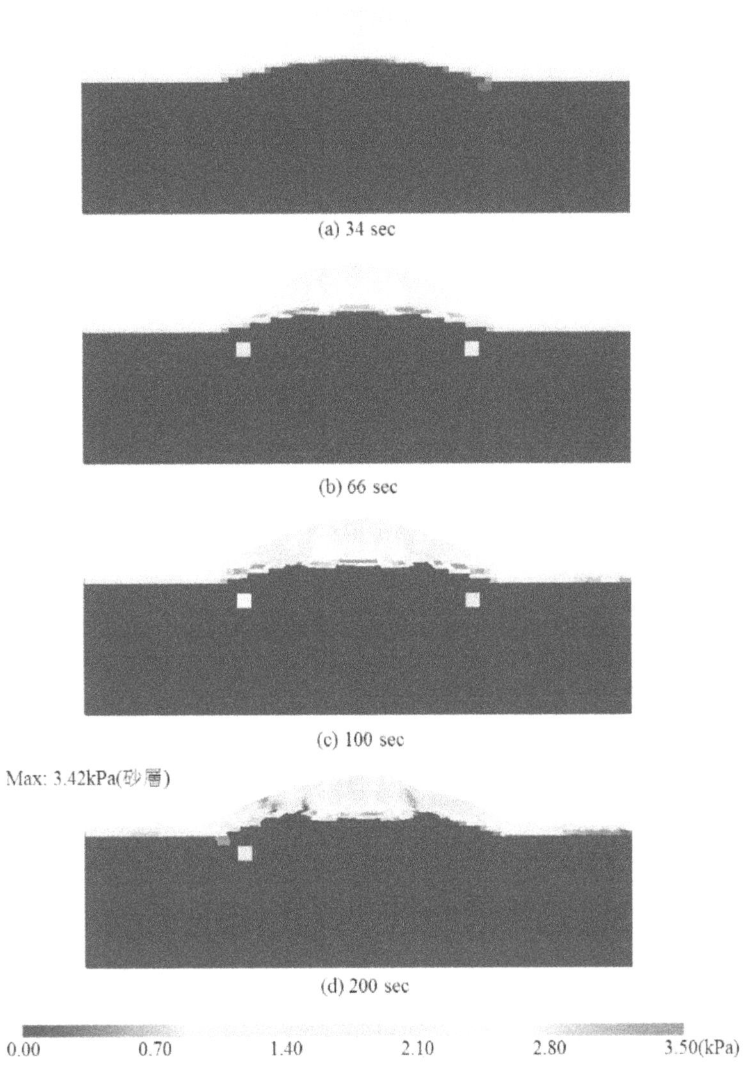

Figure 7.11 Distribution of suction (Case 2-2).

of sandy soil in the embankment. The failure mechanism of Pattern 5 in Section 7.2 was well simulated, namely, the failure of the embankment was caused by the liquefaction of the saturated soil inside the embankment body, as suggested by Sasaki et al. (1994). Comparing the results of the cases with the different thickness of the saturated sandy layer, a linear relation was obtained between the thickness of the saturated sand layer and the settlement rate.

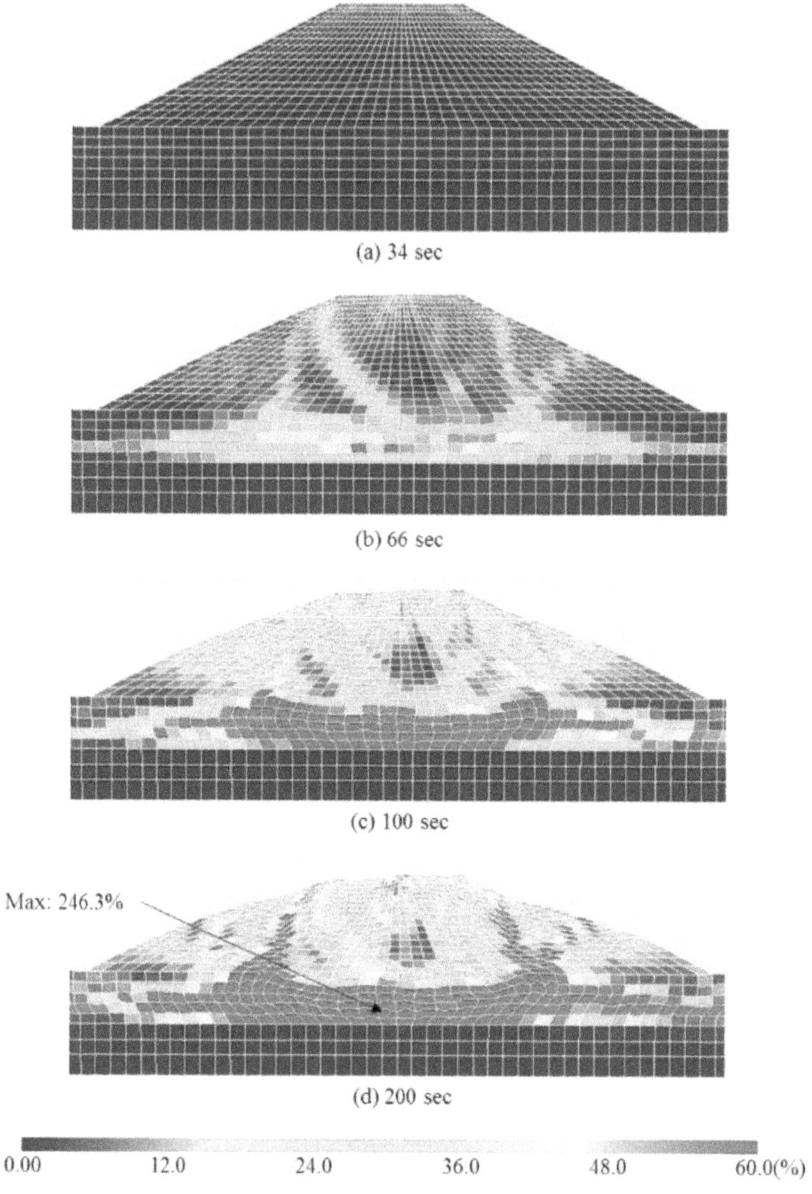

Figure 7.12 Distribution of accumulated plastic/viscoplastic deviatoric strain (Case 2-2).

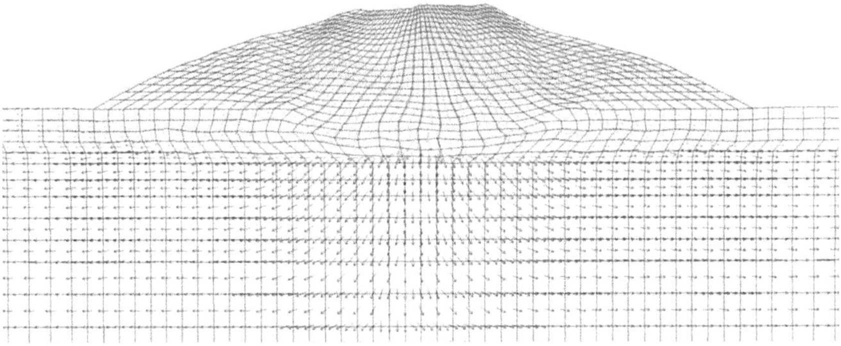

Figure 7.13 Distribution of displacement vectors (Case 2-2, after 200 s).

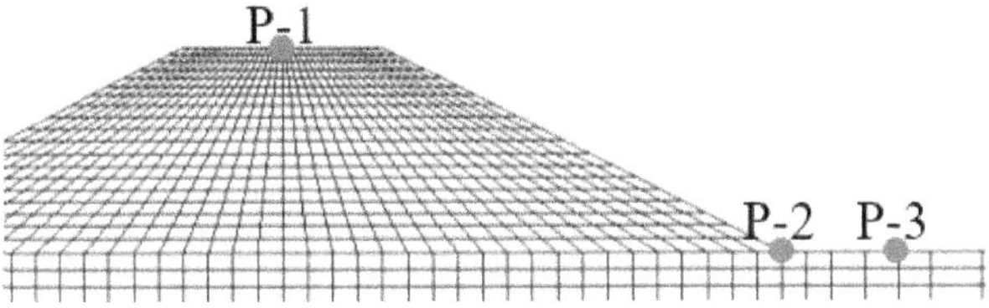

Figure 7.14 Position of the output for displacement.

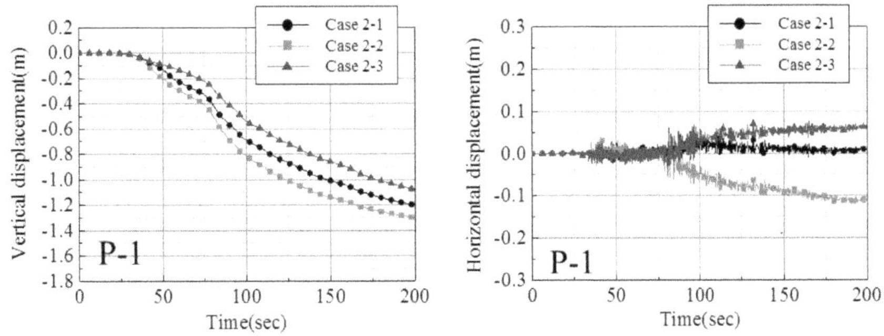

Figure 7.15 Vertical and horizontal displacement-time profile at the top (Cases 2-1, 2-2, 2-3).

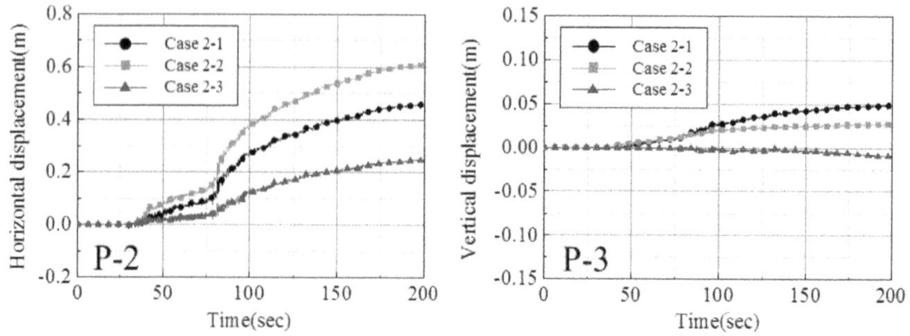

Figure 7.16 Vertical and horizontal displacement-time profile at the toe (Cases 2-1, 2-2, 2-3).

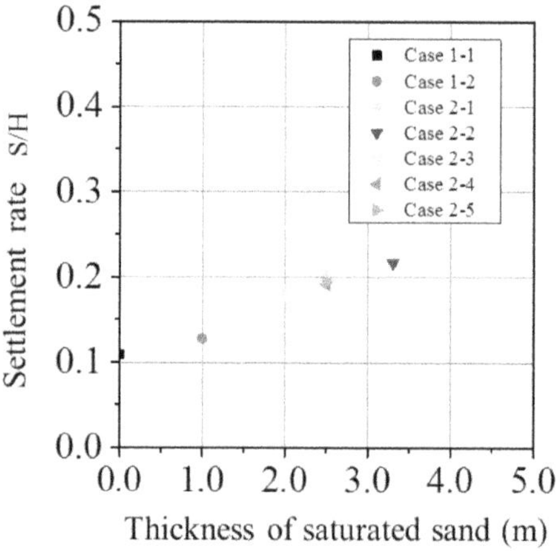

Figure 7.17 Settlement rate *S/H* and thickness of saturated sand layer.

REFERENCES

Finn, W.D.L. 1998. Estimating post-liquefaction displacements in embankment dams and priorities remediation measures. In: *Proc. International Conference on Case Histories in Geotechnical Engineering*, ed. C. Prakash, 1156–1171. Missouri University of Sci. Tech., Rolla, MI.

Harder, L.F., Leslie, F., Kelson, K.I., Kishida, T. and Kayen, R. 2011. Preliminary observations of the Fujinuma dam failure following the March 11, 2011 Tohoku Offshore Earthquake, Japan. Geotechnical Extreme Events Reconnaissance. Report No. GEER-25e.

Iai, S. and Kameoka, T. 1993. Finite element analysis of earthquake induced damage to anchored sheet pile quay walls. *Soils Found.* 33(1): 71–91.

Kaneko, M., Sasaki, Y., Nishikawa, J., Nagase, M. and Mamiya, K. 1995. River dike failure in Japan by earthquakes in 1993. In: *Proc. 3rd International*

Conference on Recent Advances in Geotechnical Earthquake Engineering and Soil Dynamics, ed. S. Prakash, 495–498. Missouri University of Sci. Tech., St. Louis, MI.

Kato, R., Oka, F. and Kimoto, S. 2014. A numerical simulation of seismic behavior of highway embankments considering seepage flow. In: *Computer Methods and Recent Advances in Geomechanics: Proc. of the 14th International Conference of International Association for Computer Methods and Recent Advances in Geomechanics*, ed. F. Oka, A. Murakami and S. Kimoto, 755–760. CRC Press/Taylor & Francis, London.

Kimoto, S., Sahbodagh, K.B., Mirjalili, M. and Oka, F. 2015. A cyclic elastovisco-plastic constitutive model for clay considering the nonlinear kinematic hardening rules and the structural degradation. *Int. J. Geomech.* 15(5): A4014005.

Kimoto, S., Matsuoka, H. and Oka, F. 2016. Liquefaction analysis of river embankment considering water-saturated region in the embankment based on finite deformation theory. Proc. of 51th Annual Meeting of JGS, D-07:1063-1064 (in Japanese).

Kimoto, S., Yui, H. and Oka, F. 2019. Dynamic analysis of partially saturated river embankment considering liquefaction using a multi-phase coupled FEM analysis method. In: Proc. 16th Asian Regional Conference of Geotechnical Engineering, IDS:SA02, SA06, TC/ATC #(Optional): ATC6.

Matsuoka, H. 2016. *Liquefaction Analysis of Levee on the Soft Clay Foundation Considering Ground Improvement*, Ms. thesis. Kyoto University (in Japanese).

Mirjalili, M. 2010. *Numerical Analysis of a Large-Scale Levee on Soft Soil Deposits Using Two-Phase Finite Deformation Theory*, PhD thesis. Kyoto University.

Oka, F., Yashima, A., Tateishi, A., Taguchi, Y. and Yamashita, S. 1999. A cyclic elasto-plastic constitutive model for sand considering a plastic-strain dependence of the shear modulus. *Geotechnique* 49(5): 661–680.

Oka, F., Tsai, P., Kimoto, S. and Kato, R. 2012. Damage patterns of river embankments due to the 2011 off the Pacific Coast of Tohoku Earthquake and a numerical modeling of the deformation of river embankments with a clayey subsoil layer. *Soils Found.* 52(5): 890–909.

Oka, F. and Uzuoka, R. 2018. Disaster of Yodo river embankments due to 2018 Osaka Earthquake. Report to the Japanese Geotechnical Society, July 25, 2018.

Oka, F. and Kimoto, S. 2012. *Computational Modeling of Multiphase Geomaterials.* Boca Raton, FL: CRC Press/Taylor & Francis.

Oka, F., Shahbodagh Khan, B. and Kimoto, S. 2019. A computational model for dynamic strain localization in unsaturated elasto-viscoplastic soils. *Int. J. Numer. Anal. Meth. Geomech.* 43(1): 138–165.

Sadeghi, H. 2014. *Dynamic Analysis of River Embankments during Earthquakes Based on Finite Deformation Theory Considering Liquefaction*, PhD thesis. Kyoto University.

Sasaki, Y., Oshiki, H. and Nishikawa, J. 1994. Embankment failure caused by the Kushiro-Oki earthquake of January 15, 1993, Performance of ground and soil structures during earthquakes. In: Proc.13th ICSMFE. JGS, 61–68.

Sasaki, Y., Towhata, I., Miyamoto, K., Shirato, M., Narita, A., Sasaki, T. and Sako, S. 2012. Reconnaissance report on damage in and around river levees caused by the 2011 off the Pacific coast of Tohoku earthquake. *Soils Found.* 52(5): 1016–1032.

Shahbodagh, B. 2011. *Large Deformation Dynamic Analysis Method for Partially Saturated elasto-Viscoplastic Soils*, PhD thesis. Kyoto University.
Shahbodagh, B., Sadeghi, H., Kimoto, S. and Oka, F. 2020. Large deformation and failure analysis of river embankments subjected to seismic loading. *Acta Geotech.* 15(6): 1381–1408.
Tohoku Regional Development Bureau of MLIT. 2011. Material-3, Interim Report by the 3rd Investigative Commission for the Restoration of the River Dikes of Kitakami river etc., July 29, 2011, 18–19 (in Japanese).
van Genuchten, M.Th. 1980. A closed-form equation for predicting the hydraulic conductivity of unsaturated soils. *Soil Sci. Soc. Am. J.* 44(5): 892–898.

Chapter 8

Numerical analysis of excavation in soft ground

8.1 INTRODUCTION

Many urban cities have been constructed over soft Holocene deposits with high groundwater levels. It is known that, in the open-cut construction method, the ground and the retaining walls are often seen as unstable if a complex construction process is not carefully followed. Many accidents during excavation works in such soft ground have been reported (e.g., Tanaka 1994; Whittle and Davies 2007). In order to prevent the failure of the ground and to reduce the deformation of the ground during excavation works, a reliable prediction method as well as high-performance construction is required. For the last three decades, numerical methods have been used to predict the deformation and the failure of the ground and earth structures, such as retaining walls. Since the ground consists of sandy and clayey soils and the boundary conditions are generally complex, appropriate modeling is required to accurately predict the response of the structure-ground system. There are two methods for such modeling. One is the imposed displacement method in which the displacement is calculated by the beam-spring model, using the subgrade reaction modulus, and then applied to the surrounding ground. The other is a method in which the stress is released at the nodes of the finite elements corresponding to the excavation of the soil. The single-phase elastic finite element method is a useful method because it requires only a small number of parameters and is effective within the range of small strains, but not in the range of plasticity. Additionally, it is known that the soil-water coupled finite element method is a powerful tool for the analysis of the ground and soil structures. In practice, however, this method has not been popular at the design stage of projects (e.g., AIJ 2006). For cases in which the deformation gradients are small, the elastic method is applicable. However, for large-scale excavation problems in soft deposits with high water levels, the deformation is elasto-plastic or elasto-viscoplastic. Particularly for soft clay deposits, creep, as well as consolidation, may occur even though the creep strain is small. Oka et al. (2016) used an elasto-viscoplastic model for their excavation analysis since experimental results showed that Osaka clay exhibits

DOI: 10.1201/9781003200031-8

strain rate sensitivity. Also, it is known that the initial volumetric strain rate is not zero in soft clay layers subjected to self-weight. For the construction of the Osaka Station building, we conducted a coupled soil-water elasto-viscoplastic finite element analysis during and after the excavation work to examine the construction method and to confirm the safety of the excavation work by comparing the measured and simulated results. A numerical analysis of the excavation by Oka et al. (2016) was presented and compared with the measured results. In the construction project, a deep mixing soil improvement method, called the "soil buttress method", was used as an additional technique for stability. Furthermore, the computer program COMVI2D-EX10 developed by Oka et al. at Kyoto University was used (Oka et al. 2002; Boonlert et al. 2007).

8.2 GEOTECHNICAL PROFILE AT THE EXCAVATION SITE

The construction was carried out in Umeda district, Osaka, Japan. The soil profile in Umeda district is as follows: the surface fill layer comprises a heterogeneous mixture of sand, gravel, crushed stone, and construction material with a thickness of 2.5 m. Underlying the fill layer, there is an alluvial sandy layer (As) with a thickness of 2.5 m containing the fine content. Below the sandy layer (As), there is a soft alluvial clay layer (Ac) with a thickness of 20 m. The standard penetration test (SPT) values for the upper layer are 1–3 and they increase just around the bottom alluvial clay layer. At a depth of 16 m, the plasticity index and the OCR of the alluvial clay layer are Ip = 24 and OCR = 1.55, respectively, while at a depth of 24 m, they are Ip = 47 and OCR = 1.56. The alluvial layers are underlain by a gravel layer (Dg1) which comprises gravel with particle sizes of 2–30 mm. The groundwater of this Dg1 layer is artesian and its SPT value is around 50. Below the Dg1 layer, there is a Pleistocene clay deposit (Dc) with a thickness of 10 m and an SPT value of 3–13. The Dc layer is underlain by a Pleistocene gravel layer (Dg2) that contains gravel with particle sizes of 2–30 mm. The SPT value is more than 50 and the groundwater is artesian.

8.3 OUTLINE OF THE EXCAVATION PROJECT

The Japan Railway Company constructed a new building just north of the station as part of a reconstruction project of Osaka Umeda Station (JR West Japan Railway Company 2011; Takada 2012). The construction started in April 2007 and ended in May 2010. The construction was carried out just beside an existing railway line. Figure 8.1 shows a plane view of the excavation works. The nearest point along the retaining walls is 1.3 m from the piers supporting the existing elevated railway tracks. The depth of

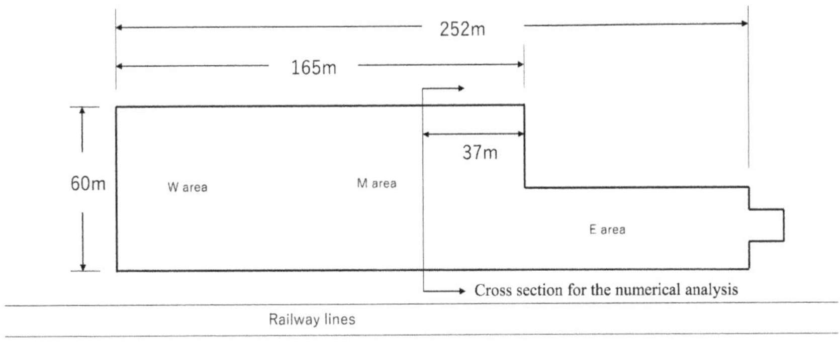

Figure 8.1 Plane view of the excavation area.

the excavation is 20 m, the width of the cross section is 29–62 m, and the length is 252 m, as shown in Figure 8.1.

In this construction, the excavation was performed inside the surrounding earth retaining walls. A plane strain two-dimensional analysis was preferable due to its shorter computation time. According to Finno's study (Finno et al. 2007), the results of a three-dimensional analysis and a two-dimensional plane strain method will be similar when the ratio of the length of the excavation over the width is more than 2.0 and the ratio of the length over the depth is more than 6.0. In this case, the length of the excavation (L) was 165–252 m, the width (B) was 29–62 m, and the depth (H) was 19.66 m, indicating that the analysis under plane strain conditions would yield similar results to the three-dimensional analysis, since L/H = 8.4–12.8 and L/B = 2.7–8.7.

The construction of the retaining walls and the excavation are described as follows. The retaining wall beside the existing elevated piers was a reinforced concrete (RC) diaphragm wall with high rigidity to prevent deformation. It had a thickness of 1.2 m and a length of 44 m, and was used as the main structure of the building. The other retaining wall was a trench cutting remixing deep wall (TRD) that can maintain continuity in the direction of the retaining wall. The excavation technique was the inverted lining method applied to effectively reduce settlement and leaning; the excavation was performed downward (top to bottom) by strutting the retaining wall.

8.4 SOIL IMPROVEMENT TECHNIQUE

The buttress type of deep mixing method was used to reduce the displacement of the retaining wall beside the railway structures in which the ground was partially improved (Building Center of Japan 2002). This method is economical and less time-consuming than the method in which the entire ground is improved. The allowable displacement limits for the safety of trains was 15 mm for both vertical and horizontal track deviations. The relative displacement limit between piers for the elevated tracks was 10 mm (secondary allowable action limit) considering the results from an analysis of elastic frames. The specifications for the improved ground were determined as follows. The mixing diameter of one circular column was 0.9 m and its equivalent effective width was 0.8 m, which was equivalent to the square column. The installation interval was 4.5 m, the improved length was 7.2–20.4 m, and the installation depth was GL −0.0 m to −28 m, which was determined by considering the trafficability and economic conditions.

8.5 EXCAVATION PROCESS

The excavation area was divided into three parts from the west side, namely, W area, M area, and E area, as shown in Figure 8.1. The inverted lining method was used for the excavation to prevent deformation; six bench beams (struts used in the step-by-step excavation) and seven bench excavations were employed. After the second, fourth, and sixth excavation steps, preloads (50%) were installed for the oblique beams to prevent deformation. The construction procedure (Steps 1–13) for the central part of the cross-sectional excavation at M area is shown in Figure 8.1. Construction area M consisted of four areas (MA, MB, MC, MD) at each step (Oka et al. 2016). After the second excavation step, the other steps were conducted starting from the area farthest from the existing pier used to support the elevated tracks.

A relief well was installed at the center of the excavation area to lower the water level of the pressurized Pleistocene gravel layer, Dg1. To reduce heaving of the ground during excavation procedures, it is common to use a relief well to decrease the water level. The analysis took account of the changes in the water level of Dg1 through the relief well. The water level of the relief well is shown in Figure 8.4.

8.6 NUMERICAL SIMULATION

8.6.1 Elasto-viscoplastic constitutive equation

The constitutive model used for clay is the elasto-viscoplastic model by Kimoto and Oka (2005), presented in section 5.7 in Oka and Kimoto (2012), which describes the degradation induced by strain, such as strain-softening behavior. The model has been applied to several geotechnical engineering problems such as the consolidation of a soft foundation (Karim et al. 2013) and the analysis of an excavation (Oka et al. 2009). The model parameters for the clay layers used in this analysis were determined through undrained triaxial compression tests with different strain rates and through consolidation tests to clarify the viscoplastic rate effect.

8.6.2 Determination of material parameters used in the analysis

For the determination of the material parameters, we used the experimental results for clay layers and the empirical relations for sandy and gravel layers, such as As, Dg1, and Dg2, and partly considered the measurement results for the permeability. In order to clarify the mechanical characteristics of Osaka Umeda clay and its material parameters, the model simulations for undrained triaxial tests were performed using the elasto-viscoplastic model by Kimoto and Oka (2005). Undrained triaxial compression tests were carried out for the soil specimens taken from the Ac2 layer at a depth of GL –16.0 m and from the Ac3 layer at a depth of GL –21.0 m. The results are shown in Figure 8.2. The undrained triaxial compression tests were conducted with different axial strain rates, namely, 0.005% min and 0.05% min, to obtain viscoplastic parameter m' for the specimen sampled at a depth of GL –21.0 m. The viscoplastic parameters of the Ac1 layer were determined using the triaxial test results.

The material parameters used in the simulation of the triaxial test results (Ac3 GL –21.0 m) were initial elastic shear modulus G_0 = 3000 kPa; compression index λ = 0.314; swelling index κ = 0.0369; initial void ratio e_0 = 1.37; consolidation yield stress $\sigma'_{mbi} = 292$ kPa; stress ratio at maximum compression $M^*_m = 1.071$; viscoplastic parameter m' = 12.22; viscoplastic parameter $C_1 = 1.5 \times 10^{-7}$ (1/s), $C_2 = 1.0 \times 10^{-7}$ (1/s); structural parameter

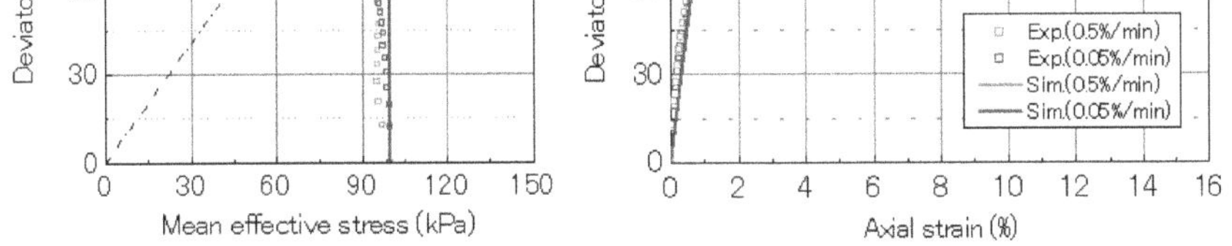

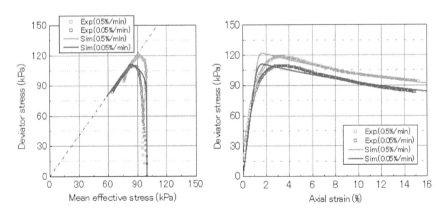

Figure 8.2 Simulation and experimental results of undrained triaxial compression tests. (After Oka, F., Takada, N., Shimono, K., Higo, Y. and Kimoto, S. 2016. A large-scale excavation in soft Holocene deposit and its elasto-viscoplastic analysis. *Acta Geotech.* 11: 625–642.)

$\sigma'_{maf} = 130\,\text{kPa}$; structural parameter $\beta = 5.0$; and initial mean effective stress $\sigma'_{m0} = 130\,\text{kPa}$.

The method of parameter determination was described in detail by Kimoto and Oka (2005) and Oka and Kimoto (2012). The compression index, λ, the swelling index, κ, and the consolidation yield stress (conventionally called preconsolidation pressure) σ'_{mbi} were determined by consolidation tests. The initial elastic shear modulus (tangential modulus at zero strain), G_0, was determined by adjusting it to the test results. The viscoplastic parameter, m', exhibited the strain rate effect. The value of the parameter is similar to the value of the Osaka soft clay in the Nakanoshima district of Osaka City (Mukai 2009; Oka et al. 2009, 2016).

The viscoplastic parameters C_{01} and C_{02} can be determined from the initial values of the strain rates, i.e., initial volumetric and deviatoric strain rates, but it is not straightforward. Hence, these values can be determined from the overall behavior of the soil. For parameters C_{01} and C_{02}, we firstly assumed that $C_{01} = C_{02}$. Then, if the dilatancy effect was not appropriately simulated, we changed C_{02} to more accurately reproduce the stress path or the volumetric strain. The structural degradation parameter, σ'_{maf}, was determined by the difference between the peak stress and the stress at a large strain. The other structural parameter, β, was determined considering the rate of strain softening. The simulated and experimental results are shown in Figure 8.2, which indicates that the strain rate sensitivity and the strain-softening behaviors are reproduced in both the stress paths and the stress–strain relations. The elastic shear modulus was determined by the stress–strain relation and the consolidation tests for the clayey layers, while the permeability coefficients were determined from the in situ pumping tests for the sandy layers and by multiphase consolidation tests for the clayey layers. For the elastic modulus, we used Young's modulus, which was a smaller value than the initial tangential value, to take account of the strain dependency considering the overall curve fitting. It is known that the viscoplastic parameters C_{01} and C_{02}, determined from laboratory tests, might be larger than those under in situ conditions because the strain rate after consolidation in the laboratory is larger than that under in situ conditions possibly caused by disturbance as indicated by Leroueil et al. (1983). And the viscoplastic parameters C_{01} and C_{02} are inversely proportional to the consolidation yield stress which depends on the strain rate (Oka and Kimoto 2012). Considering the previous application of the model to Osaka clay (Oka et al. 2009), C_{01} and C_{02} were assumed to be 10^{-2} times smaller for Ac2 and 10^{-3} times smaller for Ac3. When we used smaller values for C_{01} and C_{02}, the peak values for the deviator stress were about 25% higher than those mentioned above. Figure 8.2 indicates that the behavior is pseudo-elastic before peak stress, but the viscoplastic strain is not zero. It is worth noting that the strain level of the ground is not high in the whole domain, but may be high locally, e.g., near the excavation front. In addition, the strain at the peak stress obtained in the triaxial test is probably larger than that obtained under in situ conditions because of the sample disturbance.

The elastic shear modulus, G_0, adopted in the finite element analysis was larger than that of the triaxial test result due to the sample disturbance induced by the sampling process. For soils without experimental or measured values, such as the As (alluvial sand), Dg1 (Pleistocene gravel), and Dg2 (Pleistocene gravel) layers, we estimated the elastic coefficients and Young's modulus E, which is not a secant modulus, but includes some strain dependency based on the empirical formula using the SPT N value by the Railway Technical Research Institute of Japan (RTRI), which was first proposed by Uto (1967). RTRI (2001) proposed several methods to determine Young's modulus including the shear wave velocity. However, we adopted a method using the SPT N value because of the limited data as $E = 2500N$ kN/m^2, N: SPT value. This formula has been successfully used in the design of railway-related architectural structures for many years, confirming the agreement with the measured results (RTRI 2001). Poisson's ratio ν is given by the formula: $\nu = K_0/(1+K_0)$, where K_0 is the earth pressure coefficient at rest calculated by Jaky's formula. In the calculation, Poisson's ratio for the As, Dg1, and Dg2 layers has been estimated considering the above relation because of the lack of experimental data for these layers. RTRI (2001) proposed a correction of Young's modulus estimated by $E = 2500N$ based on the measurement of rebound for the excavation. RTRI (2001) recommended that a two to three times larger Young's modulus was appropriate when analyzing the rebounding induced by the relatively large excavation in the construction works. For the other parameters for the As and Dg layers, we used parameters obtained for the same layers at Nakanoshima near Umeda in Osaka which were reported by Oka et al. (2009) due to the lack of experimental data.

The soil improved by deep mixing for the soil buttress was assumed to be elastic and the material parameters for the soil buttress–soft soil composite material were determined by weighing the elastic constants of the soil buttress and the soft soil between the soil buttresses with respect to the widths of the materials for the interval of the installation (about 4.5 m). The composite material was assumed because of the two-dimensional modeling. The interval of the soil buttress = 4.5 m, the length of the soil buttress = 7.2–20.4 m, and the effective width of the soil buttress = 0.8 m.

The material parameters used in the analysis were determined using the above methods (Oka et al. 2016). In the finite element analysis, the sandy soil was modeled as a linear elastic model because insufficient data existed and large deformation was not expected. The soil buttress body was also modeled as linear elastic materials. Young's modulus of the soil buttress was determined by one-dimensional compression test results, and its permeability coefficient and Poisson's ratio were determined by the value recommended by the Japan Architecture Center (2002). The RC continuous wall was modeled as a linear elastic material and the strut beam was a linear elastic spring (RTRI 2004). The permeability of the RC continuous wall was determined as 1.0×10^{-10} (m/s) based on the records of the measured data.

8.6.3 Numerical modeling of excavation

For the numerical simulation, we used a finite element method for two-phase mixtures based on the finite deformation theory (Higo et al. 2006). The governing equations and the numerical method, described in Chapter 3 and in section 6.5 of Oka and Kimoto (2012), were used here for the analysis with the assumption of the saturation equal to 1.0. The mesh for the analysis comprised 2530 elements and 7793 degrees of freedom. We used the 8-node isoparametric elements for the solid skeleton and the 4-node isoparametric elements for the pore water pressure.

The excavation procedure used in this analysis was based on the method by Brown and Booker (1985). The equivalent stress due to the excavation was calculated from the excavated elements adjacent to the excavation boundary using the following equation:

$$\{\boldsymbol{R}_E\} = \int_{V_E} [\boldsymbol{B}]^{\mathrm{T}} \{\sigma\} dv - \int_{V_E} [\boldsymbol{N}]^{\mathrm{T}} \rho \{b\} dv \tag{8.1}$$

where $\{\boldsymbol{R}_E\}$ is the force vector appropriate to the nodes on the excavated surface; $[\boldsymbol{N}]$ is a matrix for the shape function of the 8-node isoparametric element; $[\boldsymbol{B}]$ is the strain-displacement matrix; $\{\sigma\}$ is the total stress vector in the elements to be excavated; ρ is the wet density; $\{b\} = \{0, g\}$ is the body force vector, g is the gravitational force; and V_E is the volume of the excavated element. The excavation process was incrementally analyzed. The finite elements, boundary conditions, and a cross section of the analysis are shown in Figure 8.3.

The areas were divided into small sections at each excavation stage. In the analysis, we adopted the two-dimensional analysis method under plane strain conditions for a cross section. Half of the cross section of the excavation area was simulated due to the symmetricity of the problem, i.e., from the diaphragm wall to the center of the excavation area. The excavation procedures for sections M of the construction area shown in Figure 8.1 were considered in the analysis. Figure 8.4 indicates the excavation-time profile.

For the relief well, shown in Figure 8.3, the water head boundary was employed. The prescribed pore water pressure, corresponding to the water head of the relief well installed in stratum Dg1 (Figures 8.3 and 8.4), was given to the boundary.

For the initial stress calculation, we only considered the dead load of 80 kN/m^2 induced by the existing pier supporting the elevated tracks as well as the self-weight. In addition, other loads induced by various activities such

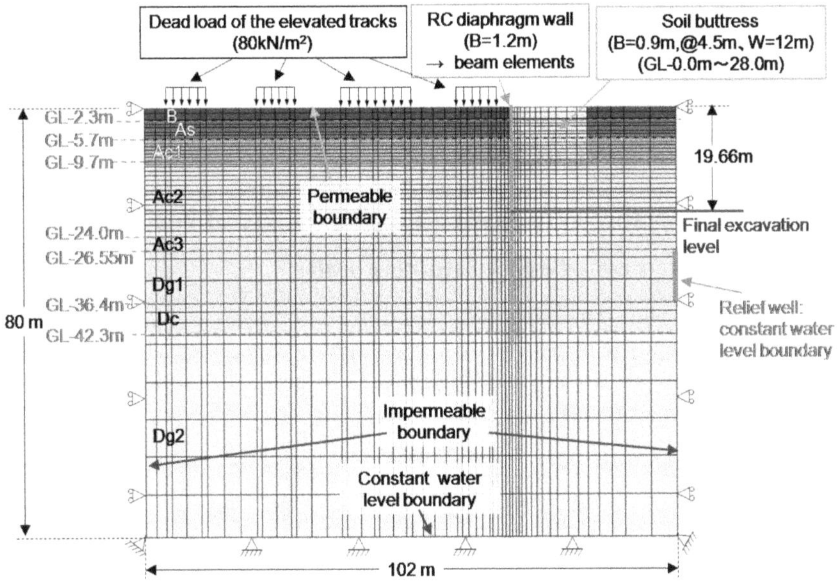

Figure 8.3 Finite element mesh and boundary conditions. (After Oka, F., Takada, N., Shimono, K., Higo, Y. and Kimoto, S. 2016. A large-scale excavation in soft Holocene deposit and its elasto-viscoplastic analysis. *Acta Geotech.* 11: 625–642.)

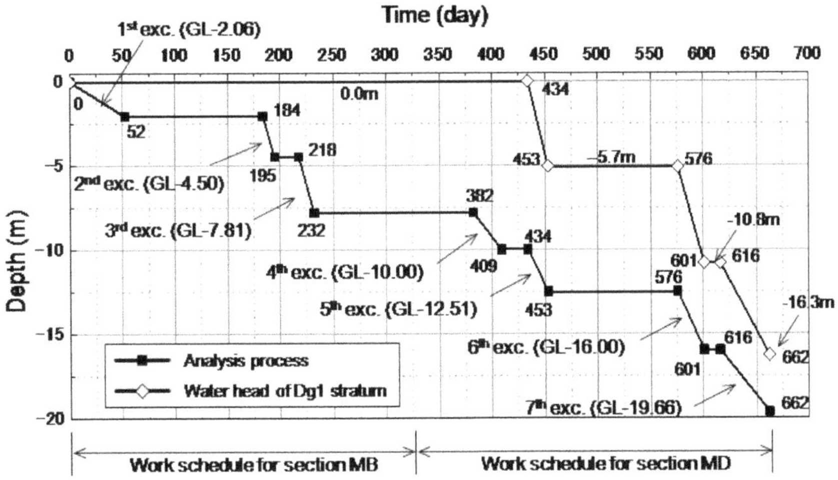

Figure 8.4 Excavation-time profile of M area (MB, MD). (After Oka, F., Takada, N., Shimono, K., Higo, Y. and Kimoto, S. 2016. A large-scale excavation in soft Holocene deposit and its elasto-viscoplastic analysis. *Acta Geotech.* 11: 625–642.)

as railway trains, cars, and a crowd of people at the station were applied to the surface as 10 kN/m^2.

8.7 NUMERICAL RESULTS AND COMPARISONS WITH MEASUREMENTS

8.7.1 Comparison of measured and simulated results

8.7.1.1 Displacement of earth retaining wall

The preloads of the strut beam were used in the analysis (Oka et al. 2016). The distribution of the measured and the simulated horizontal displacements of the retaining wall is shown in Figure 8.5. In general practice, it is often

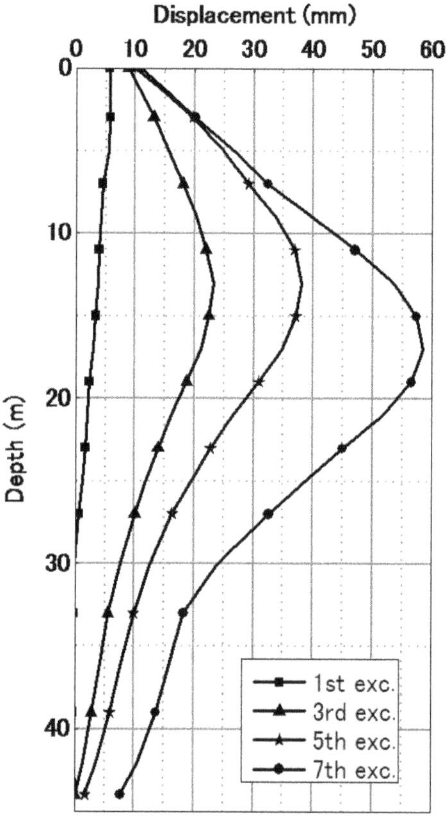

Figure 8.5a Displacement profile of retaining wall (a: observed). (After Oka, F., Takada, N., Shimono, K., Higo, Y. and Kimoto, S. 2016. A large-scale excavation in soft Holocene deposit and its elasto-viscoplastic analysis. *Acta Geotech.* 11: 625–642.)

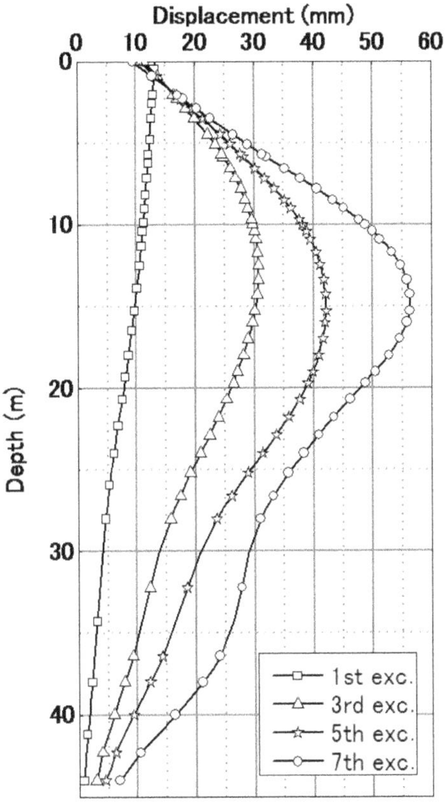

Figure 8.5b Displacement profile of retaining wall (b: simulated). (After Oka, F., Takada, N., Shimono, K., Higo, Y. and Kimoto, S. 2016. A large-scale excavation in soft Holocene deposit and its elasto-viscoplastic analysis. *Acta Geotech.* 11: 625–642.)

assumed that the deepest point of the retaining wall is a fixed point, but in the Osaka soft clay deposit, it has been reported that the deepest point of the wall, which penetrates into the diluvial gravel layer, displaces even in the deep diluvial gravel layer in the case of large-scale excavations (Oka et al. 2009). For this reason, we measured the top of the wall with respect to the fixed zero displacement point, and then inversely calculated the displacement of the deepest point of the wall. The displacement into the excavation area was positive. It was observed that the excavation induced the wall to move into the excavation area, and that the value was 14.0 mm, 751 days after the beginning of the excavation and 89 days after the end of the excavation.

Figure 8.5(a, b) shows that the trends of the displacement developments of the retaining wall agree with those of the measured ones. Until the third stage of the excavation, the simulated displacement was larger than that of the measured one, namely, by 10 mm. This may have been due to the setup of the initial stress levels. The initial value was not zero because of the

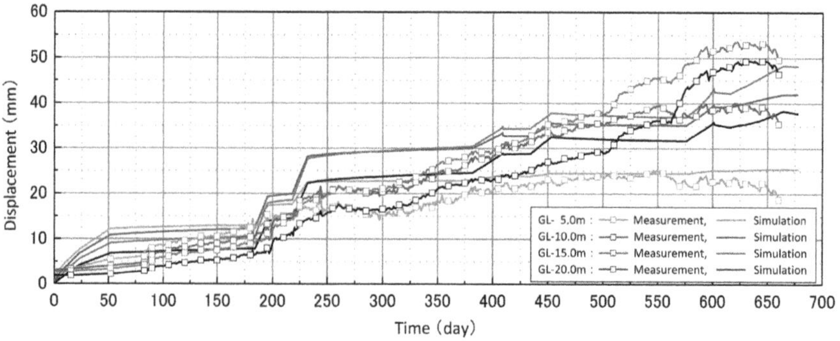

Figure 8.6 Horizontal displacement of the retaining wall-time profile. (After Oka, F., Takada, N., Shimono, K., Higo, Y. and Kimoto, S. 2016. A large-scale excavation in soft Holocene deposit and its elasto-viscoplastic analysis. *Acta Geotech.* 11: 625–642.)

excavation of the other areas, and this effect was ignored in the analysis. In order to fully take into account the effect of all the other areas, a three-dimensional analysis was required with enough data.

Figure 8.6 shows the time history of the horizontal displacements of the retaining wall (Oka et al. 2016). Although the excavation was not performed from day 52 to day 161, the deformation continued. This was partly due to the viscoplastic creep and the consolidation phenomena of the clay layer; the present analysis can reproduce the behavior of the retaining wall from day 52 to day 161, although the value is rather small.

8.7.1.2 Settlements and pore water pressure in layers

The vertical settlements were obtained at the installation points of the settlement gauges shown in Figure 8.7. Figure 8.8 shows the vertical displacement-time profile at depths of GL –2.5 m and GL –7.5 m behind the retaining wall studied, respectively. The simulated results corresponded to the first stage of the excavation until day 52. In this period, the simulated settlement was larger. This may have been due to the underestimation of the elastic modulus.

Between days 52 and 184, the excavation was stopped, and during this period, a settlement of almost zero was observed. Between days 184 and 232, i.e., during the second and third stages of the excavation, the settlement developed and corresponded to the measured results. From days 232 to 382, the excavation was stopped again; the simulated settlement was almost zero and a small settlement was seen in the measurement. This observed settlement may have been due to the excavation at the other sections. This point is partly due to the assumption of a two-dimensional analysis of the three-dimensional real behavior. Between days 382 and 453, i.e., during the

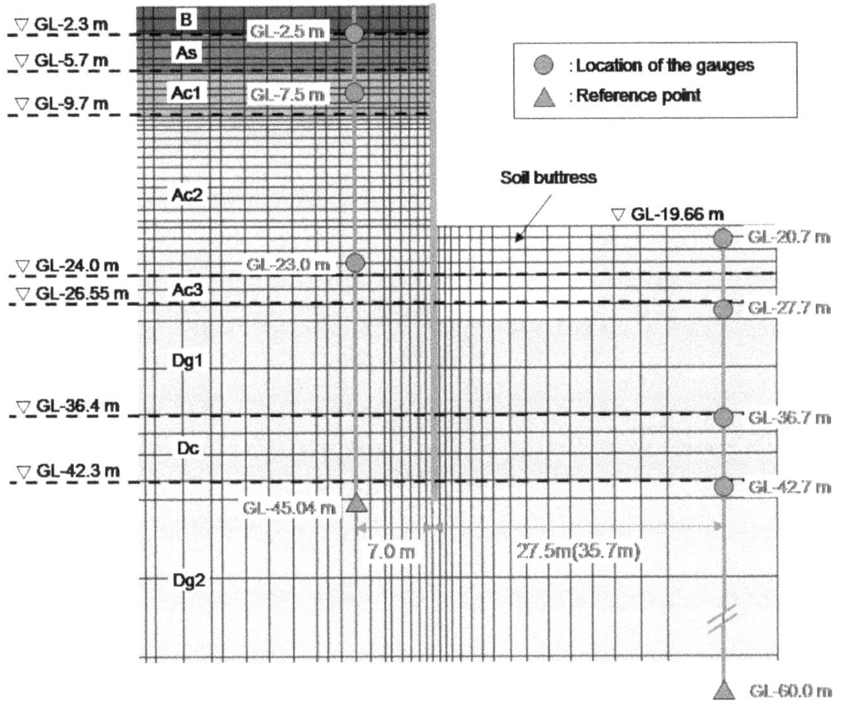

Figure 8.7 Location of settlement gauges. (After Oka, F., Takada, N., Shimono, K., Higo, Y. and Kimoto, S. 2016. A large-scale excavation in soft Holocene deposit and its elasto-viscoplastic analysis. *Acta Geotech.* 11: 625–642.)

fourth and fifth stages of the excavation, the simulated settlement was similar to the measured value. Then, the excavation was stopped again between days 453 and 576. In this period, the settlement still developed, although the magnitude was small. This was probably due to viscoplastic creep and/or consolidation.

The simulated pore pressure-time profile during the excavation was obtained. Figure 8.9 shows that the pore water pressure increased, e.g., between days 232 and 382, while the settlement was almost zero during this period. This phenomenon can be explained as follows: the pore water pressure developed due to the development of the viscoplastic volumetric strain during this period, but the dissipation of the pore pressure was not enough for the settlement of the ground. This is a feature of the consolidation of the viscoplastic ground. Then, the pore water pressure gradually decreased with time. This indicates the progress of the consolidation of the clay layer. Settlement occurred during the sixth and seventh excavation stages, i.e., between days 576 and 662. The increase in displacement just after the excavation was consistent with the displacement of the retaining

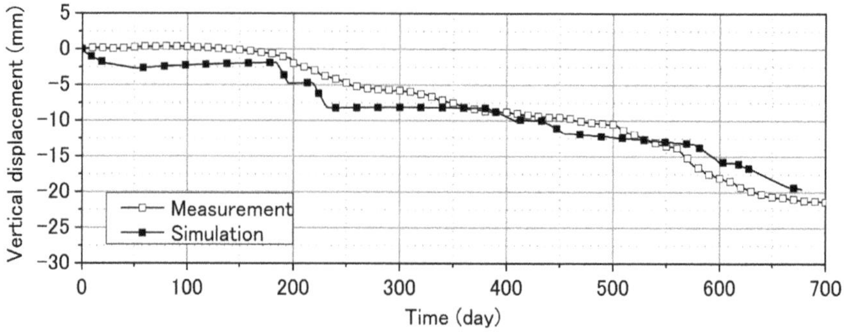

(a) GL-2.5m

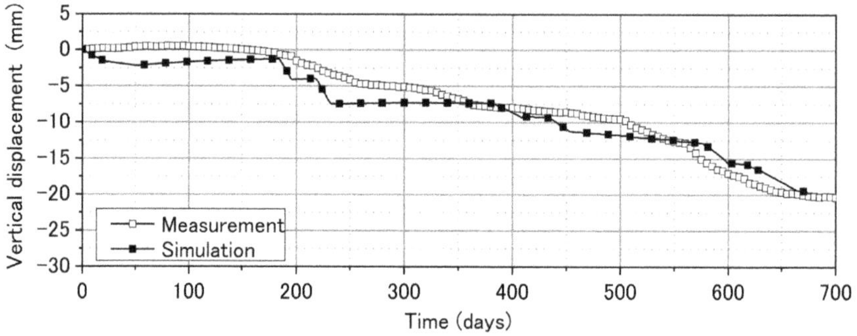

(b) GL-7.5m

Figure 8.8 Vertical settlement-time relation at a depth of (a) GL −2.5 m and (b) −7.5 m behind the retaining wall. (After Oka, F., Takada, N., Shimono, K., Higo, Y. and Kimoto, S. 2016. A large-scale excavation in soft Holocene deposit and its elasto-viscoplastic analysis. *Acta Geotech.* 11: 625–642.)

wall, shown in Figure 8.4. The settlement-time profile at a depth of 7.5 m was similar to that at a depth of 2.5 m. The overall trend of the simulated and observed settlement-time profiles at depths of 2.5 m and 7.5 m was in good agreement, except for the settlement behavior at the shallow depth in the very early stages of the excavation.

8.8 SUMMARY

1. The simulated deformation of the retaining wall was in relatively good agreement with the overall measured results, but larger than the measured one. The large displacement in the excavation area was partly due to the building piles used in the construction and partly to an underestimation of the elastic modulus. In addition, it was found that the bottom of the retaining wall moved toward the excavation area.

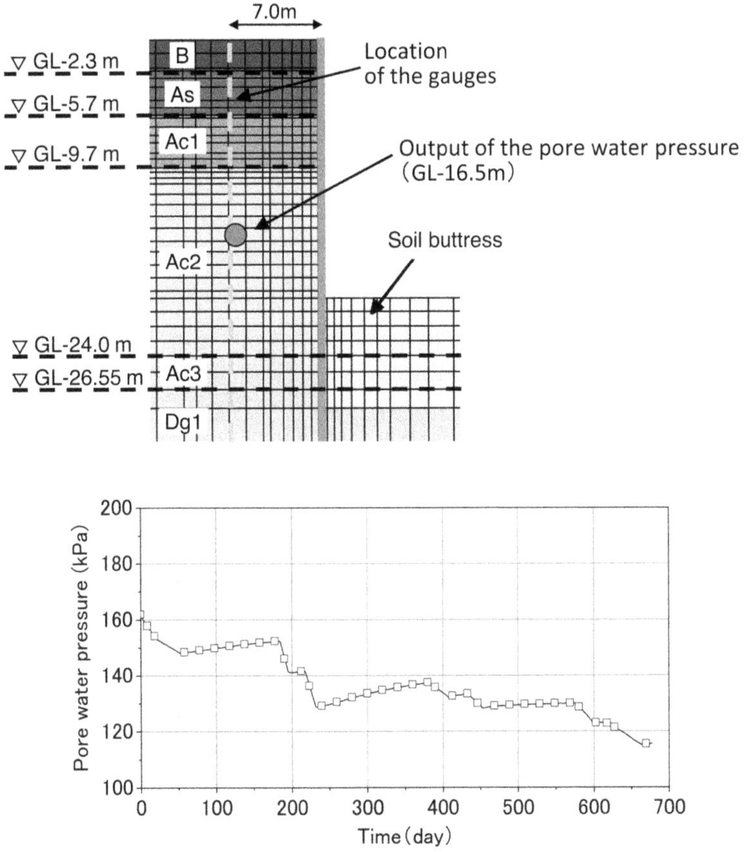

Pore water pressure – time profile

Figure 8.9 Pore water pressure-time profile. (After Oka, F., Takada, N., Shimono, K., Higo, Y. and Kimoto, S. 2016. A large-scale excavation in soft Holocene deposit and its elasto-viscoplastic analysis. *Acta Geotech.* 11: 625–642.)

2. The settlement behavior of the ground behind the wall was well simulated. The progress of the deformation, even when the excavation was stopped, indicates the possible effects of viscoplastic creep and associated consolidation.
3. The rebound behavior in the excavation area was simulated (Oka et al. 2016), but the simulated rebound value was larger. This indicates that the complexity of the construction procedure affects the results and that the estimation of the elastic modulus is of great importance.
4. The buttress type of deep mixing method was effective in reducing deformation even when the improvement area was limited.

5. Comparing the measurement and the simulation results, it was shown that the numerical simulation method reproduces the overall behavior of earth retaining walls relatively well only if the displacement at the bottom of the walls is taken into account. However, more attention must be given to the construction procedure and an accurate determination of the parameters is required in order to more accurately simulate the measured response.

REFERENCES

Architectural Institute of Japan (AIJ) 2006. Estimation of the effect of neighboring construction and countermeasure. In Symposium on Various Problems of Design of the Earth Retaining Wall, Committee Report (in Japanese).

Boonlert, S., Oka, F., Kimoto, S., Kodaka, T. and Higo, Y. 2007. Elasto-viscoplastic finite element study of the effect of degradation on bearing capacity of footing on clay ground. *Geomech. Geoeng.* 2(4): 235–251.

Brown, P.T. and Booker, J.R. 1985. Finite element analysis of excavation. *Comput. Geotech.* 1(3): 207–220.

Building Center of Japan. 2002. *Revised Version of the Guide Line for the Design of Soil Improvement of Building and Its Quality Control, Deep and Shallow Mixing Method Using Cement Type Solidification Material.* Showa Jyoho Pub. Co., Tokyo (in Japanese).

Finno, R.J., Blackburn, J.T. and Roboski, J.F. 2007. Three-dimensional effects for supported excavations in clay. *J. Geotech. Geoenviron. Eng.* ASCE, 133(1): 30–36.

Higo, Y., Oka, F., Kodaka, T. and Kimoto, S. 2006. Three-dimensional strain localization of water-saturated clay and numerical simulation using an elasto-viscoplastic model. *Philosoph. Mag.* 86(21–22): 3205–3240.

Japan Architecture Center. 2002. *Revised Version: Design of Improved Ground for Building and Guideline for Quality Control-Deep and Shallow Mixing Method Using Cement Solidification Material.* Shyowa Jyoho Pub. Co., Tokyo (in Japanese).

Karim, M.R., Oka, F., Krabbenhoft, K., Leroueil, S. and Kimoto, S. 2013. Simulation of long-term consolidation behavior of soft sensitive clay using an elasto-viscoplastic constitutive model. *Int. J. Numer. Anal. Meth. Geomech.* 37: 2801–2824.

Kimoto, S. and Oka, F. 2005. An elasto-viscoplastic model for clay considering destructuralization and consolidation analysis of unstable behavior. *Soils Found.* 45(2): 29–42.

Leroueil, S., Samson, L. and Bozozuk, M. 1983. Laboratory and field determination of preconsolidation pressures at Gloucester. *Can. Geotech. J.* 20(3): 477–490.

Mukai, H. 2009. *Deformation Analysis of Earth-Retaining Wall and Ground Induced by the Open Cut Construction*, PhD thesis. Kyoto University (in Japanese).

Oka, F. and Kimoto, S. 2012. *Computational Modeling of Multiphase Geomaterials.* CRC Press/Taylor & Francis, Boca Raton, FL.

Oka, F., Higo, Y. and Kimoto, S. 2002. Effect of dilatancy on the strain localization of water-saturated elasto-viscoplastic soil. *Int. J. Sol. Struct.* 39(13–14): 3625–3647.

Oka, F., Higo, Y., Nakano, M., Mukai, H., Izunami, T., Kakeda, S., Amano, K. and Nagaya, J. 2009. Deformation analysis of earth-retaining wall during excavation of Nakanoshima soft clay deposit by an elasto-viscoplastic finite element method. *JSCE J. Geotech. Geoenviron. Eng.* C65(2): 492–505 (in Japanese).

Oka, F., Takada, N., Shimono, K., Higo, Y. and Kimoto, S. 2016. A large-scale excavation in soft Holocene deposit and its elasto-viscoplastic analysis. *Acta Geotech.* 11(3): 625–642.

Railway Technical Research Institute of Japan. 2001. *Standard for Design of Railway Structures and Interpretation, Open-Cut Tunnel*, Maruzen Co. Tokyo: 59–64. and Referential material No. 11, *Real Rebound and its Calculation Examples*, Maruzen Co. Tokyo: 429–436 (in Japanese).

Railway Technical Research Institute of Japan. 2004. *Standard for Design of Railway Structures and Interpretation, Concrete Structures*, Maruzen Co.: 79 (in Japanese).

Takada, N. 2012. *Elasto-Viscoplastic Analysis of Large Scale Excavation of Soft Clay Deposit Considering Soil Improvement*, PhD thesis. Kyoto University (in Japanese).

Tanaka, H. 1994. Behavior of a braced excavation in soft clay and the undrained shear strength for passive earth pressure. *Soils Found.* 34(1): 53–64.

Uto, K. 1967. *Sounding of Ground, Foundations of Structures*, Kansai Branch of Japan Society of Testing Materials: 33–66 (in Japanese).

West Japan Railway Company, JR West Japan Consulting Company and Joint Venture Group for the Construction and Improvement of Osaka Station. 2011. *Report of Technical Committee of Construction and Improvement for Osaka Station*, May 2011 (in Japanese).

Whittle, A.J. and Davies, R.V. 2007. Nicoll Highway collapse: Evaluation of geotechnical factors affecting design of excavation support system. In *Proceedings of the International Conference on Deep Excavations*, 28–30 June 2006, Singapore: 1–16.

Chapter 9

Elasto-viscoplastic constitutive modeling of the swelling process of unsaturated clay

9.1 INTRODUCTION

Expansive soil is found in many areas of the world, particularly in semi-arid regions. Expansive soil (e.g., bentonite or bentonite mixtures) undergoes huge volumetric changes when exposed to water. The heaving of foundations is one of the serious consequences of founding lightly loaded structures on expansive soils. Under displacement-confined conditions, expansive soil will exhibit considerable swelling pressures, which result in serious damage to buildings and other structures. Hence, it is necessary to have an understanding of the mechanical behavior of this type of material.

Many attempts have been made in the past to understand the swelling mechanism of expansive soils. Volume changes in clay are due to the clay–water–cation interaction (Bolt 1956). The Gouy–Chapman diffuse double layer theory (Gouy 1910; Chapman 1913) has been the most widely used approach to relate clay compressibility to basic particle–water–cation interaction (Olson and Mesri 1970; Mesri and Olson 1971; Marcial et al. 2002). As shown in Figure 9.1, the swelling of bentonite is due to the absorption of water molecules into the interlayers (e.g., Butscher et al. 2016; Kurahayashi 1980). The bonding force between the negative-charged surface and the interlayer cations is lower than the interaction force between the interlayer cations and the water molecules. Consequently, the gap between the layers is widened as the interlayer cations attract the water molecules. When the interaction between the interlayer cations and the water molecules reaches its limit, the swelling stops.

During the past few decades, a number of experimental and theoretical research works have been carried out on bentonite and bentonite–soil mixtures. The relationship between swelling deformation and the distance between two montmorillonite layers was proposed (Komine and Ogata 1996). Sridharan and Choudhury (2002) proposed a swelling pressure equation for Na-montmorillonite while analyzing the compression data of slurred samples of montmorillonite reported by Bolt (1956). However, some researchers (Mitchell 1993; Tripathy et al. 2004) have shown that very little information is available on the use of the diffuse double layer

DOI: 10.1201/9781003200031-9

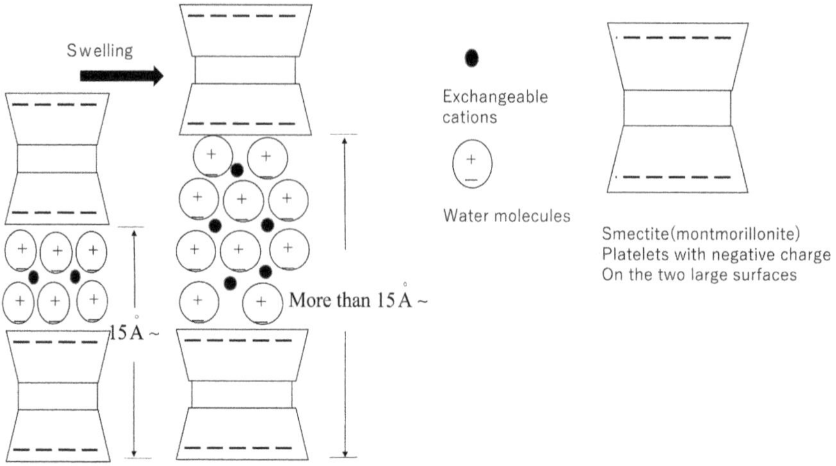

Figure 9.1 Swelling structure of smectite (montmorillonite).

theory for the determination of the swelling pressure of compacted bentonite. In addition, the microstructure of compacted expansive clays has been studied by Push (1982), who observed a double structure made up of clay aggregates and large macrostructure pores. Gens and Alonso (1992) presented the conceptual basis of a model for expansive soils. Several numerical models have also been proposed to simulate expansive soil (Alonso et al. 1991, 1995, 1999, 2005) based on the elasto-plastic theory. According to their theory, two levels of structures are considered for expansive soils: (1) a microstructural level that corresponds to the active clay minerals and their vicinity, and (2) a macrostructural level that accounts for the larger structural soil arrangements. The microstructure, namely, the swelling domain that expands when hydrated, is assumed to be water saturated even at high levels of suction. In contrast, the macrostructure is assumed to be unsaturated when subjected to suction, and its behavior may be described by a conventional framework for unsaturated soils. In addition, a theoretical model was proposed by Shuai and Fredlund (1998) to describe the volume changes during various oedometer swelling tests.

A rigorous experimental study of expansive soil presents significant challenges because the experiments need to cover a large suction range. Lloret et al. (2003) carried out experiments on highly compacted bentonite. Additionally, researchers have conducted swelling pressure tests on compacted bentonite (Push 1982; Kanno and Wakamatsu 1992; Komine and Ogata 1996, 2003).

In this chapter, an elasto-viscoplastic swelling model for unsaturated bentonite is presented based on the constitutive model developed by Feng et al. (2006), Feng (2008) and Oka et al. (2008, 2009) for unsaturated soils. An internal variable, H, which controls the growth of the absorption of water into the clay interlayer, is introduced to describe the large volumetric expansive behavior of the microstructure. This model includes the effects of suction and the swelling effect through the hardening parameter, and a van Genuchten type of soil-water characteristic curve (SWCC) (van Genuchten 1980) is adopted to describe the change in the degree of saturation with suction. Using this constitutive model, the swelling behavior of bentonite is simulated with the finite element method. Feng (2008) adopted the model for the numerical analysis of the swelling pressure of disposal barriers.

9.2 ELASTO-VISCOPLASTIC CONSTITUTIVE MODEL FOR UNSATURATED SWELLING SOIL

As presented in Section 2.4, an elasto-viscoplastic model for unsaturated soil was proposed by Oka et al. (2006), which takes into account the effect of suction on the constitutive model to describe the collapse behavior of unsaturated soil caused by a decrease in suction. The numerical results show that the behavior of unsaturated soil, such as the changes in the pore air pressure, the pore water pressure, and the volumetric strain, can be simulated well with this model. In addition, this model can predict the viscoplastic volumetric swelling phenomenon occurring during the wetting process. To capture the swelling phenomenon caused by clay particles, such as montmorillonite particles, the elasto-viscoplastic constitutive model for unsaturated soil is extended to be able to reproduce the volumetric swelling. In the model, a swelling equation is used to describe the viscoplastic volumetric swelling. Meanwhile, in order to capture soil compaction caused by swelling of the microstructure, microstructural swelling is introduced into the model.

9.2.1 Model assumptions

In order to carry out a multiphase analysis of unsaturated expansive soil, two levels of structures are considered. In the present section, the following are assumed:

1. The behavior of the macrostructure of bentonite includes the normal behavior of unsaturated soil. The initial suction level and the SWCC are considered to stand for those of the macrostructural behavior.
2. Assuming a special viscoplastic strain rate tensor, the microstructural swelling is introduced into the total strain rate tensor in addition to the elasto-viscoplastic strain rate tensor for the macrostructure.
3. In addition to the effect of suction, the effect of internal compaction caused by microstructural swelling is introduced into the constitutive equation as the shrinkage or expansion of the yield surface and the overconsolidation boundary surface.

In the model, it is assumed that the strain rate tensor consists of the elastic strain rate tensor, $\dot{\varepsilon}_{ij}^{e}$, the viscoplastic strain rate tensor, $\dot{\varepsilon}_{ij}^{vp}$, and an additional viscoplastic strain rate tensor, $\dot{\varepsilon}_{ij}^{vp(s)}$, caused by the microstructural swelling. Hence, the total strain rate tensor, $\dot{\varepsilon}_{ij}$, can be expressed as

$$\dot{\varepsilon}_{ij} = \dot{\varepsilon}_{ij}^{e} + \dot{\varepsilon}_{ij}^{vp} + \frac{1}{3}\dot{\varepsilon}_{kk}^{vp(s)}\delta_{ij} \tag{9.1}$$

where $\dot{\varepsilon}_{ij}^{e}$ is the elastic strain rate tensor given by a generalized Hooke type of law, determined by equation 5.6 in Oka and Kimoto (2012) and equation 5 in Kimoto et al. (2015).

9.2.2 Swelling equation for interparticles

From the experimental results on bentonite (Komine and Ogata 1996; Push 1982), the swelling phase is followed by an asymptotic tendency toward a constant final value. In this model, the following evolutional equation is proposed to describe the viscoplastic volumetric swelling of the swelling domain:

$$\begin{aligned} \dot{\varepsilon}_{kk}^{vp(s)} &= -\dot{H} \\ \dot{H} &= B(A - H) \end{aligned} \tag{9.2}$$

where H is an internal variable that describes the growth of the absorption of water into the clay particles, and A and B are material parameters. It is worth noting that the compression is positive.

Figure 9.2 shows the swelling equation curves at various values for parameters A and B. It can be seen that parameter A represents the potential of the absorption of water, and parameter B controls the evolution rate of H.

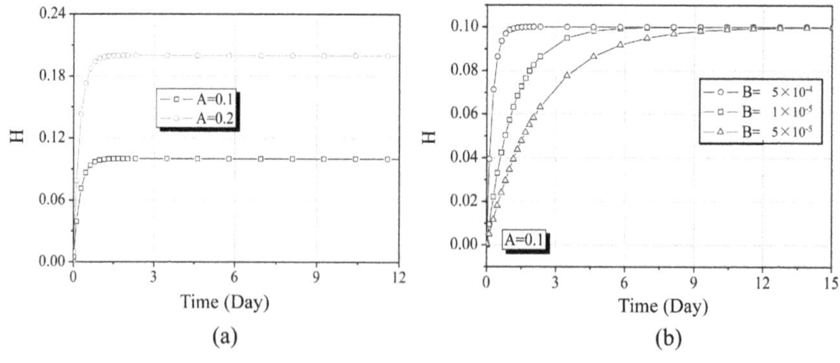

Figure 9.2 Swelling equations with different parameters for *A* and *B*.

9.2.3 Viscoplastic model including swelling effect

In the following analysis, we used a monotonic version of the viscoplastic constitutive model. According to the elasto-viscoplastic theory (Kimoto and Oka 2005), an overconsolidation boundary surface, f_b, is given as

$$f_b = \bar{\eta}^*_{(0)} + M^*_m \ln\left(\sigma'_m / \sigma'_{mb}\right) = 0 \tag{9.3}$$

$$\bar{\eta}^*_{(0)} = \left\{\left(\eta^*_{ij} - \eta^*_{ij(0)}\right)\left(\eta^*_{ij} - \eta^*_{ij(0)}\right)\right\}^{\frac{1}{2}} \tag{9.4}$$

where $\eta^*_{ij} (= S_{ij} / \sigma'_m)$ is the stress ratio tensor; S_{ij} is the deviatoric skeleton stress tensor; σ'_m is the mean skeleton stress; $\eta^*_{ij(0)}$ is the initial value of the stress ratio tensor before deformation, e.g., the value of the stress ratio at the end of consolidation; M^*_m is the value of $\eta^* = \sqrt{\eta^*_{ij}\eta^*_{ij}}$ across which dilatancy changes from negative to positive; and σ'_{mb} is a hardening parameter.

The static yield function, f_y, is given by

$$f_y = \bar{\eta}^*_{(0)} + \tilde{M}^* \ln \frac{\sigma'_m}{\sigma'^{(s)}_{my}} = 0 \tag{9.5}$$

$$\tilde{M}^* = \begin{cases} M^*_m & : \quad f_b \geq 0 \\ -\dfrac{\sqrt{\eta^*_{ij}\eta^*_{ij}}}{\ln(\sigma'_m / \sigma'_{mc})} & : \quad f_b < 0 \end{cases} \tag{9.6}$$

where σ'_{mc} denotes the mean effective stress at the intersection of the overconsolidation boundary surface with the σ'_m axis.

The viscoplastic strain rate tensor, $\dot{\varepsilon}^{vp}_{ij}$, is given by the generalized viscoplastic flow rule, which is presented in section 5.7 of Oka and Kimoto (2012) and Section 2.2.

The deviatoric and volumetric viscoplastic strain rates can be written as

$$\dot{e}^{vp}_{ij} = C_1 \exp\left[m'\left(\bar{\eta}^*_0 + \tilde{M}^* \ln \frac{\sigma'_m}{\sigma'_{mb}}\right)\right] \frac{\eta^*_{ij} - \eta^*_{ij(0)}}{\bar{\eta}^*} \tag{9.7}$$

$$\dot{\varepsilon}^{vp}_{kk} = C_2 \exp\left[m'\left(\bar{\eta}^*_0 + \tilde{M}^* \ln \frac{\sigma'_m}{\sigma'_{mb}}\right)\right]\left[\tilde{M}^* - \frac{\eta^*_{mn}\left(\eta^*_{mn} - \eta^*_{mn(0)}\right)}{\bar{\eta}^*}\right] \tag{9.8}$$

where $\dot{e}^{vp}_{ij}$ and $\dot{\varepsilon}^{vp}_{kk}$ are deviatoric and volumetric components of the viscoplastic strain rate $\dot{\varepsilon}^{vp}_{ij}$, and C_1, C_2 are viscoplastic parameters (see section 5.7 of Oka and Kimoto [2012] and Section 2.2).

Considering the suction effect by Equations 2.108 and 2.109, the hardening parameters are given as

$$\sigma'_{mb} = \sigma'_{ma} \exp\left(\frac{1+e}{\lambda-\kappa}\varepsilon_{kk}^{vp}\right)\left[1+S_I \exp\left\{-s_d\left(\frac{P_i^c}{P^c}-1\right)\right\}\right] \tag{9.9}$$

$$\sigma_{my}'^{(s)} = \frac{\sigma_{myi}'^{(s)}}{\sigma'_{mai}}\sigma'_{ma} \exp\left(\frac{1+e}{\lambda-\kappa}\varepsilon_{kk}^{vp}\right)\left[1+S_I \exp\left\{-s_d\left(\frac{P_i^c}{P^c}-1\right)\right\}\right] \tag{9.10}$$

With the absorption of water into the interlayers, the apparent volume of bentonite particles increases. In fact, the distance between the two platelets increases, as shown in Figure 9.1. Using a scanning electron microscope (SEM), Komine and Ogata (2004) reported an SEM image of the wetting process of bentonite, in which the bentonite content is 100% of mass. It can be seen that the macro voids are finally filled up by the volume increase in bentonite.

Figure 9.3 illustrates the swelling process in the case where the swelling deformation is restricted. With wetting, the void ratio of the macrostructure gradually becomes packed with swollen montmorillonite particles when the swelling deformation is restricted. As a result, the sample becomes stiffer and has a higher strength, which is similar to the soil being highly compacted. Herein, the phenomenon is called the "internal compaction effect"; it is different from traditional compaction in the changes in water

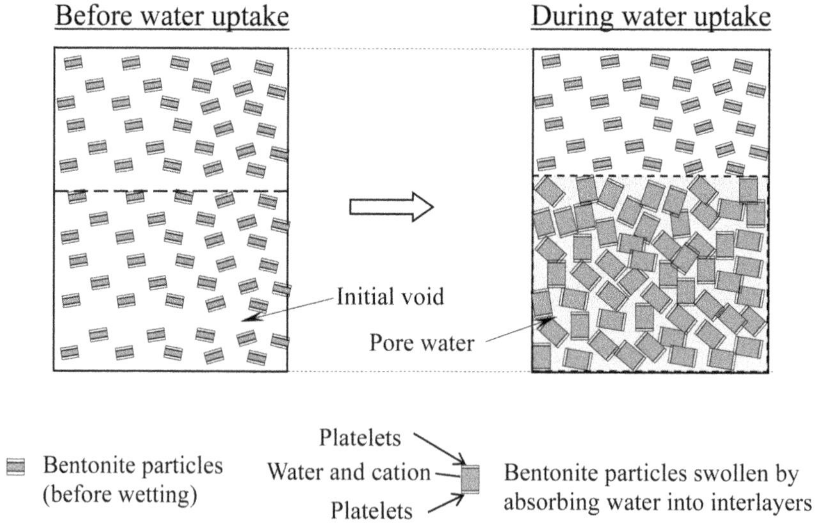

Figure 9.3 Process of swelling with restricted deformation.

content. This internal compaction effect is expressed as the expansion of the overconsolidation boundary surface, and the yield surface. Considering Equations 9.2 and 9.13, the parameters of the overconsolidation boundary surface and the static yield surface are given as

$$\sigma'_{mb} = \sigma'_{ma} \exp\left(\frac{1+e}{\lambda-\kappa}\varepsilon_{kk}^{vp^*}\right)\left[1+S_I \exp\left\{-s_d\left(\frac{P_i^c}{P^c}-1\right)\right\}\right] \tag{9.11}$$

$$\sigma_{my}^{'(s)} = \frac{\sigma_{myi}^{'(s)}}{\sigma'_{mai}}\sigma'_{ma} \exp\left(\frac{1+e}{\lambda-\kappa}\varepsilon_{kk}^{vp^*}\right)\left[1+S_I \exp\left\{-s_d\left(\frac{P_i^c}{P^c}-1\right)\right\}\right] \tag{9.12}$$

where $\varepsilon_{kk}^{vp^*}$ is the viscoplastic volumetric strain including the special swelling effect and is defined as

$$\varepsilon_{kk}^{vp^*} = \varepsilon_{kk}^{vp} + \gamma|H| \tag{9.13}$$

where γ is adopted to reflect the percentage of the swelling viscoplastic strain considered, which varies from 0 to 1; $\gamma = 0$ means that the swelling viscoplastic strain does not affect the expansion or shrinkage of the overconsolidation boundary surface and the static yield boundary surface, while $\gamma = 1$ means that all levels of the viscoplastic swelling volumetric strain have an effect on the expansion or shrinkage of the boundary surface.

9.3 SIMULATION OF SWELLING PRESSURE TESTS

9.3.1 Analysis method and the model

Numerical simulation methods are given in section 6.5 of Oka and Kimoto (2012). In the analysis, the stretching tensor is used instead of the strain rate because of the finite strain analysis. An 8-node quadrilateral isoparametric element with a reduced Gaussian (2 × 2) integration (see Figure 9.4) is used for the displacement in order to eliminate shear locking as well as to reduce the appearance of the spurious hourglass mode. The pore pressure values for air and water are defined by a 4-node quadrilateral isoparametric element. As shown in Figure 9.5, the initial thickness of the soil sample is 0.02 m with 10 elements, the bottom of which is set to be permeable for water and air by assuming that the water pressure at the bottom is –10 kPa and the air pressure is kept at 0 kPa. The other boundaries are assumed to be impermeable to water and air. The time increment is set to be 600 s when some elements are still swelling. Otherwise, it is 1200 s.

The displacements in both the X and Y directions at the top and bottom are fixed, while the displacements are fixed only in the X direction for the

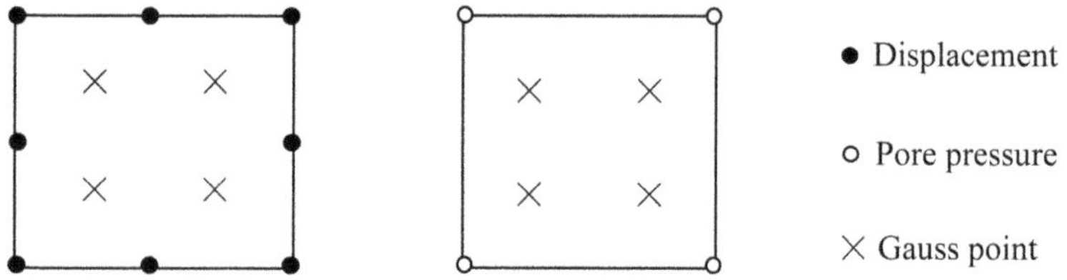

Figure 9.4 Isoparametric elements for the soil skeleton and the pore pressure.

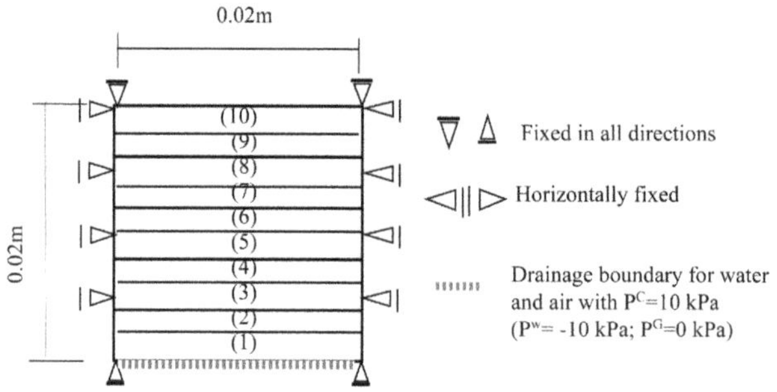

Figure 9.5 Finite element mesh and boundary conditions.

Table 9.1 Initial conditions for the simulations

Initial suction	P^C (kPa)	100
Initial void ratio	e_0	0.66
Initial saturation	S_r (%)	69.98
Initial pore water pressure	P^W (kPa)	-100
Initial pore gas pressure	P^G	0

other boundaries. The initial conditions and the material parameters for the analyses are shown in Tables 9.1 and 9.2. These parameters are measured for the Kunigel GX bentonite sample. The initial suction, which is the macrostructural suction, is assumed to be 100 kPa for all the analyses.

9.3.2 Swelling pressure with wetting process

The wetting process is shown in Figures 9.6 and 9.7. Figure 9.6 shows the changes in the degree of saturation for each element with wetting. The decrease in suction for every element with wetting is shown in Figure 9.7. In this analysis, it is assumed that each element starts to swell when the degree of saturation reaches a given value. From Figure 9.7, it can be seen that the degree of saturation of the elements, starting with the element at the bottom, reaches the onset saturation for swelling, element by element.

Table 9.2 Material parameters used in the simulation

Elastic shear modulus	G_0 (kPa)	3.74×10^4
Initial yield stress	σ'_{mbi} (kPa)	100
Swelling index	κ	0.078
Compression index	λ	0.117
Viscoplastic parameter	m'	95.4
Viscoplastic parameter	C_1 (1/s)	9.47×10^{-18}
Viscoplastic parameter	C_2 (1/s)	9.47×10^{-18}
Stress ratio at critical state	M^*_m	0.4736
Suction parameter	S_I	0.5
Suction parameter	S_d	0.25
Structural parameter	σ'_{maf} (kPa)	83
Structural parameter	β	5.0
van Genuchten parameter	α (1/kPa)	0.015
van Genuchten parameter	n	1.517
Permeability of water at $s_r = 1$	k^W (m/s)	2.0×10^{-13}
Permeability of gas at $s_r = 0$	k^G (m/s)	1.3×10^{-11}
Shape parameter	a	3.0
Shape parameter	b	2.3
Maximum saturation	S_{rmax}	1
Minimum saturation	S_{rmin}	0
Expansive parameter	A	0.1
Expansive parameter	B	0.00001
Expansive parameter	γ	0.4
Onset saturation for swelling	(%)	70.5

Consequently, the swelling starts from the bottom element and moves upward, element by element. Accordingly, the decrease in suction with wetting also starts from the bottom element and moves upward, as shown in Figure 9.7.

As mentioned previously, the predictions of the swelling pressure are affected by the following parameters: A, B, γ, permeability, and the initial swelling saturation. The first two parameters are newly introduced into the model and control the swelling equation. The last two parameters control the time needed to complete the swelling process. In the next sections, the effects of these parameters are investigated.

9.3.3 Swelling pressure with different permeabilities

As previously mentioned, the swelling of the sample depends on the degree of saturation, which in turn is strongly affected by permeability. In this section, the effect of permeability on the swelling pressure is investigated using the model proposed. In these analyses, changes in the permeability of water

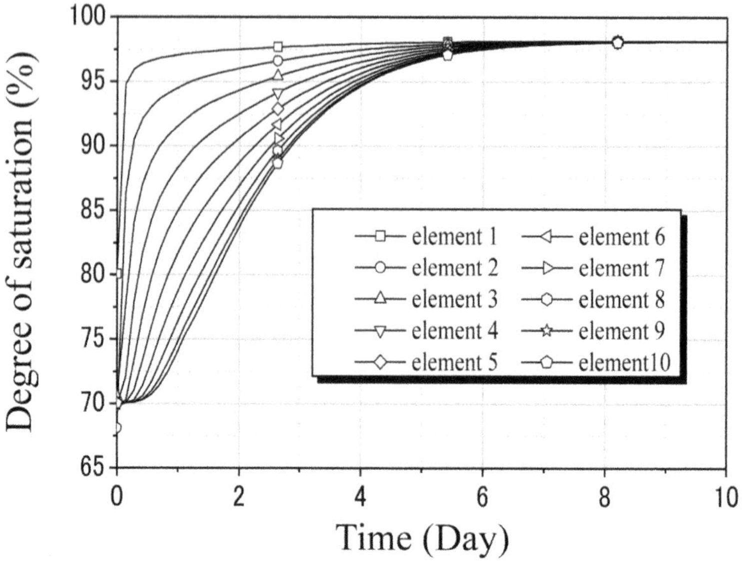

Figure 9.6 Degree of saturation for each element with wetting.

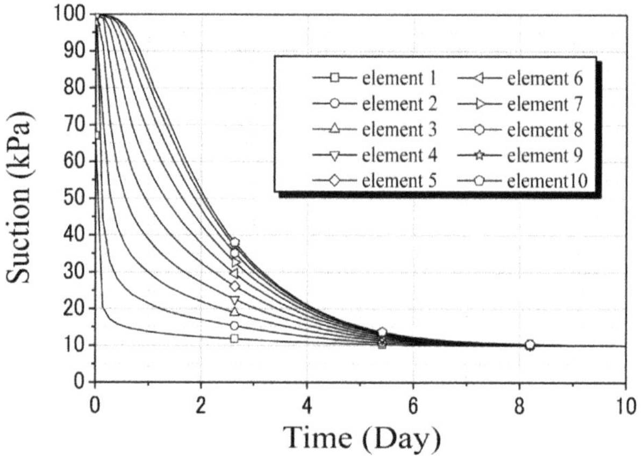

Figure 9.7 Suction for each element with wetting.

are examined. The onset saturation for swelling is set to be 75% in all calculations in this section. The remaining soil material parameters are listed in Table 9.2. Figure 9.8 provides the changes in the swelling pressure with wetting obtained from the simulations. From Figure 9.8, it is confirmed that permeability k^W delays the time needed for swelling. For higher permeability, the sample starts to swell after a short time. For lower permeability, on the contrary, the sample does not appear to start swelling until a long wetting period has passed.

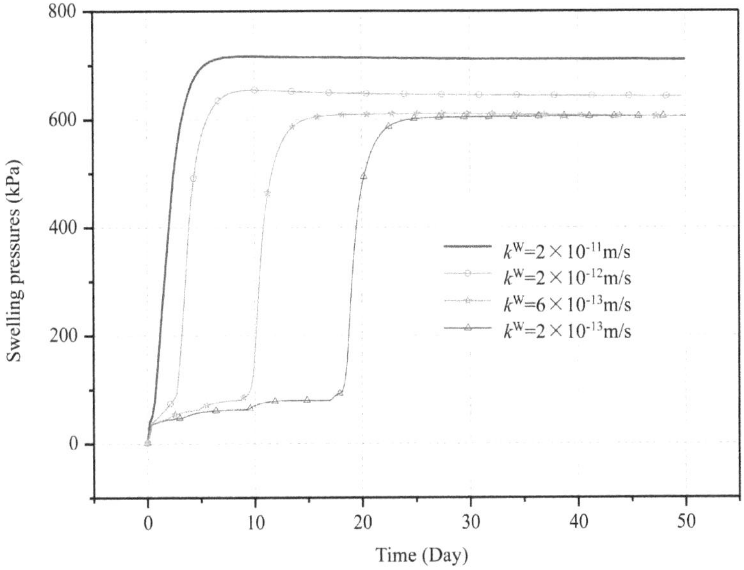

Figure 9.8 Effect of permeability on the swelling pressure.

9.3.4 Swelling pressure with different levels of onset saturation for swelling

As known, the swelling behavior is due to the absorption of water into the interlayers. In this model, the onset saturation for swelling is assumed to be the degree of saturation from which the microstructure starts to swell. In the previous section, the swelling was assumed to start when saturation had reached 75%. In this section, different levels of initial swelling saturation are investigated. The permeability of water is set to be 2.0×10^{-13} m/s, and the remaining parameters are listed in Table 9.2. Six cases are studied with different levels of onset saturation for swelling, ranging from 70% to 75%.

Figure 9.9 shows the predicted swelling pressure curves at various levels of onset saturation. Similar to the effect of permeability, the decrease in onset saturation for swelling leads to a delay in swelling. For the case in which the onset saturation for swelling is 70%, the swelling starts after a very short time, because the initial swelling saturation is very close to the initial saturation (69.98%). With an increase in the onset saturation value, a longer wetting period is required for the swelling to begin.

9.3.5 Effect of γ on the swelling pressure

In this section, the effects of the parameter γ, which is newly introduced in the model, are studied. Parametric studies on the changes in γ from 0.0 to 1.0 are performed. The other soil parameters are listed in Table 9.2. The

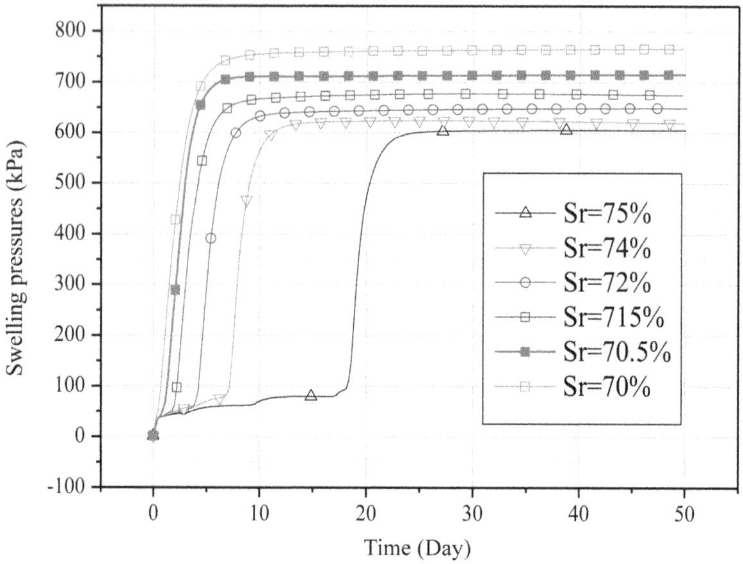

Figure 9.9 Effect of the onset saturation for swelling on the swelling pressure.

simulated results of the swelling pressures during the wetting process with different values for γ are shown in Figure 9.10. From this figure, it is seen that parameter γ ($0 \leq \gamma \leq 1$) affects not only the final pressure, but also the shape of the swelling pressure curves. In cases where no or only a small value of viscoplastic swelling volumetric strain ($\gamma < 0.3$) is considered, a time-softening behavior can be observed. However, for the case in which a higher percentage of swelling strain ($\gamma > 0.4$) is considered, time-hardening behavior can be observed. Additionally, it is found that the predicted final swelling pressure increases with an increase in γ; however, the final swelling pressure remains almost constant when $\gamma > 0.5$.

According to this elasto-viscoplastic model, viscoplastic strain increments for the overstress type of model depend on the difference between the current stress state and the static stress state. This means that a large difference between the current stress state and $\sigma_{my}^{\prime(s)}$ will lead to an obvious relaxation. Figure 9.11 presents the changes in σ'_{mb} with the wetting process. It can be seen that by considering the viscoplastic swelling strain, as in Equations 9.9 and 9.10, the value of σ'_{mb} increases quickly and reaches a higher value compared to the value of σ'_{mb} for which no swelling effect is considered. In the case of a lower percentage of microstructural swelling, we can see that compared with the mean skeleton stress, σ'_m, σ'_{mb}, or $\sigma_{my}^{\prime(s)}$ increases slowly. Therefore, the slow increase in the static yield surface and the quick increase in the mean skeleton stress are due to the increase in the

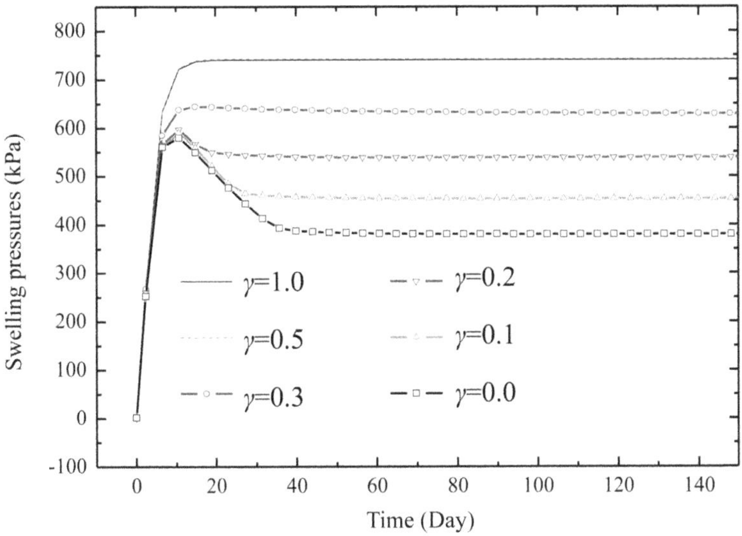

Figure 9.10 Effect of parameter γ on the swelling pressure.

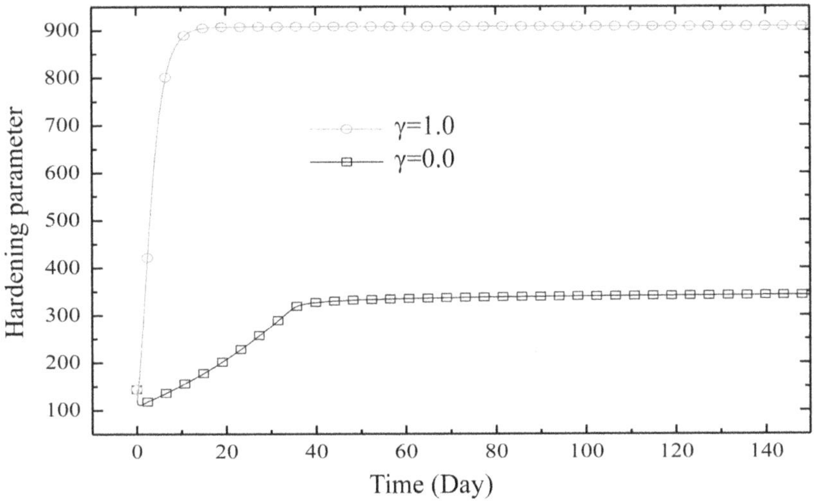

Figure 9.11 Changes in the hardening parameter σ'_{mb} with different values for γ.

viscoplastic strain. As a result, relaxation occurs and the time-softening behavior can be observed. When the effect of significant microstructural volumetric swelling is taken into account, e.g., $\gamma > 0.3$, this time-softening behavior cannot be observed. It is also found that γ markedly affects the final swelling pressure.

9.3.6 Effect of *A* on the swelling pressure

As mentioned previously, *A* is a material parameter, which represents the potential to absorb water. In this section, the effect of parameter *A* on the final swelling pressure is studied. The remaining parameters are listed in Table 9.2. Figure 9.12 provides the development of the swelling pressure with wetting under different values for parameter *A*. From Figure 9.12, it is found that *A* controls the magnitude of the swelling pressure. It is seen that the larger values of *A* result in higher levels of swelling pressure.

9.3.7 Swelling pressure considering the initial dry density and the water content

In the previous sections, the effects of the swelling parameters *A* and γ on the swelling pressure were examined. From Figure 9.12, it can be observed that at a given value for γ, the final swelling pressure will increase as parameter *A* increases. Then, as shown in Figure 9.10, by decreasing the value of γ, the time-softening behavior of the swelling pressure can be simulated. Meanwhile, we can see in Figure 9.10 that the final value of the swelling pressure also decreases with a decrease in the γ value even at the same value of *A*. According to the experimental results, we can assume that the swelling potential of the sample is affected by both the type of special mineral, such as montmorillonite or swelling chlorite, and by its concentration. This

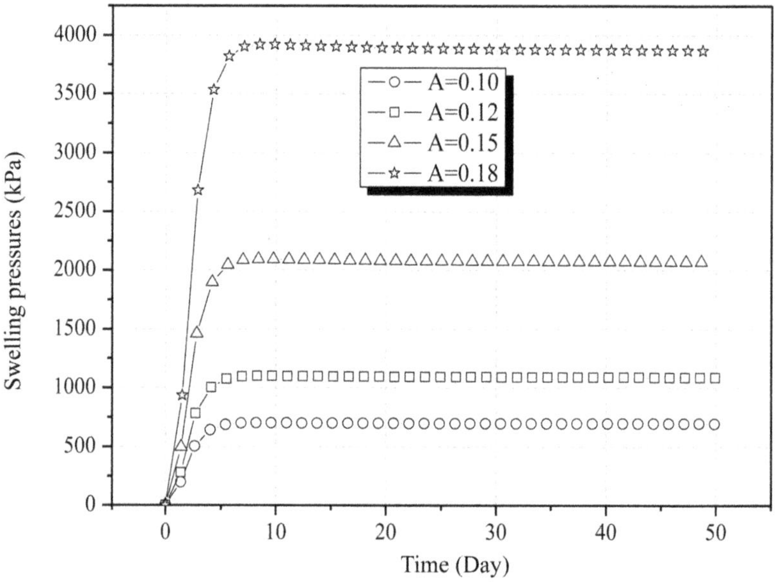

Figure 9.12 Effect of parameter *A* on the swelling pressure.

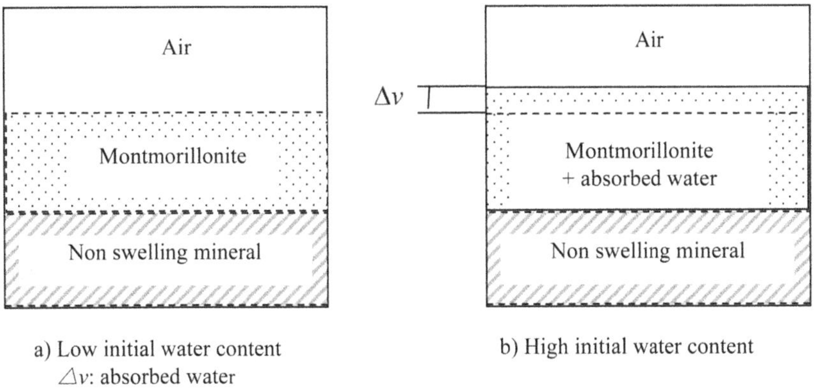

Figure 9.13 Comparison of compacted bentonite with different water content.

means that for expansive soil, parameter A is affected by the dry density (high mineral concentration).

For the swelling process, it is reasonable to think that the initial water content can also cause some degree of swelling. Figure 9.13 illustrates a comparison between two types of compacted bentonite with different levels of initial water content. As shown in the figure, a certain amount of swelling is seen within the samples with higher levels of initial water content even before the wetting process begins. In this model, parameter γ can be used to reproduce the compacted effect caused by swelling. Therefore, parameter γ can be adopted to reflect the effect of the initial water content. Theoretically, by adopting proper values for parameters A and γ, the effects of dry densities and initial water content on the swelling pressure can be captured using this model.

9.4 APPLICATION TO KUNIGEL GX BENTONITE

Swelling pressure tests were conducted for a confined situation in which changes in the volume of the specimen were not permitted. Using commercial Kunigel GX bentonite, the swelling pressure of bentonite was tested by Ono et al. (2006). The properties of this bentonite are shown in Tables 9.1 and 9.3. As shown in Table 9.4, experiments at different dry densities and initial degrees of saturation were carried out for bentonite. The experimental results are shown in Figure 9.14. The experimental results with different dry densities confirmed that the dry density controls the final swelling pressure, which means the higher the dry density of bentonite the higher the final swelling pressure. Meanwhile, it is seen that, at a given dry density, the initial water content affects the initial part of the swelling pressure curves.

Table 9.3 Properties of the Kunigel GX

	Sodium bentonite
Density (g/cm^3)	2.65
Liquid limit (%)	416
Plastic limit (%)	21
Plasticity index	395
Activity	8.53
Clay content (<2 μm) (%)	51.6

Table 9.4 Experiment cases

Case	*Dry density (Mg/m^3)*	*Water content (%)*	*Degree of saturation (%)*
SW2	1.6	6.5	27
SW3	1.8	6.5	39
SW5	1.6	21.6	91
SW6	1.8	15.3	92

Source: Data from Ono, F., Niwase, K., Tani, T., Nakgoe, A. and Chichimatsu, M. 2006. Results of permeability tests for bentonite with two-phase compaction by field compaction method. In *Proc. 61st Annual Meeting of JSCE*, 323–324. Kyoto (in Japanese).

In the case of a lower initial water content (SW2 and SW3), the swelling pressure increases to a peak value fairly quickly. Then, a time-softening phenomenon can be observed. For the cases of high initial water content (SW5 and SW6) without this type of time-softening behavior, the swelling pressure gradually reaches its final value.

As discussed in the previous section, by adjusting the parameters A and γ, finite element simulations have been carried out to simulate swelling pressure tests with different levels of dry density and initial water content. The material parameters A and γ are listed in Table 9.5. The remaining parameters and initial conditions are listed in Tables 9.1 and 9.2. The predicted swelling pressures with the wetting process are shown in Figure 9.15. For reference cases SW3 and SW6, with higher initial dry densities, higher A values (=0.18 and 0.16) are adopted. For cases with higher initial water content values (SW5 and SW6), a larger γ (=0.3) is used to represent the initial hardening effect.

As Tables 9.4 and 9.5 show, compared with the cases (S2 and S3), a slightly lower value is adopted for parameter A for high initial water content (S5 and S6) even though these simulation cases have the same dry densities respectively, i.e the dry density of S2 is the same as that of S5 and the dry density of S3 is the same as that of S6. This is because for cases with high initial water content, some amount of swelling has already taken place before wetting, so the swelling potential should be lower than in samples with lower initial water content at the same dry density.

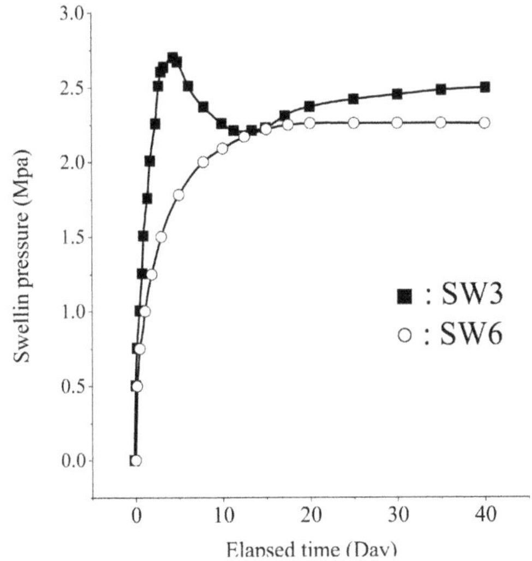

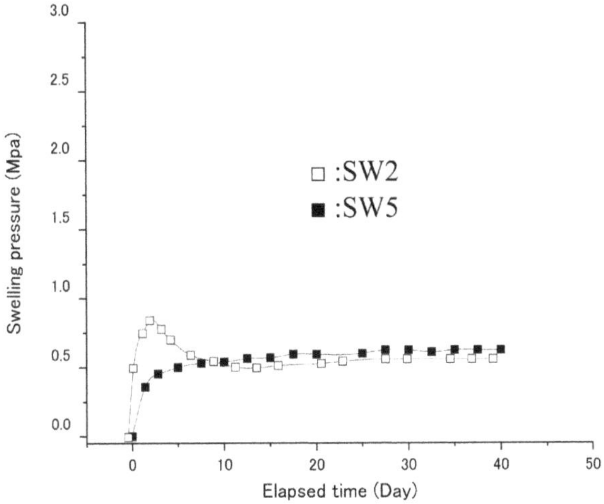

Figure 9.14 Experimental results. (Data from Ono, F., Niwase, K., Tani, T., Nakgoe, A and Chichimatsu, M. 2006. Results of permeability tests for bentonite with two-phase compaction by field compaction method, In *Proc. 61st Annual Meeting of JSCE*, 323–324: Kyoto [in Japanese].)

9.5 SUMMARY

An elasto-viscoplastic constitutive model was presented for unsaturated expansive soil. The model is an extension of the elasto-viscoplastic model by Oka et al. (2008). An internal variable H reflects the growth of the absorption of water into the interlayers of clay platelets. Due to the absorption

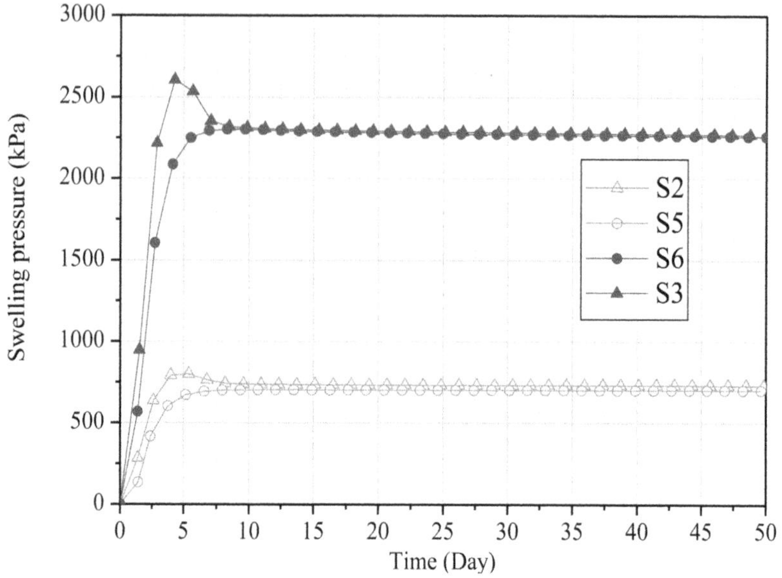

Figure 9.15 Simulated results for the swelling pressure.

Table 9.5 Simulation cases

Simulation case	***A***	γ	*No. of experiment*
S2	0.12	0.1	SW2
S3	0.18	0.1	SW3
S5	0.10	0.3	SW5
S6	0.16	0.3	SW6

of water into the interlayers, the apparent volume of the montmorillonite particles increases. Under restricted swelling deformation conditions, this process leads to a decrease in micro voids. In the present model, this internal compaction effect was expressed with the expansion of the overconsolidation boundary surface and the static yield surface.

Using the proposed model, one-dimensional finite element analyses were conducted to simulate the swelling pressure. The results showed that the proposed model can reproduce the swelling behavior of unsaturated bentonite during the wetting process. In addition, the effects of the parameters on the predicted swelling pressure were investigated.

Parameters A and γ were adopted to describe the swelling potential and the internal compaction effect of bentonite. The proposed model was applied to swelling pressure tests on bentonite (Kunigel GX). Compared with the experimental results, it was found that the model can well reproduce the effect of dry density and initial water content on the swelling pressure.

REFERENCES

Alonso, E.E., Gens, A. and Lloret, A. 1991. Double structure model for the prediction of long-term movements in expansive materials. In *Proceedings of the 7th International Conference on Computer Methods and Advances in Geomechanics*, Cairns, eds. J.R. Booker, J.P. Carter, G. Beer, 54–548. Balkema.

Alonso, E.E., Lloret, A., Gens, A. and Yong, D.Q. 1995. Experimental behaviour of highly expansive double-structure clay. In *Proceedings of the 1st International Conference on Unsaturated Soils*, UNSAT'95, Paris, eds. E.E. Alonso and P. Delage, 11–16. Balekema.

Alonso, E.E., Vaunat, J. and Gens, A. 1999. Modeling the mechanical behaviour of expansive clays. *Eng. Geol.* 54(1–2): 173–183.

Alonso, E.E., Romero, E., Hoffmann, C. and Garcia-Escudero, E. 2005. Expansive bentonite-sand mixtures in cyclic controlled-suction drying and wetting. *Eng. Geol.* 81(3): 213–226.

Bolt, G.H. 1956. Physico-chemical analysis of the compressibility of pure clays. *Geotecnique* 6(2): 86–93.

Butscher, C., Mutschler, T. and Blum, P. 2016. Swelling of clay-sulfate rocks: A review of processes and controls. *Rock Mech. Rock Eng.* 49(4): 1533–1549.

Chapman, J.D. 1913. A contribution to the theory of electrocapillary. *The London, Edinburgh and Dublin Philosophical Magazine and J. of Sci.* 25(148): 475–481.

Feng, H. 2008. *Multi-Phase Deformation Analysis of elasto-Viscoplastic Unsaturated Soil and Modeling of Bentonite*, PhD thesis. Kyoto University.

Feng, H., Kimoto, S., Oka, F., Kodaka, T. and Suzuki, H. 2006. Three dimensional multiphase analysis of elasto-viscoplastic unsaturated soil. In *Proceedings of the 19th KKCNN Symposium on Civil Engineering*, Kyoto, 449–452.

Gens, A. and Alonso, E.E. 1992. A framework for the behaviour of unsaturated expansive clays. *Can. Geotech. J.* 29(6): 1013–1032.

Gouy, G. 1910. Electric charge on the surface of an electrolyte. *J. Phys.* 4(9): 457.

Kanno, T. and Wakamatsu, H. 1992. Water-uptake and swelling properties of unsaturated bentonite buffer materials. *Can. Geotech. J.* 29(6): 1102–1107.

Kimoto, S. and Oka, F. 2005. An elasto-viscoplastic model for clay considering destructuralization and consolidation analysis of unstable behavior. *Soils Found.* 45(2): 29–42.

Kimoto, S., Shahbodagh, B., Mirjalili, M. and Oka, F. 2015. A cyclic elastoviscoplastic constitutive model for clay considering nonlinear kinematic hardening rules and structural degradation. *Int. J. Geomechanics ASCE* 15(5): A4014005-1, A4014005-14.

Komine, H. and Ogata, N. 1996. Prediction for swelling characteristics of compacted bentonite. *Can. Geotech. J.* 33(1): 11–22.

Komine, H. and Ogata, N. 2003. New equations for swelling characteristics of bentonite-based buffer materials. *Can. Geotech. J.* 40(2): 460–475.

Komine, H. and Ogata, N. 2004. Predicting swelling characteristics of bentonites. *J. Geotech. Geoenviron. Eng.* 130(8): 818–829.

Kurahayashi, S. 1980.*Clay and Life.* Tokai Gakujyutsu Sennsyo (in Japanese).

Lloret, A., Villar, M.V., Sanchez, M., Gens, A., Pintado, X. and Alonso, E.E. 2003. Mechanical behaviour of heavily compacted bentonite under high suction changes. *Geotechnique* 53(1): 27–40.

Marcial, D., Delage, P. and Cui, Y.J. 2002. On the high stress compression of bentonite. *Can. Geotech. J.* 39(4): 812–820.

Mesri, G. and Olson, R.E. 1971. Consolidation of characteristics of montmorillonite. *Geotechnique* 21(4): 341–352.

Mitchell, J.K. 1993. *Fundamentals of Soil Behaviour*, 2nd ed. John Wiley & Sons, New York.

Oka, F., Kodaka, T., Kimoto, S., Kim, Y.-S. and Yamasaki, N. 2006. A multi-phase coupled FE analysis using an elasto-viscoplastic model for unsaturated soil. In *Proceedings of the 2nd US-Japan Workshop on Geomechanics*, Kyoto, Geomechanics II, Geotechnical Special Publication, ASCE, 124–131.

Oka, F., Feng, H., Kimoto, S. and Higo, Y. 2008. An elasto-viscoplastic numerical analysis of the swelling process of unsaturated bentonite. *J. Appl. Mech., JSCE* 11: 369–376.

Oka, F., Feng, H., Kimoto, S., Tinet, A.J. and Higo, Y. 2009. Numerical modeling of swelling and the osmotic consolidation of bentonite. In *Proceedings of the 1st International Symposium on Computational Geomechanics*, ed. S. Pietruszczak, G.N. Pande, C. Tamagnini and R. Wan, 532–543. IC2E, International Center for Computational Engineering, Rhodes, Greece.

Oka, F. and Kimoto, S. 2012. *Computational Modeling of Multiphase Geomaterials*. CRC Press, Taylor & Francis Group, Boca Raton.

Olson, R.E. and Mesri, G. 1970. Mechanism controlling the compressibility of clay. *J. of the Soil Mechanics of Foundation Div.* ASCE. 96(SM6): 1860–1878.

Ono, F., Niwase, K., Tani, T., Nakgoe, A. and Chichimatsu, M. 2006. Results of permeability tests for bentonite with two-phase compaction by field compaction method. In *Proceedings of the 61st Annual Meeting of JSCE*, 323–324. Kyoto (in Japanese).

Push, R. 1982. Mineral-water interactions and their influence on the physical behaviour of highly compacted Na bentonite. *Can. Geotech. J.* 19(3): 289–299.

Shuai, F. and Fredlund, D.G. 1998. Model for the simulation of swelling-pressure measurements on expansive soil. *Can. Geotech. J.* 35(1): 96–114.

Sridharan, A. and Choudhury, D. 2002. Swelling pressure of sodium montmorillonites. *Geotechnique* 52(6): 459–462.

Tripathy, S., Sridharan, A. and Schanz, T. 2004. Swelling pressures of compacted bentonites from diffuse double layer theory. *Can. Geotech. J.* 41(3): 437–450.

van Genuchten, M.Th. 1980. A closed-form equation for predicting the hydraulic conductivity of unsaturated soils. *Soil Sci. Soc. Am. J.* 44(5): 892–899.

Chapter 10

Numerical analysis of hydrate-bearing subsoil during dissociation

10.1 INTRODUCTION

Recently, methane hydrates have been viewed as a potential energy resource because a large amount of gas is trapped within ocean sediments and regions of permafrost. However, we have little knowledge about the performance of sediments induced by the dissociation of hydrates in the ground. Some geologists have discussed their possible triggering of landslides in the seabed ground during and after the last deglaciation. Sultan et al. (2004) indicated that the dissociation of gas hydrates is a possible reason for the giant Storegga Slide on the Norwegian margin where the slide mass involved was about 3000 km^3 and affected an area of about 90000 km^2. Kvalstad et al. (2005) mentioned that temperature changes after the last deglaciation may have triggered dissociation, but it could only have affected a small part of the slide area, whereas pore pressure generation with strain softening and rapid deposition was considered to have higher potential. Wong et al. (2003) showed that one of the possible triggers for mass wasting in the northwestern Sea of Okhotsk is gas hydrate instability. These phenomena occurred in the geological period, and the extreme change in environment during gas production may induce similar behavior. The dissociation process follows the phase changes from solids to fluids, i.e., from hydrates to water and gas, and hence, the ground will be under unsaturated conditions. In addition, heat transfer becomes important during the dissociation process, since the phase equilibrium is controlled by temperature and pressure, and the dissociation reaction is an endothermic reaction. Some researchers conducted experimental studies to investigate the characteristics of hydrate-bearing sediments. Following their experimental study, Wu and Grozic (2008) indicated that dissociation of even a small percentage of gas hydrates can lead to failure. It is essential to develop a numerical framework to rigorously study the ground deformation associated with the dissociation of methane hydrate.

Hydrate-bearing sediments are composed of soil particles, hydrates, water, and gas during the dissociation process. Sakamoto et al. (2008) conducted experimental studies on the dissociation behavior of hydrate

DOI: 10.1201/9781003200031-10

sediments and showed that the increase in the effective stress at the initial stage of depressurization is a dominant factor for deformation. Hyodo et al. (2005) conducted a series of triaxial tests on hydrate–sediments mixture. The sediments used in the mixture were obtained from the deep-sea bottom at Nankai Trough. Miyazaki et al. (2008) experimentally showed that the strength of hydrate sediments depends on the hydrate saturation and the strain rate. Iwai et al. (2018) carried out triaxial tests on carbon dioxide hydrate-bearing sand and found that the material is rate dependent.

For the numerical study, several numerical simulators have been developed to evaluate gas production values. Masuda et al. (2002) and Ahmadi et al. (2004) developed numerical models using the finite difference method to predict the flow of gas and water that accompanies hydrate dissociation. While they considered both the fluid flow and heat transfer, the solid phase was assumed to be immobile. Although other numerical simulators were developed, the solid phase was assumed to be rigid in most of them (e.g., Bejan et al. 2002; Tsypkin 2000; Bondarev et al. 1999).

The behavior of multiphase materials can be described within the framework of a macroscopic continuum approach through the use of the theory of porous media (de Boer 1998). The theory is considered to be a generalization of Biot's two-phase mixture theory for saturated soil (Biot 1962). Proceeding from the general geometrically non-linear formulation, the governing balance relations for multiphase materials can be obtained (e.g., Boer 1998; Loret and Khalili 2000; Ehlers et al. 2004). Mass conservation laws for the gas phase as well as for the liquid phase are considered in the analyses. In the field of geotechnics, the air pressure is frequently assumed to be zero in many research works (e.g., Sheng et al. 2003) since unsaturated geomaterials usually exist near the surface of the ground. Considering hydrate dissociation, however, we have to deal with the high level of gas pressure that exists deep in the ground; this means that the mass balance for both fluid phases must be considered in the analysis. Oka et al. (2006, 2009, 2010) proposed an air–water–soil coupled finite element model in which the skeleton stress is used as a stress variable with the effect of suction considered through the parameters in the constitutive equation for soil. The compression behavior of unsaturated soils was simulated even under impermeable conditions for both water and gas flow.

The conservation of energy is required when there is a considerable change in temperature during the deformation process. Oka et al. (2005a,b) and Kimoto et al. (2007a,b,c) numerically simulated the thermal consolidation process of deformable solids using a coupled thermo-hydro-mechanical finite element analysis method with a thermo-elasto-viscoplastic constitutive model. Since hydrate dissociation is an endothermic reaction, heat transfer plays an important role in both gas production and ground deformation. Rutqvist and Moridis (2008) and Rutqvist et al. (2009) proposed a numerical simulator for analyzing the geomechanical performance of hydrate-bearing permafrost. They used the staggered technique to couple the hydraulic and mechanical models. Kimoto et al. (2007a) extended the previously developed theory considering the dependency of the permeability coefficients for water and gas on hydrate saturation.

In the following section, we present a numerical analysis of methane gas hydrate dissociation in hydrate-bearing sediments in the seabed ground. The simulation method has been developed based on a coupled chemo-thermo-mechanically analysis, taking into account the phase changes from solids to fluids, i.e., water and gas, the flow of water and gas, heat transfer, and the ground deformation (Kimoto et al. 2007a,b,c, 2010; Oka et al. 2011). The following assumptions are adopted in the formulation:

1. Soil particles and water are incompressible.
2. Methane gas is treated as an ideal gas and the effect of dissolution into water is disregarded.
3. The flows of gas and water are independent and they follow Darcy's law.
4. The soil and hydrates behave as a solid mass, so that the velocity of hydrates is equal to the velocity of soil.
5. The acceleration term is disregarded in the equation of motion.

10.2 MULTIPHASE MIXTURE THEORY FOR SOIL CONTAINING HYDRATE

Geomaterials generally fall into the category of multiphase materials. They are basically composed of soil particles, water, and air. The behavior of multiphase materials can be described within the framework of a macroscopic continuum approach through the use of the theory of porous media (e.g., de Boer 1998; Oka and Kimoto 2012).

Oka et al. (2006) proposed an air–water–soil coupled finite element model in which the skeleton stress is used as a stress variable with the effect of suction considered through the parameters in the constitutive equation for soil.

Since the hydrate dissociation is an endothermic reaction, heat transfer plays an important role in both the gas production and the ground deformation. The multiphase porous theory has been extended to the sediment containing methane hydrate by Kimoto et al. (2007a, b, 2010).

10.2.1 General setting

The multiphase material, ψ, is composed of four phases, namely, soil (S), water (W), gas (G), and hydrates (H), which are continuously distributed throughout the medium.

$$\psi = \sum_{\alpha} \psi^{\alpha} \quad (\alpha = S, W, G, H) \tag{10.1}$$

in which S, W, G, and H indicate the soil, water, gas, and hydrate phases, respectively. For simplicity, we assume that hydrates move with soil particles; in other words, the solid phase, denoted by SH, is composed of soil and hydrates which exist around the soil particles. Total volume, V, is obtained from the sum of the partial volumes of the constituents, i.e.

$$\sum_{\alpha} V^{\alpha} = V \ (\alpha = S, W, G, H) \tag{10.2}$$

The total volume of fluids, V^F, is given by

$$V^F = \sum_{\gamma} V^{\gamma} \ (\gamma = W, G) \tag{10.3}$$

The volume fraction, n^{α}, is defined as the local ratio of the volume element with respect to the total volume as

$$n^{\alpha} = \frac{V^{\alpha}}{V}, \ \sum_{\alpha} n^{\alpha} = 1 \ (\alpha = S, W, G, H) \tag{10.4}$$

The volume fraction of the void, n, is written as

$$n = \sum_{\beta} n^{\beta} = \frac{V - V^S}{V} = 1 - n^S \ (\beta = W, G, H) \tag{10.5}$$

The volume fraction of the fluid, n^F, is given by

$$n^F = n - n^H \tag{10.6}$$

In addition, the fluid saturation is given by

$$s^{\gamma} = \frac{V^{\gamma}}{V^F}, \ \sum_{\gamma} s^{\gamma} = 1 \ (\gamma = W, G) \tag{10.7}$$

Water saturation, s^W, is denoted by S_r in the following:

$$S_r = \frac{V^W}{V^F} = \frac{n^W}{n^F} \tag{10.8}$$

The density of each material, ρ^{α}, and the total phase, ρ, are denoted by

$$\rho^{\alpha} = \frac{M^{\alpha}}{V^{\alpha}}, \ \rho = \sum_{\alpha} n^{\alpha} \rho^{\alpha} \ (\alpha = S, W, G, H) \tag{10.9}$$

10.2.2 Definition of stress

In the theory of porous media, the concept of the effective stress tensor is related to the deformation of the soil skeleton and plays an important role. The effective stress tensor has been defined by Terzaghi (1943) for water-saturated soil; however, the effective stress needs to be redefined if the fluid is made of compressible materials. For unsaturated soil, the skeleton stress tensor, σ'_{ij}, is used for the stress variable in the constitutive relation for the soil skeleton as well as suction. The total stress tensor, σ_{ij}, is obtained from the sum of the partial stress values, σ^{α}_{ij}, namely,

$$\sigma_{ij} = \sum_{\alpha} \sigma^{\alpha}_{ij} \quad (\alpha = S, W, G, H) \tag{10.10}$$

where the partial stresses for different phases are given by

$$\sigma^{W}_{ij} = -n^{W} P^{W} \delta_{ij} \tag{10.11}$$

$$\sigma^{G}_{ij} = -n^{G} P^{G} \delta_{ij} \tag{10.12}$$

$$\sigma^{SH}_{ij} = \sigma^{S}_{ij} + \sigma^{H}_{ij} = \sigma'_{ij} - n^{SH} P^{F} \delta_{ij} \tag{10.13}$$

in which the volume fraction, n^{SH}, of the solid phase and the average pressure of the fluids, P^F, are given by

$$n^{SH} = n^{S} + n^{H} \tag{10.14}$$

$$P^{F} = S_r P^{W} + (1 - S_r) P^{G} \tag{10.15}$$

From the above equations, the skeleton stress, σ'_{ij}, is obtained as

$$\sigma'_{ij} = \sigma_{ij} + P^{F} \delta_{ij} \tag{10.16}$$

Notice that the solid phase denoted by the superscript "*SH*" is assumed to be composed of soil and hydrates. Tension is positive in the equations. The skeleton stress is consistent with the effective stress when the water saturation, S_r, equals 1.0.

10.2.3 Conservation of mass

The conservation of mass for the soil, water, gas, and hydrate phases, $\alpha(= S, W, G, H)$, is given as follows:

$$\frac{\partial}{\partial t}(n^\alpha \rho^\alpha)+(n^\alpha \rho^\alpha v_i^\alpha)_{,i}-\dot{m}^\alpha=0 \quad (\alpha=S,W,G,H) \tag{10.17}$$

where ρ^α is the material density for each phase; $\dot{m}^\alpha$ is the mass increasing ratio per unit volume generated by dissociation; and v_i^α is the velocity of the α phase.

When soil particles, hydrates, and water are assumed to be incompressible, $\dot{\rho}^S=0$, $\dot{\rho}^W=0$, $\dot{\rho}^H=0$, and $\dot{m}^S=0$ and $\dot{m}^H+\dot{m}^W+\dot{m}^G=0$ are assumed. Neglecting the space derivative of the volume fractions, Equation 10.17 yields

$$-\dot{n}\rho^S+(1-n)\rho^S v_{i,i}^{SH}=0 \tag{10.18}$$

$$\dot{n}^F S_r \rho^W+n^F \dot{S}_r \rho^W+n^F S_r \rho^W v_{i,i}^W-\dot{m}^W=0 \tag{10.19}$$

$$\dot{n}^F(1-S_r)\rho^G-n^F\dot{S}_r\rho^G+n^F(1-S_r)\dot{\rho}^G+n^F(1-S_r)\rho^G v_{i,i}^G-\dot{m}^G=0 \tag{10.20}$$

$$\dot{n}^H\rho^H-\dot{m}^H=0 \tag{10.21}$$

where we assume $v_i^{SH}=v_i^S$ and the term $n^H\rho^H v_{i,i}^H$ is disregarded.

The stretching tensor for each phase, D_{ij}^α, is defined using the velocity vector, v_i^α, as

$$D_{ij}^\alpha=\frac{1}{2}(v_{i,j}^\alpha+v_{j,i}^\alpha) \quad (\alpha=S,W,G) \tag{10.22}$$

and $v_{i,i}^{SH}$ is assumed to be given by

$$v_{i,i}^{SH}=D_{ii}^{SH}=D_{ii} \tag{10.23}$$

The mass conservation equation for the water phase is obtained by adding Equation 10.18 multiplied by $S_r(\rho^W/\rho^S)$ and Equation 10.19 as

$$\left(1-n^H\right)S_r v_{i,i}^{SH}+\dot{S}_r n^F-\dot{n}^H S_r+n^F S_r(v_{i,i}^W-v_{i,i}^{SH})-\frac{\dot{m}^W}{\rho^W}=0 \tag{10.24}$$

in which the soil and hydrates are assumed to behave as solid mass, we suppose v_i^{SH} $(=v_i^S=v_i^H)$. Similarly, the mass conservation equation for the gas phase is obtained from the sum of Equation 10.18 multiplied by $(1-S_r)$ (ρ^G/ρ^S) and Equation 10.20 as

$$\begin{aligned}&\left(1-n^{H}\right)\left(1-S_r\right)v_{i,i}^{SH}-\dot{S}_r n^{F}-\dot{n}^{H}\left(1-S_r\right)\\&\quad+n^{F}(1-S_r)(v_{i,i}^{G}-v_{i,i}^{SH})+n^{F}\left(1-S_r\right)\frac{\dot{\rho}^{G}}{\rho^{G}}-\frac{\dot{m}^{G}}{\rho^{G}}=0\end{aligned} \tag{10.25}$$

For describing changes in the gas density, the constitutive equation of ideal gases is used, i.e.,

$$\rho^{G}=\frac{M^{G}P^{G}}{R\theta} \tag{10.26}$$

where M^G is the molecular weight of gas; R is the gas constant; θ is the temperature; and tension is positive in the equation. Considering the time derivative of Equation 10.26, we have

$$\frac{\dot{\rho}^{G}}{\rho^{G}}=\frac{\dot{P}^{G}}{P^{G}}-\frac{\dot{\theta}}{\theta} \tag{10.27}$$

The mass changing rates, $\dot{m}^{\alpha}$, for the water and gas phases are related to the dissociation rate which will be discussed in Section 10.3.

As for the mass conservation of hydrate phase given by Equation 10.21, $\dot{n}^{H}$ is negative when dissociation occurs, and the mass increasing ratio of hydrates, $\dot{m}^{H}$, is calculated from the dissociation ratio. The flux vector of hydrate is neglected in the above equations.

10.2.4 Conservation of linear momentum

Momentum balance is required for each phase, namely,

$$n^{\alpha}\rho^{\alpha}\dot{v}_i^{\alpha}=\sigma_{ji,j}^{\alpha}+\rho^{\alpha}n^{\alpha}\bar{F}_i-\tilde{P}_i^{\alpha}\quad(\alpha=S,W,G,H) \tag{10.28}$$

in which $\bar{F}_i$ is the gravity force and $\tilde{P}_i^{\alpha}$ is related to the interaction term given by

$$\tilde{P}_i^{\alpha}=\sum_{\gamma}D^{\alpha\gamma}(v_i^{\alpha}-v_i^{\gamma}),\quad D^{\alpha\gamma}=D^{\gamma\alpha}\quad(\alpha,\gamma=S,W,G,H) \tag{10.29}$$

where $D^{\alpha\gamma}$ are parameters that describe the interaction between each two phases and are given by

$$D^{WS}=\frac{(n^{W})^{2}\rho^{W}g}{k^{W}},\;D^{GS}=\frac{(n^{G})^{2}\rho^{G}g}{k^{G}} \tag{10.30}$$

The momentum balances for the soil (S) and hydrates (H) phases are given from Equation 10.28 as

$$n^S \rho^S \dot{v}_i^S = \sigma_{ji,j}^S + \rho^S n^S \overline{F}_i - \tilde{P}_i^S \tag{10.31}$$

$$n^H \rho^H \dot{v}_i^H = \sigma_{ji,j}^H + \rho^H n^H \overline{F}_i - \tilde{P}_i^H \tag{10.32}$$

Assuming $v_i^{SH} (= v_i^S = v_i^H)$ and adding Equations 10.31 and 10.32, we obtain

$$\left(n^S \rho^S + n^H \rho^H\right)\dot{v}_i^{SH} = \left(\sigma_{ji}^S + \sigma_{ji}^H\right)_{,j} + \left(n^S \rho^S + n^H \rho^H\right)\overline{F}_i - \tilde{P}_i^S - \tilde{P}_i^H \tag{10.33}$$

Considering Equations 10.13 and 10.14, the momentum balance for the solid phase (SH) is given as

$$n^{SH} \rho^{SH} \dot{v}_i^{SH} = \sigma_{ji,j}^{SH} + n^{SH} \rho^{SH} \overline{F}_i - \tilde{P}_i^{SH} \tag{10.34}$$

in which $n^{SH} \rho^{SH} = n^S \rho^S + n^H \rho^H$, $\tilde{P}_i^{SH} = \tilde{P}_i^S + \tilde{P}_i^H$.

Neglecting the acceleration term, the sum of the conservation of momentum for the solid (SH), water (W), and gas (G) phases gives the equilibrium equation as

$$\sigma_{ji,j} + \rho^E \overline{F}_i = 0 \tag{10.35}$$

$$\rho^E = \sum_{\alpha} n^\alpha \rho^\alpha \quad (\alpha = SH, W, G) \tag{10.36}$$

Assuming that the space derivative of the volume fraction $n_{,i}^\alpha$ is negligible and the interaction between the water and gas phases D^{GW} and D^{WG} is zero, the Darcy law–type of equations for the water and gas phases are obtained from the momentum balance and are, respectively, expressed as

$$w_i^W = n^W\left(v_i^W - v_i^{SH}\right) = -\frac{k^W}{\rho^W g}\left(P_{,i}^W - \rho^W \overline{F}_i\right) \tag{10.37}$$

$$w_i^G = n^G\left(v_i^G - v_i^{SH}\right) = -\frac{k^G}{\rho^G g}\left(P_{,i}^G - \rho^G \overline{F}_i\right) \tag{10.38}$$

in which k^W and k^G are the permeability coefficients for the water phase and the gas phase, respectively.

10.2.5 Weak form of the total balance of the linear momentum

Since we will use an updated Lagrangian method in the numerical analysis, we will take the rate type of conservation for the momentum as

$$\frac{D}{Dt}\int_S t_i dS = 0 \tag{10.39}$$

in which D/Dt is the material time derivative and t_i is the surface traction vector.

Using Cauchy's stress theorem, Equation 10.39 provides

$$\frac{D}{Dt}\int_V \sigma_{ji,j} dV = 0 \tag{10.40}$$

Adopting Nanson's law and the Gauss theorem, Equation 10.40 gives the rate form of the equilibrium equation in the current configuration as

$$\int_V \dot{S}^t_{ji,j} dV = 0 \tag{10.41}$$

in which $\dot{S}^t_{ij}$ is the total nominal stress rate tensor with respect to the current configuration defined as

$$\dot{S}^t_{ij} = \dot{\sigma}_{ij} + L_{pp}\sigma_{ij} - L_{ip}\sigma_{pj} \tag{10.42}$$

in which L_{ij} is the velocity gradient tenor.

10.2.6 Conservation of energy

The following energy conservation equation is adopted in order to consider the heat conductivity and the heat sink rate associated with hydrate dissociation:

$$(\rho c)^E \dot{\theta} = D^{vp}_{ij}\sigma'_{ij} - q_{Hi,i} + \dot{Q}^E \tag{10.43}$$

$$(\rho c)^E = \sum_{\alpha} n^\alpha \rho^\alpha c^\alpha \quad (\alpha = S, W, G, H) \tag{10.44}$$

$$\dot{Q}^E = \sum_{\alpha} Q^\alpha \quad (\alpha = S, W, G, H) \tag{10.45}$$

where c^α is the specific heat for the α phase; θ is the temperature for the total phase; D^{vp}_{ij} is the viscoplastic stretching tensor; q_{Hi} is the heat flux;

and $\dot{Q}^E$ is the dissociation heat rate per unit volume due to hydrate dissociation. We assume that the temperature is uniform for all the phases in Equation 10.43.

The heat flux q_{Hi} is given by

$$q_{Hi} = -\lambda^E \theta_{,i} \tag{10.46}$$

where λ^E is the thermal conductivity for the total phase:

$$\lambda^E = \sum_{\alpha} n^\alpha \lambda^\alpha \quad (\alpha = S, W, G, H) \tag{10.47}$$

in which λ^α is the thermal conductivity for the α phase.

The dissociation heat rate per unit volume $\dot{Q}^E$ is given by

$$\dot{Q}^E = \frac{\dot{N}^H Q}{V}, \; Q = 56599 - 16.744\theta \tag{10.48}$$

where Q (kJ/kmol) is the dissociation heat per unit kilomole which changes with temperature (K) (Kuustraa et al. 1983; Masuda et al. 2002).

10.2.7 Soil-water characteristic curve

The relation between suction and saturation is given in the following equation proposed by van Genuchten (1980):

$$S_{re} = \left\{1 + \left(\alpha' P^C\right)^{n'}\right\}^{-m} \tag{10.49}$$

where α', m, and n' are the material parameters, and the relation $m = 1 - 1/n'$ is assumed; $P^C (= P^G - P^W)$ is the suction; and S_{re} is the effective saturation, i.e.,

$$S_{re} = \frac{S_r - S_{r\min}}{S_{r\max} - S_{r\min}} \tag{10.50}$$

where $S_{r\max}$ and $S_{r\min}$ are the maximum and minimum values of the saturation degree, respectively. The material parameters are determined to be $\alpha' = 0.0112$ and $n' = 1.7$ in the present study based on the experimental data of sandy soil. The soil-water characteristic curve used in the analysis is shown in Figure 10.1.

The permeability coefficients for water and gas, k^W (m/s) and k^G (m/s), are dependent on hydrate saturation S_r^H (Masuda et al. 2002):

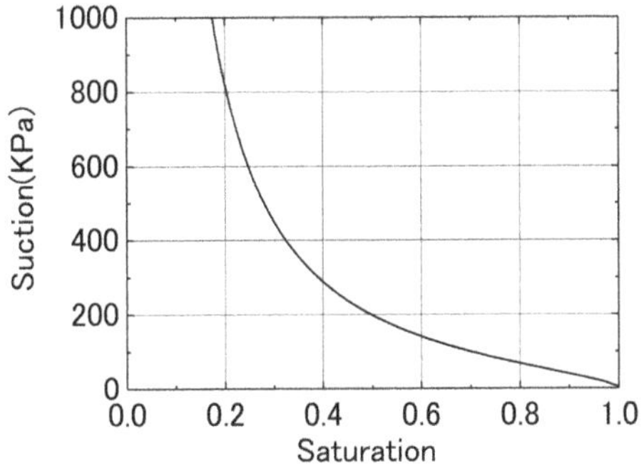

Figure 10.1 Soil-water characteristic curve.

$$k^\alpha = k^\alpha_{S_r^H=0}\left(1 - S_r^H\right)^N \quad (\alpha = W, G) \tag{10.51}$$

$$k^\alpha_{S_r^H=0} = k_0^\alpha \exp\left(\frac{e - e_0}{2}\right) \tag{10.52}$$

$$S_r^H = \frac{V^H}{V^v} \tag{10.53}$$

where V^v is the volume of the void; V^H is the volume of the hydrates; e is the current void ratio; e_0 is the initial void ratio; and $k^\alpha_{S_r^H=0}$ is the permeability when hydrate saturation S_r^H is equal to zero. From the experimental observations by Sakamoto et al. (2004), the permeability decreases to around 1/30 when $S_r^H = 0.4$ compared to that of $S_r^H = 0$. Subsequently, the material parameter N is determined to be 7. The relationship between the permeability ratio $k^\alpha / k^\alpha_{S_r^H=0}$ defined as $\left(1 - S_r^H\right)^N$ which varies from 1 to 0 when the hydrate saturation increases from 0 to 1 is shown in Figure 10.2.

10.2.8 Dissociation of hydrates

The stability of hydrates depends on the temperature and the pressure. The methane hydrate is stable under low-temperature and high-pressure conditions, which is above the equilibrium curve in Figure 10.3. If the conditions for the pore pressure and temperature shift to the unstable region given in Equation 10.54 (Bejan et al. 2002), gas hydrates dissociate into water and gas with the reaction expressed in Equation 10.55:

$$P \le c\exp\left(a - \frac{b}{T}\right) \text{ (Unstable region)} \tag{10.54}$$

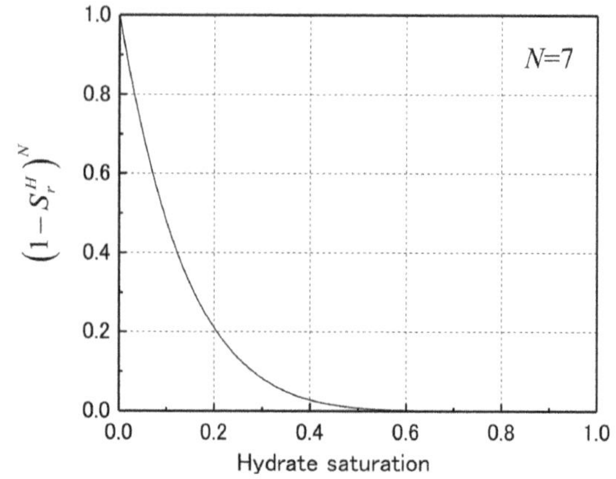

Figure 10.2 Dependency of hydrate saturation on permeability.

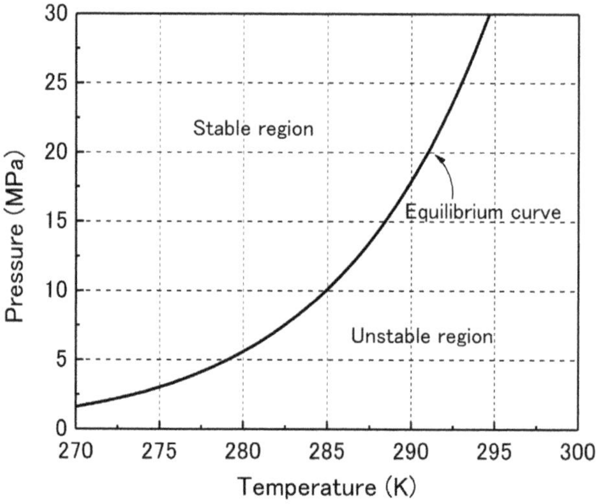

Figure 10.3 Equilibrium curve for methane hydrate.

$$CH_4 \cdot nH_2O\,(\text{hydrate}) \rightarrow nH_2O\,(\text{water}) + CH_4\,(\text{gas}) \tag{10.55}$$

where a, b, and c are material parameters and n is a hydrate number and is assumed to be equal to 5.75. The dissociation ratio, $\dot{N}^H$ (kmol/s), is given by the following Kim–Bishnoi equation (Kim et al. 1987):

$$\dot{N}^H = -5.85\times10^{12}\times\exp^{-\frac{9400}{\theta}}\left(P^e - P\right)N^{H_0\frac{1}{3}}N^{H\frac{2}{3}} \tag{10.56}$$

where N^H (kmol) is the moles of hydrates in volume V (m^3); N^{H_0} (kmol) is the moles of hydrates at the initial state; P (MPa) is the average pore pressure; and P^e is the equilibrium pressure at temperature θ (K). In Equation 10.56, it is assumed that the dissociation rate is proportional to a driving force, i.e., the difference in the fugacity of the methane at the equilibrium pressure and the fugacity of the methane at the present state. The dependency of the reaction rate on temperature follows the Arrhenius-type equation. When dissociation occurs, the dissociation ratio is negative, i.e., $\dot{N}^H < 0$. The water- and gas-generating ratios, $\dot{N}^W$ and $\dot{N}^G$ (kmol/s), are given by

$$\dot{N}^W = -5.75\dot{N}^H \tag{10.57}$$

$$\dot{N}^G = -\dot{N}^H \tag{10.58}$$

Mass increasing ratios (t/s/m^3) for the hydrates, water, and gas phases required in the continuity law can be calculated from the above equations. The reaction described in Equation 10.55 is reversible, i.e., methane gas rehydrates are in the stable region. Only the dissociation process is considered in the present section.

10.3 ELASTO-VISCOPLASTIC MODEL FOR UNSATURATED SOIL CONTAINING HYDRATE

The elasto-viscoplastic model considering soil structure degradation developed for saturated soils has been extended to unsaturated soils using the skeleton stress and the newly introduced suction effect within the constitutive model (Oka et al. 2006, 2010). Natural hydrates exist between soil particles and are considered to have a bonding effect which increases the soil strength. From this point of view, we introduced the effect of hydrate saturation in the elasto-viscoplastic model (Kimoto et al. 2004) by means of the shrinkage or the expansion of the overconsolidation (OC) boundary surface and the static yield surface.

It is assumed that the total stretching tensor consists of the elastic stretching tensor, D_{ij}^e, and the viscoplastic stretching tensor, D_{ij}^{vp}, as

$$D_{ij} = D_{ij}^e + D_{ij}^{vp} \tag{10.59}$$

The elastic stretching is given by a generalized Hooke type of law, i.e.,

$$D_{ij}^e = \frac{1}{2G}\dot{S}_{ij} + \frac{\kappa}{3(1+e)}\frac{\dot{\sigma}'_m}{\sigma'_m}\delta_{ij} \tag{10.60}$$

where S_{ij} is the deviatoric skeleton stress tensor; σ'_m is the mean skeleton stress; G is the elastic shear coefficient; e is the initial void ratio; κ is the swelling index; and the superimposed dot denotes the time differentiation.

In this model, it is assumed that there is an overconsolidation boundary surface that delineates the normally consolidated (NC) region, $f_b \geq 0$, and the overconsolidated region, $f_b < 0$, as follows:

$$f_b = \bar{\eta}^*_{(0)} + M^*_m \ln\left(\sigma'_m / \sigma'_{mb}\right) = 0 \tag{10.61}$$

$$\bar{\eta}^*_{(0)} = \left\{\left(\eta^*_{ij} - \eta^*_{ij(0)}\right)\left(\eta^*_{ij} - \eta^*_{ij(0)}\right)\right\}^{\frac{1}{2}}, \quad \eta^*_{ij} = S_{ij} / \sigma'_m \tag{10.62}$$

where M^*_m is the value of $\eta^* = \sqrt{\eta^*_{ij}\eta^*_{ij}}$ when the volumetric strain increment changes from compression to dilation, which is equal to the stress ratio at the critical state, and σ'_{mb} is the hardening/softening parameter. The stress ratio tensor at the end of the anisotropic consolidation is given by $\eta^*_{ij(0)}$.

The suction effect is incorporated by relating it to the value of $\sigma'_{ma(s)}$ which controls the size of the OC boundary surface (Oka et al. 2006). In addition, the effect of hydrate saturation on strength is introduced. It has been experimentally revealed that the strength of soils containing methane hydrates depends on the hydrate saturation in the void, since hydrates have a bonding effect between soil particles. Once hydrates dissociate, degradation of the soil structure occurs with decreasing hydrate saturation. Considering the effect of the suction and the hydrate saturation, the hardening-softening rule of σ'_{mb} is given as

$$\sigma'_{mb} = N_m N_s \sigma'_{ma}(z) \exp\left(\frac{1+e}{\lambda - \kappa}\varepsilon^{vp}_{kk}\right) \tag{10.63}$$

where N_m and N_s denote the effects of hydrate saturation and suction, respectively; $\sigma'_{ma}(z)$ refers to structural degradation with increasing viscoplastic strain; λ is the compression index; and κ is the swelling index.

The term $\sigma'_{ma}(z)$ describes a strain softening that controls the degradation of the material caused by structural changes, i.e.,

$$\sigma'_{ma} = \sigma'_{maf} + \left(\sigma'_{mai} - \sigma'_{maf}\right)\exp(-\beta z) \tag{10.64}$$

$$z = \int_0^t \dot{z}\,dt, \quad \dot{z} = \sqrt{D^{vp}_{ij} D^{vp}_{ij}} \tag{10.65}$$

where σ'_{mai} and σ'_{maf} are the initial and the final value of σ'_{ma}; β is a material parameter that controls the rate of structural changes; and z is the accumulation of the second invariant of the viscoplastic stretching, D^{vp}_{ij}.

The parameter N_s is defined as

$$N_s = 1 + S_I \exp\left\{-s_d\left(\frac{P_i^C}{P^C} - 1\right)\right\} \tag{10.66}$$

where S_I is the strength ratio of unsaturated soils when the value of suction P^C equals P_i^C, and s_d controls the decreasing ratio of strength with decreasing the suction. The term P_i^C is set to be the maximum value for suction. At the initial state when $P^C = P_i^C$, the strength ratio of the unsaturated soil to the saturated soil is $1 + S_I$ and decreases with a decrease in suction.

The parameter N_m describes the dependency of the hydrate saturation on the stress–strain behavior:

$$N_m = 1 + n_m \exp\left\{-n_d\left(\frac{S_{ri}^H}{S_r^H} - 1\right)\right\} \tag{10.67}$$

where S_r^H is the hydrate saturation in the void. In Equation 10.67, n_m is the material parameter that describes the ratio of the strength when the hydrate saturation S_r^H is equal to S_{ri}^H to that when the hydrate saturation S_r^H is equal to zero, and n_d is the stress decreasing ratio with decreasing hydrate saturation. The material parameters that describe the effect of hydrate saturation are determined to be $S_{ri}^H = 0.51$, $n_m = 0.6$, and $n_d = 0.75$, based on the experimental data on carbon dioxide hydrate mixtures (Matsui et al. 2005).

The static yield function is given by

$$f_y = \bar{\eta}_{(\chi)}^* + \tilde{M}^* \ln\left(\sigma'_m / \sigma'^{(s)}_{my}\right) = 0 \tag{10.68}$$

$$\bar{\eta}_\chi^* = \left\{\left(\eta_{ij}^* - \chi_{ij}^*\right)\left(\eta_{ij}^* - \chi_{ij}^*\right)\right\}^{\frac{1}{2}} \tag{10.69}$$

In Equations 10.68 and 10.72, $\tilde{M}^*$ is the dilatancy coefficient; $\sigma'^{(s)}_{my}$ denotes the static hardening parameter; and χ_{ij}^* is a so-called back stress parameter which has the same dimensions as the stress ratio tensor η_{ij}^*.

The dilatancy coefficient, $\tilde{M}^*$, is assumed to be constant in the NC region and to vary with the current stress in the OC region as

$$\tilde{M}^* = \begin{cases} M_m^* & \text{for NC region } (f_b \geq 0) \\ -\dfrac{\sqrt{\eta_{ij}^* \eta_{ij}^*}}{\ln\left(\sigma'_m / \sigma'_{mc}\right)} & \text{for OC region } (f_b < 0) \end{cases} \tag{10.70}$$

where M_m^* is the value of $\sqrt{\eta_{ij}^* \eta_{ij}^*} / \sigma'_m$ at the critical state, and σ'_{mc} denotes the mean skeleton stress at the intersection of the OC boundary surface and the σ'_m axis.

In the same way as for the OC boundary surface, the effects of suction and hydrate saturation are introduced in the value of $\sigma'^{(s)}_{my}$:

$$\sigma'^{(s)}_{my} = \frac{N_m N_s \sigma'_{ma}(z)}{\sigma'_{mai}} \sigma'^{(s)}_{myi} \exp\left(\frac{1+e}{\lambda - \kappa} \varepsilon^{vp}_{kk}\right) \tag{10.71}$$

where $\sigma'^{(s)}_{myi}$ is the initial value of $\sigma'^{(s)}_{my}$ when $N_m = N_s = 1$.

The viscoplastic potential surface is described as

$$f_p = \bar{\eta}^*_{(\chi)} + \tilde{M}^* \ln\left(\sigma'_m / \sigma'_{mp}\right) = 0 \tag{10.72}$$

The evolution equation for the non-linear kinematic hardening parameter χ^*_{ij} is given by

$$\dot{\chi}^*_{ij} = B^*\left(A^* D^{vp}_{ij} - \chi^*_{ij} \dot{\gamma}^{vp}\right) \tag{10.73}$$

where A^* and B^* are the material parameters; D^{vp}_{ij} is the viscoplastic deviatoric stretching; $\dot{\gamma}^{vp} = \sqrt{D^{vp}_{ij} D^{vp}_{ij}}$; and A^* is related to the stress ratio at failure, namely, $A^* = M^*_f$.

The viscoplastic stretching tensor is expressed by the following equation which is based on Perzyna's viscoplastic theory:

$$D^{vp}_{ij} = C_{ijkl} \left\langle \Phi_1(f_y) \right\rangle \frac{\partial f_p}{\partial \sigma'_{kl}} \tag{10.74}$$

$$C_{ijkl} = a\delta_{ij}\delta_{kl} + b\left(\delta_{ik}\delta_{jl} + \delta_{il}\delta_{jk}\right),\ C_1 = 2b,\ C_2 = 3a + 2b \tag{10.75}$$

where $\langle\ \rangle$ are Macaulay's brackets; $\langle f(x) \rangle = f(x)$, if $x > 0$, $=0$, if $x \le 0$; C_1 and C_2 are the viscoplastic parameters for the deviatoric and the volumetric components, respectively; and Φ_1 indicates the material function for the strain rate sensitivity (Kimoto et al. 2010). The dependency of the viscoplastic property of soils on temperature is also introduced in the viscoplastic parameters C_1 and C_2. Figure 10.4 shows simulations of the undrained triaxial test for saturated soil without hydrates by the elasto-viscoplastic model. The experimental data were obtained for the soil specimen sampled from the seabed ground at Nankai Trough (Oka and Kimoto, 2008).

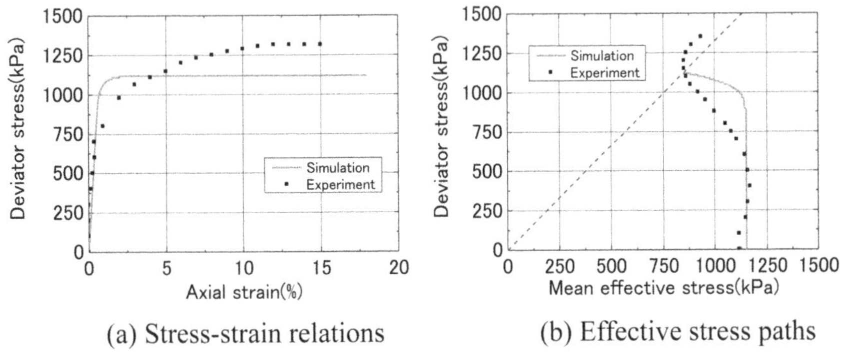

Figure 10.4 Simulation of undrained triaxial test by the elasto-viscoplastic model.

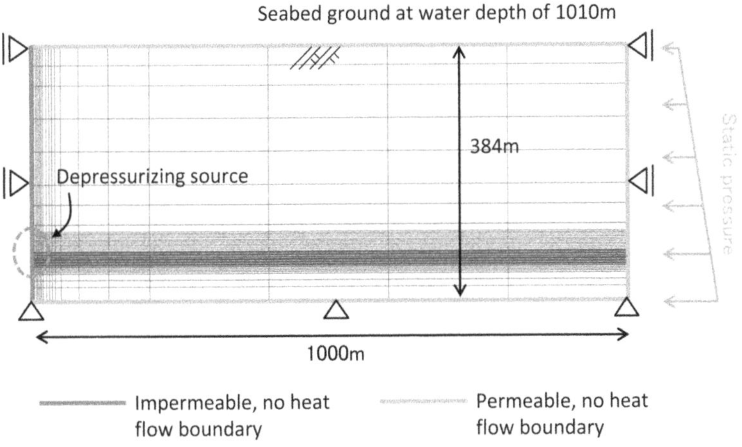

Figure 10.5 Finite element mesh.

10.4 NUMERICAL MODEL AND SIMULATION

10.4.1 Simulation model

Weak forms of the conservation of mass for water and gas, the conservation of momentum, and the conservation of energy are discretized in space and solved by the finite element method. For the finite element method, an updated Lagrangian method with the objective Jaumann rate of Cauchy's stress is used (Kimoto et al. 2010; Oka et al. 2006).

We have simulated the production process by depressurizing in the seabed ground at a water depth of around 1010 m. The finite element mesh and the boundary conditions for the simulation are shown in Figure 10.5, in which the plane strain condition is assumed. The ground is assumed to consist of silty soil, and hydrate-bearing sediment exists at a ground depth

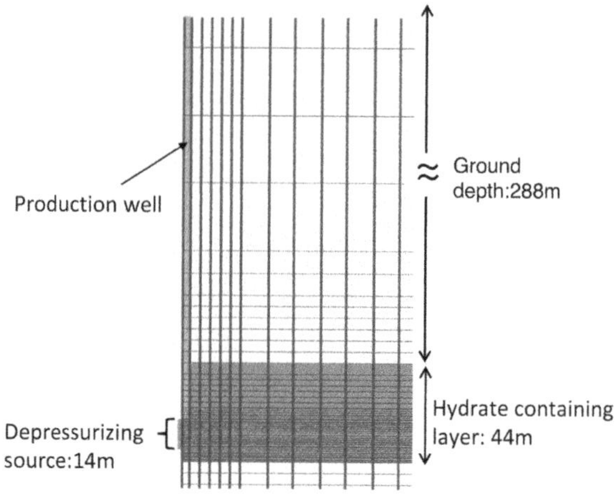

Figure 10.6 Finite element meshes around the depressurizing source.

Table 10.1 Material parameters of the production well

Elastic shear modulus G (kPa)	11.67×10^6
Elastic bulk modulus K (kPa)	15.56×10^6
Permeability coefficient of water k^W (m/s)	1.0×10^{-11}
Permeability coefficient of gas k^G (m/s)	1.0×10^{-10}

Table 10.2 Parameters for the initial stress analysis

Young's modulus E_0 (kPa)	150640.0
Poisson ratio (kPa)	0.4
Internal friction angle (degree)	30
Effective unit weight (kN/m^3)	8.82079

of 288–332 m. Figure 10.6 schematically shows the elements around the depressurizing source. The production well is assumed to exist on the left side of the model. The elements for the production well are modeled as an elastic material and the material parameters are shown in Table 10.1. The initial stress is calculated by the static gravity analysis and the parameters for the initial stress analysis are shown in Table 10.2. The initial conditions for the dissociation analysis are shown in Table 10.3. The pressure is depressurized from 13 MPa to 6 MPa in 25 h, as shown in Figure 10.7. The material parameters for the soil skeleton are given in Table 10.4.

10.4.2 Simulation results

Figure 10.8 shows the distribution of the hydrate saturation ratio around the depressurizing source. The hydrate saturation ratio is determined as

Table 10.3 Initial conditions

Initial volume fraction of void n_0	0.47
Initial hydrate saturation S_{ri}^H	0.51
Internal water saturation S_{ri}	1.0

Table 10.4 Material parameters used in the simulation

Parameters for the flow conductivity	
Permeability coefficient of water k^W (m/s)	1.0×10^{-5}
Permeability coefficient of gas k^G (m/s)	1.0×10^{-4}
Permeability parameter of hydrate N	7
Van Genuchten parameter α'	0.012
Van Genuchten parameter n'	1.7
Parameters of the soil skeleton	
Elastic shear modulus G (kPa)	53800
Compression index λ	0.169
Swelling index κ	0.017
Quasi-overconsolidation ratio $\sigma'_{mbi} / \sigma'_{m0}$	1.0
Stress ratio at the critical state M^*	1.08
Viscoplastic parameter m'	23.0
Viscoplastic parameter C_0 (1/s)	1.0×10^{-12}
Structural parameter β	0.0
Structural parameter $\sigma'_{maf} / \sigma'_{mai}$	1.0
Thermo-viscoplastic parameter α	0.15
Suction parameter S_I	0.2
Suction parameter s_d	0.25
Suction parameter P_i^C	100.0
Hydrate parameter n_m	0.6
Hydrate parameter n_d	0.75
Hydrate parameter S_{ri}^H	0.51

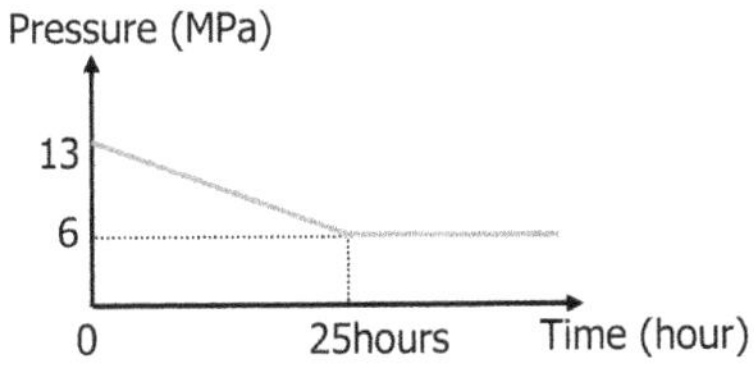

Figure 10.7 Depressurizing condition at the source.

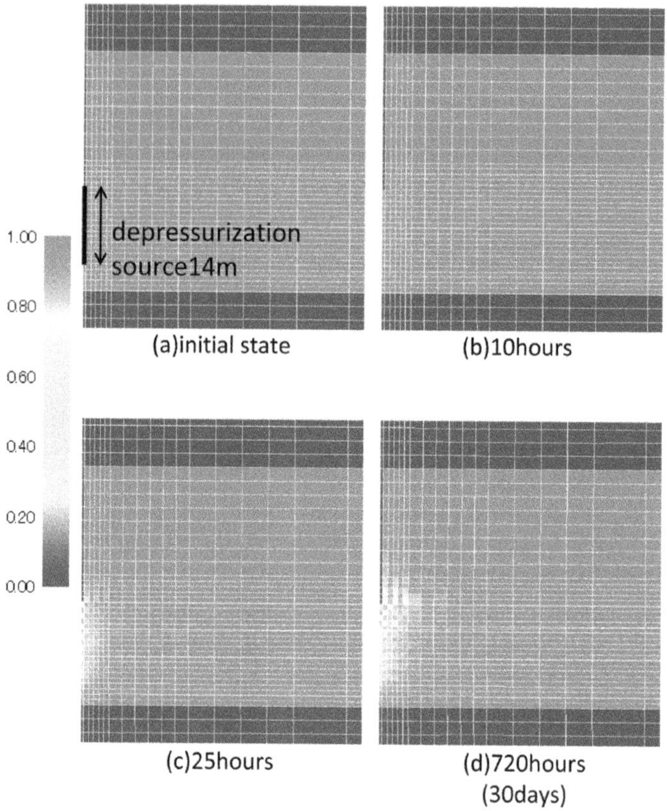

Figure 10.8 Distributions of hydrate saturation ratio around the depressurizing source.

the remaining hydrate moles divided by the initial hydrate moles in each element. From Figure 10.8, dissociation progresses around the depressurizing source; however, about 80% of hydrates remain as shown. This is because of the self-preservation effect of methane hydrate, i.e., the decrease in temperature due to the endothermic reaction shown in Figure 10.9. Figures 10.10 and 10.11 show the distributions of the pore water pressure and the pore gas pressure around the depressurizing source. When depressurizing is finished after 25 h, the pore water pressure decreases around the source. The pore gas pressure is generated in the dissociated area and the value is around 6 MPa. Figure 10.12 shows the distribution of the volumetric strain around the source. The volumetric compression occurs around the source and the maximum value is about 8%. The vertical settlement on the seabed surface is shown in Figure 10.13. Rapid settlement occurs in the initial stage of production and then gradually increases, and the total settlement becomes 0.12 m. The skeleton stress paths in the elements around the source are shown in Figure 10.14. The mean skeleton stress increases

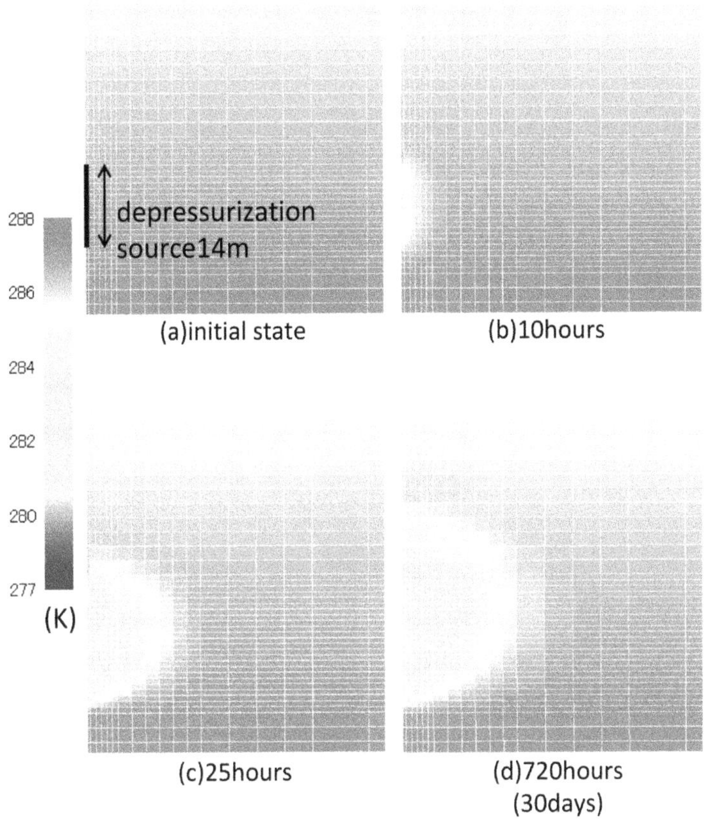

Figure 10.9 Distributions of temperature around the depressurizing source.

in all the elements except in the production well, which causes volume compression.

10.5 SUMMARY

In this chapter, we presented a coupled chemo-thermo-mechanical analysis method to predict the ground deformation caused by the dissociation of gas hydrates. The method was developed based on the coupled analysis, taking into account the phase changes from solids to fluids, i.e., water and gas, the flow of fluids, heat transfer, and the ground deformation. A two-dimensional dissociation analysis of the seabed ground of hydrate-bearing sediments was conducted. The results show that deformation occurs around the depressurizing source during the dissociation process.

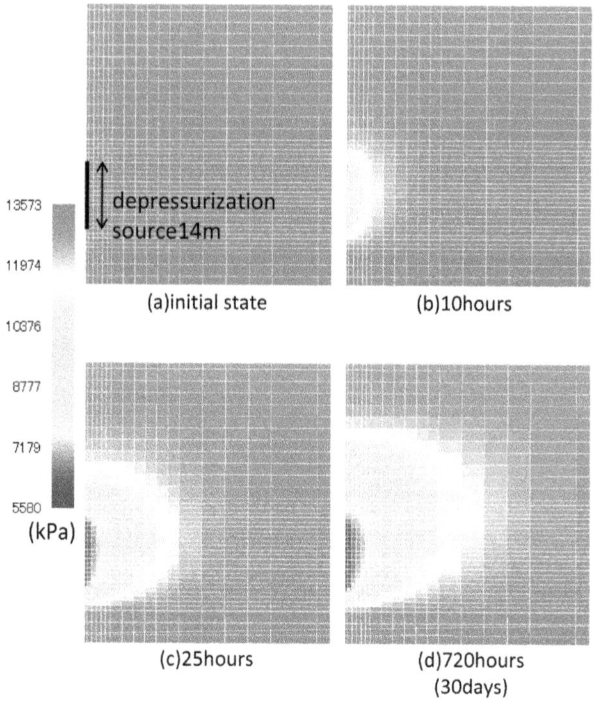

Figure 10.10 Distributions of pore water pressure around the depressurizing source.

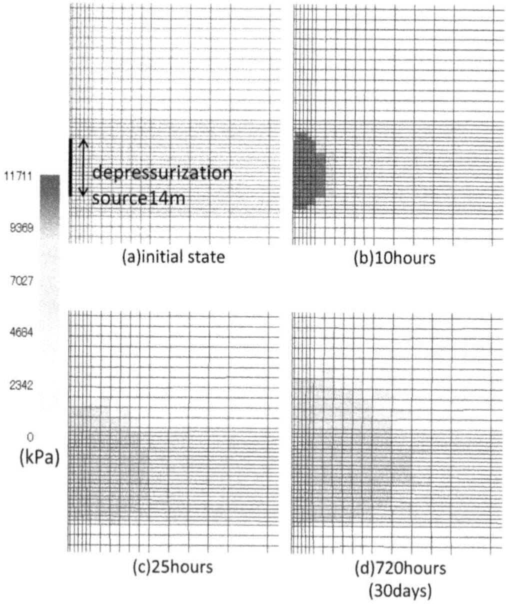

Figure 10.11 Distributions of pore gas pressure around the depressurizing source.

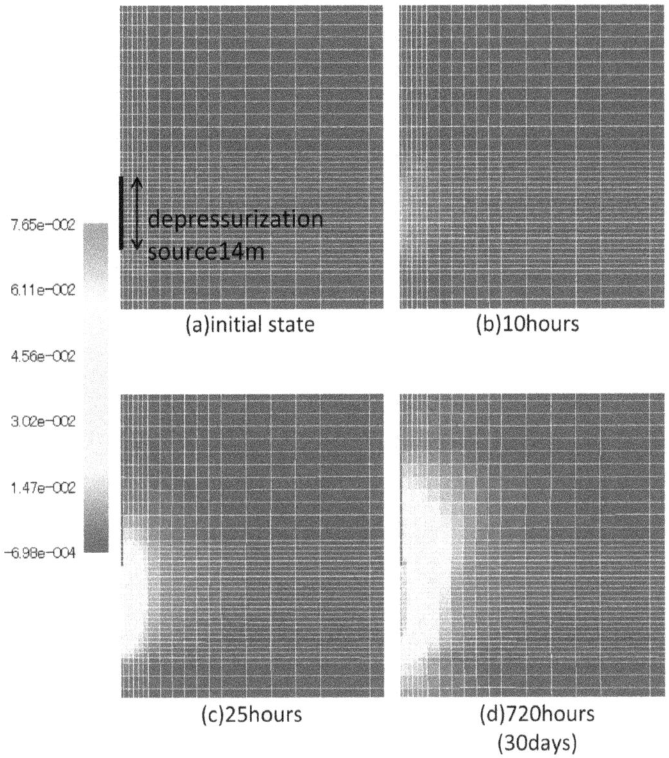

Figure 10.12 Distributions of volumetric strain around the depressurizing source.

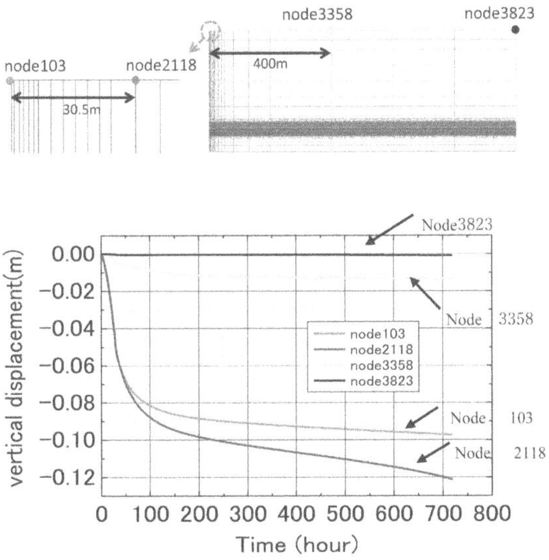

Figure 10.13 Vertical displacement around the depressurizing source.

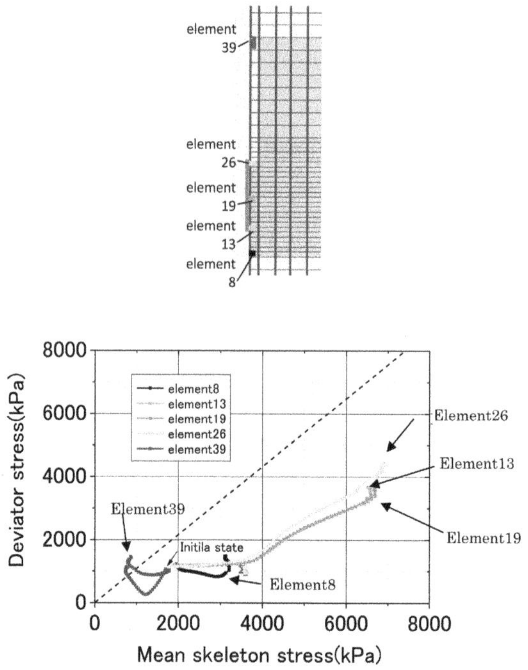

Figure 10.14 Skeleton stress paths around the depressurizing source.

REFERENCES

Ahmadi, G., Ji, C. and Smith, D. 2004. Numerical solution for natural gas production from methane hydrate dissociation. *J. Petrol. Sci. Eng.* 41(4): 269–285.

Bejan, A., Rocha, L.A.O. and Cherry, R.S. 2002. Methane hydrates in porous layers: Gas formation and convection. In: *Transport Phenomena in Porous Media II*, ed. B.D. Ingham and I. Pop, 365–396. Pergamon, Oxford.

Biot, M.A. 1962. Theory of propagation of elastic waves in fluid-saturated porous media. *J. Acoust. Soc. Am.* 28(2): 168–191.

Bondarev, E.A. and Kapitonova, T.A. 1999. Simulation of multiphase flow in porous media accompanied by gas hydrate formation and dissociation. *Russ. J. Eng. Thermophys.* 9(1): 2.

de Boer, R. 1998. Theory of porous media-past and present. *Angew Z. J. Math. Technol.* 78(7): 441–466.

Ehlers, W., Graf, T. and Ammann, M. 2004. Deformation and localization analysis of partially saturated soil. *Comp. Meth. Appl. Mechan. Eng.* 193(27–29): 2885–2910.

Hyodo, M., Nakata, Y., Yoshimoto, N. and Ebinuma, T. 2005. Basic research on the mechanical behavior of methane hydrate-sediments mixture. *Soils Found.* 45(1): 75–85.

Iwai, H., Konishi, Y., Saimyou, K., Kimoto, S. and Oka, F. 2018. Rate effect on the stress-strain relations of synthetic carbon dioxide hydrate-bearing sand and its dissociation test by thermal stimulation. *Soils Found.* 58(5): 1113–1132.

Kim, H.C., Bishnoi, P.R., Heidemann, R.A. and Rizvi, S.S.H. 1987. Kinetics of methane hydrate decomposition. *Chem. Eng. Sci.* 42(7): 1645–1653.

Kimoto, S., Oka, F. and Higo, Y. 2004. Strain localization analysis of elastoviscoplastic soil considering structural degradation. *Comput Methods Appl Mech Eng*: 193: 2854–2866.

Kimoto, S., Oka, F., Fushita, T. and Fujiwaki, M. 2007a. A chemo-thermo-mechanically coupled numerical simulation of the subsurface ground deformation due to methane hydrate dissociation. *Comput. Geotech.* 34(4): 216–228.

Kimoto, S., Oka, F., Fushita, T. and Fujiwaki, M. 2007b. Numerical simulation of the ground deformation due to methane hydrate dissociation by a chemo-thermo-mechanically coupled analysis. In: *Proceedings of the 9th International Symposium on Numerical Models in Geomechanics*, Rodos, ed. G.N. Pande and S. Pietruszczak, 303–309. Bakema/Taylor & Francis, Rotterdam.

Kimoto, S., Oka, F., Kim, Y.-S., Takada, N. and Higo, Y. 2007c. A finite element analysis of the thermo-hydro-mechanically coupled problem of a cohesive deposit using a thermo-elasto-viscoplastic model. *Key Eng. Mater.* 340–341: 1291–1296.

Kimoto, S., Oka, F. and Fushita, T. 2010. A chemo-thermo-mechanically coupled analysis of ground deformation induced by gas hydrate dissociation. *Int. J. Mech. Sci.* 52(2): 365–376.

Kuustraa, V.A., Hammershaimb, E.C., Holder, G.D. and Sloan, E.E. 1983. *Handbook of Gas Hydrate Properties and Occurrence.* Lewin & Associates, Washington DC, DOE/MC/19239-1546: 94.

Kvalstad, T.J., Andresen, L., Forsberg, C.F., Berg, K., Bryn, P. and Wangen, M. 2005. The Stregga slide: Evaluation of triggering sources and slide mechanics. *Mar. Petrol. Geol.* 22(1–2): 245–256.

Loret, B. and Khalili, N. 2000. A three-phase model for unsaturated soils. *Int. J. Numer. Anal. Meth. Geomech.* 24(11): 893–927.

Masuda, Y., Kurihara, M., Ohuchi, H. and Sato, T. 2002. A field-scale simulation study on gas productivity of formations containing gas hydrates. In: *Proceedings of the 4th International Conference on Gas Hydrate*, 40–46, Yokohama, Japan, May 2002.

Matsui, A., Hyodo, M., Nakata, Y., Yoshimoto, N. and Taketomi, K. 2005. Formation and mechanical properties of carbon dioxide hydrate in sand. In: *Proc. of the 40th JGS Symposium.* Hakodate, Japan: ID(212): 40–46. (in Japanese)

Miyazaki, K., Masui, A., Yamaguchi, T., Sakamoto, Y., Haneda, H., Ogata, Y., Aoki, K. and Okubo, S. 2008. Strain rate dependency of peak and residual strength of sediment containing synthetic methane hydrate. *J. MMIJ* 124(10/11): 619–625 (in Japanese).

Oka, F., Kimoto, S., Kim, Y.-S., Takada, N. and Higo, Y. 2005a. A finite element analysis of the thermo-hydro-mechanically coupled problem of cohesive deposit using a thermo-elasto-viscoplastic model. In: *Poromechanics-Biot-Centennial, Proceedings of the 3rd Biot Conference on Poromechanics*, Norman, ed. Abousleiman, Y.N., Cheng, A.H.-D. and Ulm, F.-J., 383–388. Balkema, Rotterdam.

Oka, F., Kimoto, S., Kodaka, T., Takada, N., Fujita, Y. and Higo, Y. 2005b. A finite element analysis of the deformation behavior of a multiphase seabed ground due to the dissociation of natural gas hydrates. In: *Proceedings of 11th International Conference of IACMAG*, Torino, ed. G. Barla and M. Barla, 127–134. AGI, Patron Editore, Rome.

Oka, F., Kodaka, T., Kimoto, S., Kim, Y.-S. and Yamasaki, N. 2006. An elasto-viscoplastic model and multiphase coupled FE analysis for unsaturated soil. In: *Proceedings of 4th International Conference on Unsaturated Soils*, April 2–6, 2006, Carefree, AZ: Geotechnical. Special Publication: 147, ASCE, 2: 2039–2050.

Oka, F. and Kimoto, S. 2008. An elasto-viscoplastic constitutive model and its application to the sample obtained from seabed ground at Nankai Trough. *J. Soc. Mater. Sci., Jpn.* 57(3): 237–242 (in Japanese).

Oka, F., Kimoto, S., Takada, N. and Higo, Y. 2009. A multiphase elasto-viscoplastic analysis of an unsaturated river embankment associated with seepage flow. In: *Proceedings of the International Symposium on Prediction and Simulation Methods for Geohazard Mitigation*, ed. F. Oka, A. Murakami and S. Kimoto, 128–132. Taylor & Francis, Boca Raton, FL.

Oka, F., Kimoto, S., Takada, N., Gotoh, H. and Higo, Y. 2010. A seepage-deformation coupled analysis of an unsaturated river embankment using a multiphase elasto-viscoplastic theory. *Soils Found.* 50(4): 483–494.

Oka, F., Kimoto, S., Kitano, T., Iwai, H. and Higo, Y. 2011. A numerical analysis of hydrate-bearing subsoil during dissociation using a chemo-thermo-mechanically coupled analysis method. In: *4th International Conference GeoProc2011: Cross Boundaries through THMC Integration*, Perth, Australia, 6–9 July, 2011, Paper No: GP060.

Oka, F. and Kimoto, S. 2012. *Computational Modeling of Multiphase Geomaterials.* CRC Press/Taylor & Francis, Boca Raton, FL.

Rutqvist, J. and Moridis, G.J. 2008. Development of a numerical simulator for analyzing the geomechanical performance of hydrate-bearing sediments. In: *42nd US Rock Mechanics Symposium. and 2nd U.S.-Canada Rock Mechanics Symposium.* Paper No. ARMA-08-139, American Rock Mechanics Society.

Rutqvist, J., Moridis, G.J., Grover, T. and Collett, T. 2009. Geomechanical response of permafrost-associated hydrate deposits to depressurization-induced gas production. *J. Petrol. Sci. Eng.* 67(1–2): 1–12.

Sakamoto, Y., Komai, T., Kawabe, T., Tenma, N. and Yamaguchi, T. 2004. Formation and dissociation behavior of methane hydrate in porous media. Estimation of permeability in methane hydrate reservoir, Part 1. *J. Min. Mater. Proc. Inst. Jpn.* 120: 85–90 (in Japanese).

Sakamoto, Y., Shimokawara, M., Ohga, K., Miyazaki, K., Komai, T., Aoki, K. and Yamaguchi, T. 2008. Experimental study on consolidation behavior and permeability characteristics during dissociation of methane hydrate by depressurization process. Estimation of permeability in methane hydrate reservoir, Part 6. *J. Min. Mater. Proc. Inst. Jpn.* 124(8): 498–507 (in Japanese).

Sheng, D., Sloan, W., Gens, A., Smith, D.W. 2003. Finite element formulation and algorithms for unsaturated soils, Part I: Theory. *Int. J. Numer. Anal. Meth. Geomech.* 27(9): 745–765.

Sultan, N., Cochonat, P., Foucher, J.-P. and Mienert, J. 2004. Effect of gas hydrates melting on seafloor slope instability. *Mar. Geol.* 213(1–4): 379–401.
Terzaghi, K. 1943. *Theoretical Soil Mechanics*. John Wiley & Sons, New York.
Tsypkin, G.G. 2000. Mathematical models of gas hydrates dissociation in porous media. *Ann. NY Acad. Sci.* 912(1): 428–436.
van Genuchten, M.T. 1980. A closed-form equation for predicting the hydraulic conductivity of unsaturated soils. *Soil Sci. Soc. Am. J.* 44(5): 892–898.
Wong, H.K., Ludmann, T., Baranov, B.V., Ya Karp, V., Konerding, P. and Ion, G. 2003. Bottom current-controlled sedimentation and mass wasting in the northwestern Sea of Okhotsk. *Mar. Geol.* 201(4): 287–305.
Wu, L. and Grozic, J.L.H. 2008. Laboratory analysis of carbon dioxide hydrate-bearing sands. *J. Geotech. Geoenviron. Eng.* 134(4): 547–550.

Chapter 11

A numerical model for the internal erosion of geomaterials

11.1 INTRODUCTION

Soil particles in the earth structures are moved and transported by seepage flow when the hydraulic flow is sufficiently intense. This is called internal erosion. Progressive degradation of the soil structure due to internal erosion may lead to local failure, such as piping, and may result in cavities in the earth structures, such as earth dams, dikes, and levees (Fell and Fry 2013). This is one of the main causes of the failure of geomaterials. Internal erosion can be classified as concentrated leak erosion, backward erosion, contact erosion, and suffusion (Fell and Fry 2013).

In this section, first we present the governing equations to simulate the hydro-mechanical behavior of soil subjected to internal erosion within the framework of the multiphase mixture theory. Then, we formulate the constitutive equations governing internal erosion, i.e., the erosion criteria and the rate equation of mass transfer following the works by Kimoto et al. (2017), Akaki et al. (2017), Akaki (2017), Loret (2015, 2019). We assume that erosion takes place when the hydro-dynamical driving force acting on a small volume of the mixture is larger than the resistance force over the area. Finally, laboratory suffusion tests using a gap-graded sandy soil are simulated by the erosion model and its validity of the proposed model is discussed with respect to the rate of eroded soil mass and the particle size distribution after the erosion test.

11.2 EQUATIONS OF MOTION AND MASS BALANCE EQUATIONS

11.2.1 General setting for the mixture

Figure 11.1 shows the volume fractions associated with the phase transition; *S*, *W*, and *FS* indicate the solid particle, water, and fluidized soil particle phases, respectively.

DOI: 10.1201/9781003200031-11

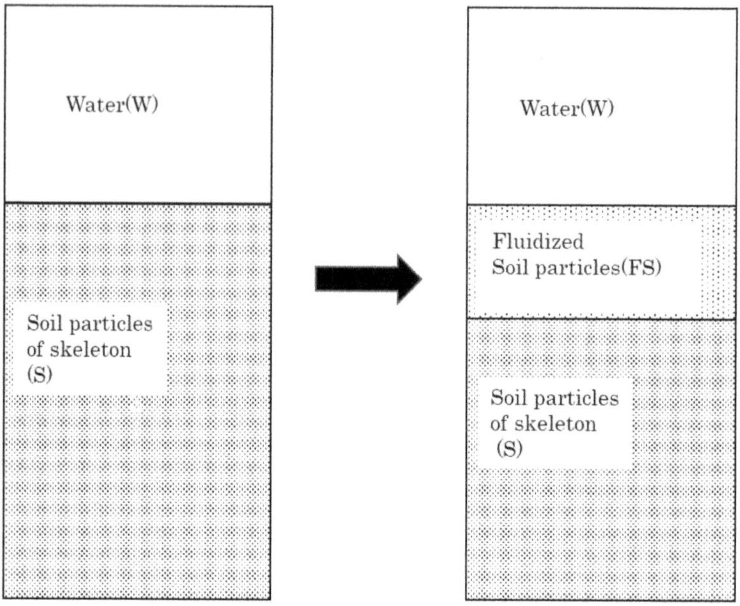

Figure 11.1 Phase transition: Soil particles of skeleton to fluidized soil particles.

The volume fractions, n^α, are defined as

$$n^\alpha = \frac{V^\alpha}{V}, \sum_\alpha n^\alpha = 1; \quad (\alpha : S, W, FS) \tag{11.1}$$

where V is the volume of the mixture; V^α is the volume of the material of phase alpha; ρ^α is the density of the constituent of the phase; n is the porosity; and c is the concentration defined as

$$n = n^W + n^{FS} = 1 - n^S \tag{11.2}$$

$$c = \frac{n^{FS}}{n} \tag{11.3}$$

11.2.2 Partial stress and effective stress

Partial stresses are defined as

$$\sigma_{ij}^S = \sigma'_{ij} - (1-n)P\delta_{ij} \tag{11.4}$$

$$\sigma_{ij}^L = \sigma_{ij}^W + \sigma_{ij}^{FS} = -nP\delta_{ij} \tag{11.5}$$

where σ_{ij}^{S}, σ_{ij}^{W}, and σ_{ij}^{FS} are the partial stresses for the S, W, and FS phases; σ'_{ij} is the effective stress; and P is the pore pressure. The partial stress of the L phase, σ_{ij}^{L}, is given by the summation of those of the W and FS phases. The tension is taken as positive for sigma, whereas the pressure is considered positive in compression.

Summing up the partial stresses yields the total stress, σ_{ij}, as

$$\sigma_{ij} = \sigma_{ij}^{S} + \sigma_{ij}^{L} = \sigma'_{ij} - P\delta_{ij} \tag{11.6}$$

11.2.3 Conservation of linear momentum

Although the derivative with respect to the space can be taken with respect to that of each phase in general, the space derivative is now taken with respect to the solid frame.

The conservation of linear momentum for each phase, i.e., for the S, W, and FS phases, is given by

$$\sigma_{ji,j}^{\alpha} + n^{\alpha}\rho^{\alpha}b_i + \sum_{\alpha} P_i^{\alpha\beta} = 0, \; (\alpha, \beta = S, W, FS) \tag{11.7}$$

where σ_{ij}^{α} is the partial stress of the α phase; n^{α} is the volume fraction; ρ^{α} is the density of the constituent of each phase; b_i is the body force; and $P_i^{\alpha\beta}$ is the interaction between the α and β phases and $P_i^{\alpha\beta} = -P_i^{\beta\alpha}$. Summing up the conservation of momentum equations for the three phases yields the conservation law of the total mixture:

$$\sigma_{ji,j} + \rho b_i = 0 \tag{11.8}$$

where σ_{ij} is the total stress and ρ is the density of the mixture ($= \Sigma_{\alpha} n^{\alpha}\rho^{\alpha}$).

We assume that the velocity of the fluidized soil particles is equal to that of the pore water.

The conservation of linear momentum of the L phase is given by the summation of those of the W and FS phases. As mentioned above, the partial stress of the L phase is given by $-nP\delta_{ij}$. The interaction term given by $P_i^{SL} = P_i^{SW} + P_i^{SFS}$ is assumed to be proportional to the relative velocity as

$$P_i^{SL} = -\frac{n^2\rho^L g}{k^L}(v_i^L - v_i^S) \tag{11.9}$$

where n is the porosity; $v_i^L (= v_i^W = v_i^{FS})$ is the velocity of the L phase; v_i^S is the velocity of the soil skeleton; k^L is the permeability; ρ^L is the density of the L phase as $\rho^L = c\rho^S + (1-c)\rho^W$; and g is the gravitational force.

Substituting Equation 11.9 into the conservation of linear momentum equation of the L phase, we have

$$w_i^L = n(v_i^L - v_i^S) = \frac{k^L}{\rho^L g}(-P_{,i} + \rho^L b_i) \tag{11.10}$$

11.2.4 Conservation of mass

The mass conservation law for each phase is given by

$$\frac{d(n^\alpha \rho^\alpha)}{dt} + n^\alpha \rho^\alpha v_{i,i}^\alpha = \dot{m}^\alpha, (\alpha = S, W, FS) \tag{11.11}$$

where $\dot{m}^\alpha$ is the velocity of the mass transfer per unit volume of the α phase, $\dot{m}^S + \dot{m}^{FS} = 0$ and $\dot{m}^W = 0$. Assuming the incompressibility of soil particles and substituting $n^S = 1 - n$ into Equation 11.11, we obtain

$$-\dot{n} + (1-n)v_{i,i}^S = \dot{m}^S / \rho^S \tag{11.12}$$

Assuming the incompressibility of soil particle and water and summing up the mass conservation laws for the W and FS phases, we have

$$\dot{n} + w_{i,i}^L + nv_{i,i}^S = -\dot{m}^S / \rho^S \tag{11.13}$$

where $\rho^{FS} = \rho^S$ is used.

Summing up Equations 11.12 and 11.13 gives

$$v_{i,i}^S + w_{i,i}^L = 0 \tag{11.14}$$

Then, we derive the transport equation of fluidized soil particles. Substituting $n^{FS} = cn$ into the mass conservation laws, we obtain

$$\dot{c}n + c\dot{n} + (cw_i^L)_{,i} + cnv_{i,i}^S = -\dot{m}^S / \rho^S \tag{11.15}$$

Eliminating $\dot{n}$ in Equation 11.15 by using Equation 11.13 and substituting Equation 11.14 into the resulting equation, the following transport equation of fluidized soil particles is obtained:

$$n\dot{c} + w_i^L c_{,i} = -(1-c)\dot{m}^S / \rho^S \tag{11.16}$$

11.3 EVOLUTIONAL EQUATION OF THE INTERNAL EROSION

Firstly, we derive the onset conditions of the internal erosion of uncemented soil. It is assumed that erosion occurs when the driving force that acts on a small volume of soil is larger than the resistance force (see Figure 11.2). In

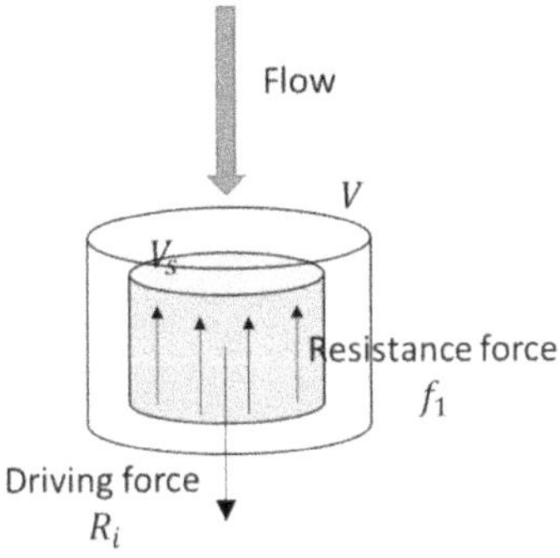

Figure 11.2 Driving force and resistance force of erosion for volume V of a set of soil particles in direction n_i.

the model, the driving force of erosion is given by the interaction term P_i^{SL} in Equations 11.7 and 11.9. The onset condition in the flow direction, n_i, is given by

$$\begin{aligned} &P_i^{SL} n_i \bar{V}_g - f_1(\sigma'_{kl}, \varepsilon_{kl}^P, S) S_g \geq 0 \\ &(\text{driving force}) - (\text{resistance force}) \geq 0 \end{aligned} \tag{11.17}$$

where $\bar{V}_g = V_g / n^s$ is the total volume around the soil particles; V_g is the volume of one soil particle; S_g is the surface area over which the resistance force acts; f_1 is the resistance force against erosion per unit surface that may depend on σ'_{kl}, i.e., the effective stress acting in the direction n_i; ε_{kl}^p is the inelastic strain tensor; and S is the other factor such as the saturation of gas in the case of the existence of gas, the initial fine content, clay content, chemical action, etc. In Figure 11.2, $S_g / \bar{V}_g = 4n^S / D$ for the cylindrical volume; $S_g / \bar{V}_g = \pi D^2 / (\pi D^3 / 6n^S) = 6n^S / D$ for the spherical volume; and D is the diameter of the soil particle.

Since the direction of P_i^{SL} is the same as that of the relative velocity direction n_i, we obtain

$$P_i^{SL} n_i = \frac{n\rho^L g}{k^L} w_i^L \cdot \frac{w_i^L}{\left\| w_i^L \right\|} = \frac{n\rho^L g}{\tilde{k} D^2} \left\| w_i^L \right\| \tag{11.18}$$

in which the permeability is assumed to be dependent on the diameter of the particles:

$$k^L = \tilde{k} D^2 \tag{11.19}$$

where D is the diameter of the particles and $\tilde{k}$ is a parameter.

If k^L is constant, the resistance force from Equation 11.17 is inversely proportional to the diameter of the particles and this is not consistent with the experimental evidence that the small particle is eroded first.

$$\frac{n\rho^L g}{\tilde{k}D^2}\left\|w_i^L\right\|\frac{\pi D^3}{6n^S} - f_1\left(\sigma'_{ij},\varepsilon_{ij}^p,S\right)\pi D^2 \geq 0$$
$$\Leftrightarrow q \equiv \left\|w_i^L\right\| \geq f_1\left(\sigma'_{ij},\varepsilon_{ij}^p,S\right)\frac{6n^S\tilde{k}}{n\rho^L g}D \tag{11.20}$$

Since soil is composed of many particles with different diameters, the onset condition of erosion is given for each particle group with different diameters D_i $(i = 1, 2,\ldots, N)$.

$$\dot{m}^S(D_i) = -<\alpha(q - q_{cr})> \frac{m^S(D_i) - Rm_0^S(D_i)}{m_T^S},\ (i = 1,2,\ldots,N) \tag{11.21}$$

where $< >$ is Macaulay's bracket as $<x> = 0$ if $x \leq 0$ and $<x> = x$ if $x > 0$.

$$q_{cr} = f_1\left(\sigma'_{ij},\varepsilon_{ij}^p,S\right)\frac{6n^S\tilde{k}}{n\rho^L g}D_i,\ (i = 1,2,\ldots,N) \tag{11.22}$$

in which $\dot{m}^S(D_i)$ and $\dot{m}_0^S(D_i)$ are the mass per unit volume of the soil group with a representative diameter D_i and its initial value, respectively; m_T^S is the mass of all the particles per unit volume; α is a coefficient of the erosion rate; and R is the ratio of the mass of remaining soil particles to that of the initial mass of all particles. In the following, $f_1 = 1$ for simplicity.

11.4 PERMEABILITY COEFFICIENT

If the effects of the resedimentation of the fluidized particles and the clogging are relatively small, the permeability increases due to the increase in the volume of the void. The void ratio dependency of the permeability is assumed to be controlled by the Kozeny–Carman equation (Carman 1937; Taylor 1948):

$$K = C_{KD}\tilde{D}^2\frac{e^3}{1+e} \tag{11.23}$$

where K is the intrinsic permeability; C_{KD} is the Kozeny–Carman constant; $\tilde{D}$ is the representative diameter; and e is the void ratio.

The relationship between the permeability, k^L, and the intrinsic permeability, K, is given by

$$k^L = \frac{\rho^L g}{\mu^L} K \tag{11.24}$$

where $\mu^L = \mu^W$ (viscosity of water).

In this section, we assume that the fluidized soil particles do not affect the viscosity of water and that the effects of the variations of C_{KD}, $\tilde{D}$, K, and ρ^L from the initial state are smaller than that of the void ratio. Then, we can derive

$$k^L = k_0^L \frac{1+e_0}{1+e} \frac{e^3}{e_0^3} \tag{11.25}$$

11.5 NUMERICAL ANALYSIS METHOD

The elasto-viscoplastic constitutive equation of the soil skeleton proposed by Kimoto and Oka (2005) was used in the analysis. Since the finite element analysis based on the updated Lagrangian method has been adopted, the strain rate is replaced by the stretching and the Jaumann stress rate of the Cauchy stress tensor is used. The constitutive equation and the finite element formulation are presented in sections 5.7 and 6.5 of Oka and Kimoto (2012). For the discretization of Equation 11.16, the streamline-upwind Petrov–Galerkin (SUPG, Brooks and Hughes 1982) method is used. The unknown variables are the velocity of the skeleton, the pore pressure, and the concentration of the soil particles. The eight-node isoparametric finite elements and the four-node isoparametric elements are used for the pore pressure and the concentration, respectively. The backward Euler scheme is used for the time integration. The governing equations are solved by the finite element method (FEM) and the unknowns are the velocity of the soil skeleton, the pore fluid pressure, and the concentration c of the eroded soil particles defined by $c = n^{FS}/n$.

11.6 NUMERICAL RESULTS

In order to validate the analysis method, we have simulated the experiments of the internal erosion tests, i.e., suffusion for the gap-graded sands of a mixture of silica sand No. 3 and No. 8 with a downward flow conducted by Ke and Takahashi (2014). The suffusion is defined by washing out the fine content during seepage (Fell and Fry 2013; Skempton and Brogan 1994). Figure 11.3 shows the finite element mesh and the boundary conditions of the analysis model. Tables 11.1 and 11.2 list the material parameters. The grain size distribution curves before and after erosion are shown in Figure 11.4, and the simulated and experimental results of the eroded soil volume for the different values of the coefficient α_{er} of the erosion rate are illustrated in Figure 11.5. As observed in Figure 11.5, the suffusion (the washing out) of the fine sand (silica sand No. 8) is well reproduced.

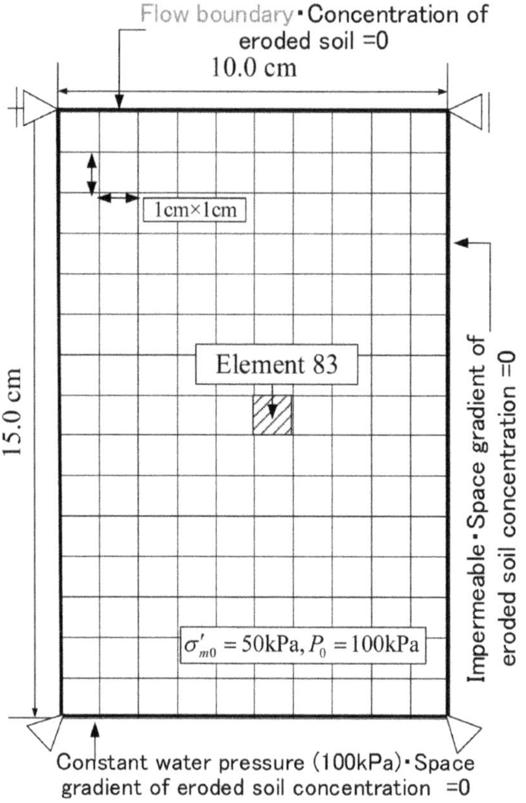

Figure 11.3 Finite element mesh and boundary conditions.

Table 11.1 Material parameters of the soil skeleton

Parameter	*Value*
Compression index, swelling index λ, κ	0.053, 0.005
Initial shear modulus G_0 (kPa)	2000.0
Initial void ratio e_0	0.59
Quasi-overconsolidation ratio OCR^*	1.0
Viscoplastic parameters m', C_0 (1/s)	40.0, 1.0×10^{-12}
Structural parameters $\sigma'_{maf}/\sigma'_{mai}, \beta$	1.0, 0.0
Stress ratio at critical state M^*_m	1.34
Initial permeability coefficient k^L_0 (m/s)	6.0×10^{-5}

**OCR* is a sumbol of quasi-overconsolidation ratio.

Table 11.2 Parameters for erosion characteristics

Parameter	*Value*
Particle diameter dependency of fluidization $\tilde{k}$ (1/m/s)	4.0
Coefficient of fluidization rate α (kg/m^4)	1.0, 2.0, 3.0
Remaining rate of soil particles R	0.3

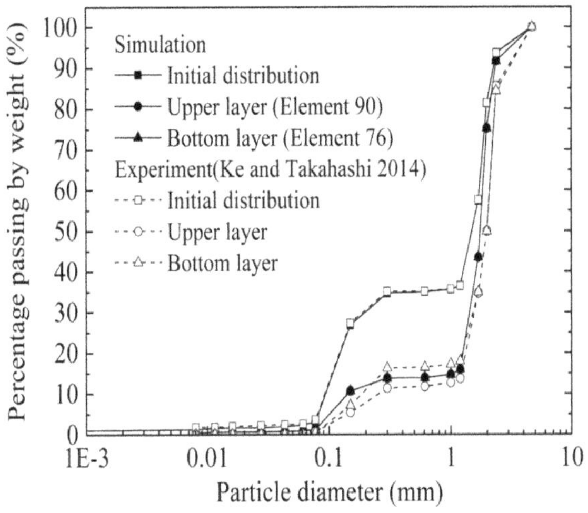

Figure 11.4 Simulated and experimental results of grain size distribution before and after erosion.

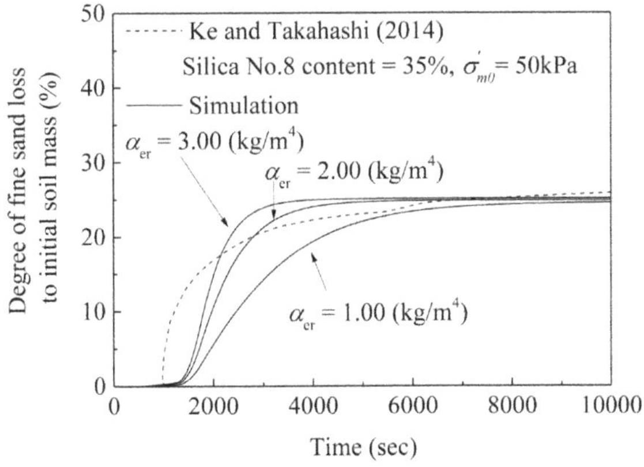

Figure 11.5 Simulated and experimental results of the eroded soil.

11.7 SUMMARY

In this chapter, the constitutive equations governing internal erosion, i.e., the erosion criteria and the rate equation of mass transfer, were formulated. Subsequently, the field equations governing the hydro-mechanical behavior of soil subjected to internal erosion were derived in the framework of the multiphase mixture theory. The validity of the proposed model was discussed with respect to the rate of eroded soil mass and the particle size distribution after the erosion test. The main conclusions obtained were as follows:

1. The coupled behavior of the soil–water interaction and the internal erosion followed by the transportation of eroded particles can be simulated by the erosion model. The change of grain size distribution due to the outflow of fine particles observed in the experiments (Ke and Takahashi 2014) was well reproduced.
2. The simulated mass of eroded soil was compared to that obtained in the experiments. The tendency that the rate of eroded soil increases with the increasing flow rate and then decreases to zero was well captured, even though the behavior at an early stage was slightly different from that obtained in the experiments.

REFERENCES

Akaki, T. 2017. *Numerical Analysis of Earthquakes and Internal Erosion during Gas Production for Hydrate-Bearing Sediments*, PhD thesis. Kyoto University.

Akaki, T., Aota, S., Kimoto, S. and Oka, F. 2017. Numerical simulation of internal erosion by multiphase coupled analysis method. *J. Jpn. Soc. Civil Eng. A2 (Appl. Mech.)* 72(2): I-75–I-86 (in Japanese).

Brooks, A.N. and Hughes, T.J.R. 1982. Streamline upwind/Petrov–Galerkin formulations for convection dominated flows with particular emphasis on the incompressible Navier–Stokes equations. *Com. Meth. Appl. Mech. Eng.* 32: 199–259.

Carman, P.C. 1937. Fluid flow through granular beds. *Trans. Inst. Chem. Eng.* 15: 150–157.

Fell, R. and Fry, J.J. 2013. State of art on the likelihood of internal erosion of dams and levees by means of testing. In: *Erosion in Geomechanics Applied to Dams and Levees*, ed. S. Bonelli, 1–99. ISTE Ltd. and Wiley, Hoboken, NJ.

Ke, L. and Takahashi, A. 2014. Experimental investigations on suffusion characteristics and its mechanical consequences on saturated cohesionless soil. *Soils Found.* 54(4): 713–730.

Kimoto, S., and Oka, F. 2005. An elasto-viscoplastic model for clay considering destructuralization and consolidation analysis of unstable behavior. *Soils Found.* 45(2): 29–42.

Kimoto, S., Akaki, T., Loret, B. and Oka, F. 2017. A numerical model of internal erosion for multiphase geomaterials. In: *Bifurcation and Degradation of Geomaterials with Engineering Applications: Proceedings of the 11th International Workshop on Bifurcation and Degradation in Geomaterials, 2017*, ed. E. Papamichos, P. Papanastasiou, E. Pasternak and A. Dyskin, 125–131. Springer, Cham.

Loret, B. 2015. *Sand Production during Hydrate Dissociation: Constitutive and Field Equations*, Personal Note, August 30, 2015.

Loret, B. 2019. *Fluid Injection in Deformable Geological Formations*, p. 387. Springer, Cham, Switzerland.

Oka, F. and Kimoto, S. 2012. *Computational Modeling of Multiphase Geomaterials*. CRC Press/Taylor & Francis, Boca Raton, FL.

Skempton, A.W. and Brogan, J.M. 1994. Experiments on piping in sandy gravels. *Geotechnique* 44(3): 449–460.

Taylor, D.W. 1948. *Fundamentals of Soil Mechanics*. Wiley, New York.

Index

U

V

W

Y

Milton Keynes UK
Ingram Content Group UK Ltd.
UKHW031142141024
449569UK00024B/1132

9 781032 059570